HETEROGENEOUS ENANTIOSELECTIVE HYDROGENATION

Catalysis by Metal Complexes

Volume 31

The titles published in this series are listed at the end of this volume.

HETEROGENEOUS ENANTIOSELECTIVE HYDROGENATION

Theory and Practice

by

EVGENII KLABUNOVSKII

Laboratory of Asymmetric Catalysis. N.D. Zelinskii Institute of Organic Chemistry Russian Academy of Sciences, Moscow, Russian Federation

GERARD V. SMITH

Southern Illinois University, Carbondale, IL, U.S.A.

and

ÁGNES ZSIGMOND

University of Szeged, Szeged, Hungary

Springer

A C.I.P. Catalogue record for this book is available from the Library of Congress.

ISBN-10 1-4020-4294-9 (HB)
ISBN-13 978-1-4020-4294-2 (HB)
ISBN-10 1-4020-4296-5 (e-book)
ISBN-13 978-1-4020-4296-6 (e-book)

Published by Springer,
P.O. Box 17, 3300 AA Dordrecht, The Netherlands.

www.springer.com

Printed on acid-free paper

Printed in the Netherlands.

CONTENTS

PREFACE

At the present time we observe the second birth of asymmetric heterogeneous catalysis. During the past 50 years asymmetric heterogeneous catalysis has developed from very modest results, to highly effective catalytic systems similar to enantioselective efficiencies of chiral metal-complex catalysts. This has been particularly true in the hydrogenation of prochiral compounds containing double bonds, in which were demonstrated only the principal possibility of realizing *in vitro* reactions similar to those occurring in nature by enzymes.

The first attempts to carry out hydrogenation using heterogeneous catalysts involved metals supported on dissymmetric materials of natural or synthetic origin, such as crystals of optically active quartz and the chiral fibers of silk or wool. These showed small asymmetric effects. Consequently, interest in such efforts died away; moreover, these early results were sometimes irreproducible and even were criticized as artifacts. Later however, the early data were reproduced and confirmed as true, especially during the 1960's, when many effective heterogeneous catalytic systems were being elaborated using nickel catalysts modified with optically active compounds.

At the end of the 1960's the new effective rhodium-phosphine catalyst, the Wilkinson catalyst, was combined with chiral phosphine ligands resulting in very effective metal-complexes which catalyze hydrogenations of prochiral compounds (mainly C=C double bond) with optical yields close to 100%. This success of the elaboration of chiral metal-complex catalysts mimics the idea taken from heterogeneous asymmetric catalysis that asymmetric reactions on the surface of metal catalysts occur through catalytic intermediate complexes involving the molecule of chiral modifier, the molecule of substrate, and the metal surface atoms. Application of this concept produced homogeneous metal complex catalysts with broad practical uses. Over one hundred novel chiral ligands were developed to produce effective chiral homogeneous catalytic systems with high enantioselectivities for many reactions besides hydrogenation; for example hydrosilylation, hydroformylation, epoxidation, and oxidation.

The present interest in asymmetric catalysis was demonstrated by awarding Nobel prizes to three winners: W. S. Knowles (USA) for elaboration of rhodium complex catalysts effective in asymmetric synthesis of anti-Parkinson medicine, R. Noyori (Japan) for elaboration of a new catalytic system based on chiral ruthenium-phosphine complex catalysts that are very effective in hydrogenation reactions, and B. Sharpless (USA) for elaboration of epoxidation and other reactions under the action of chiral titanium complexes.

These developments were achieved despite the known technological difficulties associated with the practical use of homogeneous chiral metal-complex catalysts. They have low durabilities and repeatabilities, the chiral phosphine ligands are difficult to synthesize, the catalytic complexes are not recoverable from reaction mixtures, and the most effective catalysts based on rhodium complexes are very expensive to synthesize.

The latter difficulty could be avoided by using non-precious metals and by immobilizing complexes onto carrier materials. But this last method did not always give good results. All these facts encouraged a redoubling of the efforts to further develop pure heterogeneous chiral catalytic systems.

In the 1960's works of the groups of Izumi (Osaka) and Klabunovskii (Moscow) elaborated enantioselective heterogeneous catalytic systems (mainly Raney nickel catalysts and other hydrogenating metal catalysts, such as Cu, Co, Ru, Pd, bimetals, and multimetalic systems) based on modification with chiral compounds (mainly with hydroxycarboxylic acids or amino acids). These new catalysts, especially modified Raney nickel, were studied in detail and the resulting works became triggers for many studies both in heterogeneous and homogeneous enantioselective catalytic reactions.

The optical yields obtained in heterogeneous catalysis were at first not very high and much lower than those from chiral metal complex catalysts; however, as understanding improved, the ensuing reports, such as in the session on enantioselective heterogeneous catalysis at the 1978 Gordon Conference on Catalysis and in the similar session at the 1980 International Congress on Catalysis, were beginning to show significant progress in heterogeneous asymmetric catalysis. During that and subsequent periods promising model compounds were developed to test the effectiveness of newly emerging enantioselective heterogeneous catalytic systems. These were based mainly on esters of acetoacetic acid, which, when the beta-carbonyl group is hydrogenated, produce optically active hydroxybutyrate esters. Eventually, some of these catalytic systems produced optical yields up to 90%. In fact, in the case of the Ni based catalytic system modified with *R,R*-tartaric acid, beta-ketoesters were hydrogenated to optical yields up to 98%, and it should be noted that this accomplishment, which is close to the effectiveness of enzymic systems, does not use precious metals

Then during the 1990's a second very effective process of enantioselective heterogeneous hydrogenation was developed. Using platinum supported on alumina and modified with optically active alkaloids of the cinchona group, the keto group of alpha-ketocarboxylic acid esters were hydrogenated to lactates with optical yields above 95%. Thus in a short time two types of enantioselective heterogeneous catalytic hydrogenation systems were developed, each effective only for hydrogenation of mainly beta- or alpha-keto esters.

The subsequent studies of the properties of substrate specificity and structure sensivity allowed for elucidation of reaction mechanisms and of modification processes as well as for the study of the phenomenon of ligand rate acceleration for producing high enantioselectivity

This book contains many publications which represent analyses of the steps of elaboration of effective heterogeneous enantioselective hydrogenation catalysts, of their significant role in the theory of catalysis, and of their role in the practice of asymmetric catalysis. In addition to reviewing the first works on catalytic hydrogenation of C=C double bond in prochiral compounds on metal catalysts supported on chiral carriers, which admittedly have only historical interest, the Chapters 1-3 review data on asymmetric adsorption of enantiomers and separation of racemic mixtures on organic and inorganic adsorbents.

In analyzing more than 250 publications Chapter 4 covers effective metal catalysts, mainly nickel, but also bimetal- and multimetal-systems, and their best modifiers, amino acids and tartaric acids. It is noted that it took more than 25 years to improve the modified nickel catalysts from their original poorly efficient systems into the modern excellent heterogeneous catalysts that hydrogenate carbonyl compounds with enantioselectivities of 96-98%.

The studies of platinum supported aluminum oxide modified with optically active alkaloids of the cinchona group are considered in Chapter 5, where more than 200 communications on this topic are summarized. The studies of Pt-alumina catalysts modified with cinchonidine allowed for the preparation of a number of very important chiral synthons and medicines. These publications give substantial information for better understanding of the mechanism of heterogeneous catalysts and of the structure of intermediate complexes in enantioselective hydrogenation reactions.

Relatively few publications are considered in chapter 6, which is devoted to electrochemical enentioselective reactions using electrode-catalysts modified with optically active compounds. This method presents significant opportunities for practical use in the synthesis of chiral compounds.

Chapter 7 contains most of the important data on practical enantioselective processes catalyzed by both heterogeneous enantioselective metal catalysts and by homogeneous metal-complex catalysts.

LIST OF ABBREVIATIONS

α^t_D	=	observed optical rotation at t ^oC and 589 nm (D-line)
$[\alpha]^t_D$	=	specific optical rotation of scalemic product, enantiomer mixture in solution, determined by α^t_D / *l* x *c*, where *l* is the length of tube (dm) and *c* is the sample concentration (g/ml)
$[\alpha]^t_l$	=	specific optical rotation of pure optical enantiomer at t ^oC and *l* nm wavelength
AA	=	amino acid
acac	=	acetylacetone (pentane-2,4-dione)
AcOH	=	acetic acid
Ala	=	alanine
BCPM	=	(2*S*,4*S*)-*N*-(*t*-butoxycarbonyl)-2-[(diphenylphosphino)methyl]-4-(dicyclohexyl phosphino)pyrrolidine
BCPP	=	(2*S*,4*S*)-*N*-(*t*-butoxycarbonyl)-2-[(dicyclohexylphosphino)methyl]-4-(diphenylphosphino)pyrrolidine
BINAP	=	2,2'-bis(diphenylphosphino)-1,1'-binaphthyl
BPPM	=	(2*S*,4*S*)-*N*-(*t*-butyloxycarbonyl)-4-(diphenylphosphino)-2-[(diphenylphosphino)methyl)]pyrrolidine
CAMP	=	cyclohexyl-(*o*-anisyl)-methylphosphin
COD	=	cycloocta-1,5-dione
Cn	=	cinchonine
Cnd	=	cinchonidine
Cy	=	cyclohexyl
CyCAPP	=	*N*-(*N'*-R-carbamoyl)-4-dicyclohexylphosphino-2-[(dicyclohexylphosphino) methyl]pyrrolidine
d	=	dextrorotatory, (+)

dl	=	racemic mixture
D	=	configurational symbol of saccharides and amino acids, the configuration is related to D-(+)-glyceraldehyde
DBU	=	1,8-diazabicyclo[5.4.0]undec-7-ene
DHCnd	=	10,11-dihydrocinchonidine
de (%)	=	diastereomeric excess (%)
DHVin	=	dihydrovinpocetine
Diol	=	pentane-2,4-diole
DIOP	=	*O,O*-isopropylidene-2,3-dihydroxy-1,4-bis(diphenylphosphino)butane
S-DOPA	=	(*S*)-3,4-dihydroxyphenylalanine
EAA	=	ethyl acetoacetate
ee (%)	=	enantiomeric excess (%) (optical yield or asymmetric yield) defined as ([R]- [S])/ ([R] + [S]) x 100
EtBf	=	ethyl benzoylformate
EtMn	=	(*R*)-(-) ethyl mandelate
EtLa	=	(*R*)-(-) ethyl lactate
EtPy	=	ethyl pyruvate,
EHB	=	(*S*)-ethyl *beta*-hydroxybutyrate; $[\alpha]^{20}_{D} = + 23.95^{o}$ (neat)
HD	=	hexa-1,5-diene
IMC	=	intermetallic compounds
KPL	=	ketopantolactone (dihydro-4,4-dimethylfurane-2,3-dione)
l	=	levorotatory (-)
L	=	configurational symbol of saccharides and amino acids, the configuration is related to L-(-)-glyceraldehyde
LEED	=	low-energy electron diffraction
MePy	=	methyl pyruvate
MeBf	=	methyl benzoylformate

MeLa	=	(*R*)-(-)-methyl lactate
MeMn	=	(*R*)-(-)-methyl mandelate
MeO-DHCnd	=	9-MeO-10,11-dihydrocinchonidine
MAA	=	methyl acetoacetate
MHNi	=	modified nickel catalyst, prepared by decomposition and reduction of nickel salts (nickel formate, nickel nitrate or Ni $(CO)_4$)
MHB	=	(*R*)-methyl *beta*-hydroxybutyrate, $[\alpha]^{20}_D$= - 22.95° (neat)
MHB	=	(*S*)-methyl *beta*-hydroxybutyrate, $[\alpha]^{20}_D$ = + 22.95° (neat)
		(all the *ee* values of MHB cited in this book were corrected according to the above specific rotations)
MNDO	=	modified neglect of diatomic overlap
MRNi	=	modified Raney Nickel catalyst
MRNi-TA	=	Raney Nickel catalyst modified with (2*R*,3*R*)-tartaric acid
MRNi-TA-NaBr	=	Raney Nickel catalyst modified with (2*R*,3*R*)-tartaric acid and NaBr
Naproxen	=	(*S*)-(+)-2-(6-methoxy-2-naphthyl)propanoic acid
NBD	=	norbornadiene (bicyclo[2,2,1]hepta-2,5-diene)
NORPHOS	=	(2*S*,3*S*)-2,3-bis(diphenylphosphino)bicyclo[2,2,1]hept-5-ene
PPM	=	(2*S*,4*S*)-2,4-bis[(diphenylphosphino)methyl]pyrrolidine
PAMP	=	methyl-anysil-phenylphosphine
Phe	=	phenylalanine
Pl	=	(*R*)-(-)-pantolactone, (D-(-)-dihydro-3-hydroxy-4,4-dimethyl-2-(3H)-furanone
Pt-alumina-Cnd	=	platinum catalysts, supported on alumina and modified with cinchonidine
Pt-silica-Cnd	=	platinum catalysts, supported on silica and modified with cinchonidine
RT	=	room temperature
Qn	=	quinine
Qnd	=	quinidine

'treo-amine' = *treo*-2-amino-1-(*p*-nitrophenyl)propane-1,3-diol

Tf = triflat (trifluoroacetate)

Tyr = tyrosine

Trp = triptophane

Val = valine

y = chemical yield (% of conversion)

Chapter 1

ASYMMETRIC ADSORPTION ON MINERALS

Abstract

Chapter 1 considers the possible relationships of earthly clays and other minerals to the origin of chirality in organic molecules. Attempts to establish experimental evidence of asymmetric adsorption on clays were unsuccessful, but the search for chirality did find naturally occurring enantiomorphic crystals like quartz. Asymmetric adsorption of organic molecules on quartz crystals such as separation of racemic mixtures, like Co or Cr complexes, alcohols and other compounds, allowed for the conclusion that quartz crystals can serve as possible sources of chirality but not of homochirality. This latter conclusion results from the finding that all studied locations of quartz crystals contain equal amounts of *d*- and *l*-forms. The preparations of synthetic adsorbents such as imprinting silica gels are also considered. More than 130 references are analyzed.

1.1. Introduction

There are many natural minerals and salts that posses optical activity in their crystalline state owing to their chiral lattices, such as quartz, cinnabar, mica, chlorates, bromates, and iodates. Crystal chirality of other minerals, like aluminosilicates, such as zeolites, were not investigated, but these minerals are considered by many investigators as possible sources of chirality and the origin of homochirality in our biosphere [1,2,3]. The optical activities of clays have not revealed reliable evidence of chirality [4] and therefore they do not play any positive role in our understanding of the origin or of the amplification of homochirality in nature.

On the other hand, if aluminosilicates are used as supports for chiral metal complex catalysts, they might reveal a contribution to asymmetric action as was found in the case of the hydrogenation catalyst [CoSalen] complex supported on hectorite [5].

Even though clays have no chirality associated with their crystal structure, Julg and other authors [6,7,8,9,10], claim asymmetric effects. As stated by Julg [10]: "...owing to the weak interactions, the two enantiomeric forms of kaolinite exhibit different energies. Experiments on synthesis and the condensation of L-amino acids is a supplemental argument for the role played by kaolinite on the appearance of the first proteins and their L-homochirality...".

1.2. Attempts of asymmetric adsorption on clays

The role of clays in chemical evolution was suggested by Bernal [1] in 1951 based on preferential adsorption of optical isomers of amino acids by optically active crystals of quartz and possibly also by clays. Cairns-Smith [2] suggested that the surfaces of minerals, probably clays, played important roles in biogenesis. Field and Spencer [11] showed that in the presence of alumina the formation of glycine from water, CO_2, and NH_3; fructose from water and CO; and other sugars from HCHO and hydrogen, can be accomplished. Gabel and Ponnamperuma [12] may have confirmed these data in studies of the condensation of HCHO on alumina. According to Bernal's suggestion,[1] such processes provide the mechanism for the concentration of homochirality in nature. See also other authors positions on this subject [13,14,15].

The first results on selective adsorption of individual enantiomers by clay minerals were claimed by Degens [6] and Jackson [16]. Degens [6] found 25% conversion of the polymerization of L-Asp after heating on kaolinite for 32 days but only 3% conversion of the polymerization of D-Asp and 14% of DL-Asp. These data are considered to result from the chiral character of the kaolinite surface favoring a chiral condensate made of L-Asp. Large changes in the extent of copolymerization on clays at shorter reaction times were found by Jackson [16]. He found a 25% preferential adsorption of L-phenyl-alanine (Phe) compared to D-Phe at pH 5.8. But both of these results are very doubtful because of experimental uncertainties. Nevertheless, based on the ideas of Cairns-Smith [2] and Degens [6], theories of the origin of homochirality based on adsorption on chiral minerals were elaborated.

Degens [6] proposed that the surface of kaolinite contains centers which adsorbed mainly L-amino acids. In contrast, Bonner and colleagues [17,18,19] found that kaolinite is not an active chiral promoter neither in adsorption of enantiomers of Phe, nor in stereoselective polymerization of Asp. Flores and Bonner [19] incubated enantiomers of amino acids with kaolinite for 8 days but with negative results for stereoselective reactions. McCullough and Lemmon [20] tried to confirm those results [6] by using labeled aspartic acid (Asp) enantiomers and paper chromatography analysis for polymer formation, but they obtained only negative results. Jackson, [21] based on earlier data of Thomson and Tsukashina [22], criticized the data of Bonner and coworkers [17,18,19] claiming that their negative results may be caused by uncertainties in different experiments and by traces of promoters of the process. Bondy and Harrington [7,23] emphasized the need to distinguish the adsorption of "natural" and "unnatural" enantiomers on clays. They studied diluted solutions of D-

and L-^{3}H-labeled leucine (Leu), Asp, and glucose contacted with colloidal clay, bentonite and montmorillonite, and observed the radioactivity of samples bounded by the clay. They claimed that "natural" compounds are bound for 6.5-11.3 times more effective then "unnatural" enantiomers: D-Leu, D-Asp, and L-glucose do not exhibit any selective adsorption on bentonite. They connected these data with the origin of chirality on Earth because the primordial clays can be disposed of once there exists more effective stereoselective samples, which posses chiral crystal structures. This difference could account for life forms possessing a great preponderance of L-amino acids and D-glucose. But Jackson et al.[21] objected to the possible chirality of the clays used and the validity of experimental data in these experiments. Wellner [4] indicated that the chiral structure of bentonite is not confirmed yet and the differences of adsorption of optical isomers could not be responsible for differential binding. Moreover it is necessary to note that the bentonite structure differs from that of kaolinite. Friebele and colleagues [24] carefully treated a sample of montmorillonite and studied its adsorption of enantiomers of Val, NorVal, and *alpha*-aminobutyric acid. They found that at pH 10 the difference in the excess of adsorption of L-amino acids is equal to 0.5-2.0 %. Their standard deviation of gas chromatographic (GLC) analysis was 0.01%, which was much better than that claimed earlier by Bonner and colleagues (0.2%) [25]. In conclusion, Friebele and colleagues [24] noted that "...clay minerals have no known chirality; therefore, to rationalize a small asymmetric effects due to adsorption of amino acid is difficult, but a very small effect can not be excluded...". Later authors [26] rejected the idea of asymmetric adsorption on clays. Yuatt and Brown [27] showed that stereoselective binding of amino acids by clays does not occur and the earlier results can be attributed to adsorption of the radiochemical decomposition products formed during the experiments [7,23]. Therefore, it will be more correct to measure the adsorption by the amount of adsorbate on the surface of the adsorbents rather than by the decrease in the concentration of the substrate in solution. This criticism belongs to Bonner's experiments too, where he [28,29] announced differential adsorption of up to 12-20 %. Friebele and colleagues [24], using GLC methods, seemed to show that there is stereoselective adsorption of enantiomers of amino acids on Na-montmorillonite in solution at pH 3, 7, and 10. L-amino acids were 0.5-2.0% more adsorbed in the clay than were their D-enantiomers. But later it was noted that the first results were erroneous and it was shown, using enantiomeric analysis of the same samples, that the results were not consistent in all cases with the GLC data (data in Table 1.1 with asterix) and no differential adsorption was found. Thus these results show that apparently clays have no significant ability to sort out L- and D-amino acids under the conditions studied and that clays have not played an important role in the enantiomeric selection of amino acids in the processes of chemical evolution.

Amino acids	pH of solutions 3	7	10
Alanine	1.022	0.976	0.935
		1.001*	0.983*
α-Aminobutyric acid	1.005	0.996	0.925
NorValine	0.986	0.979	0.986
Valine	0.977	0.996	1.015

* Determined using amino acid analyzer.
Adapted modified data from Friebele et al.[24]

Table 1.1. Ratio of *D/L* of amino acids adsorbed on Na-Montmorillonite as determined by GLC [24].

Ignoring all these negative experimental data Julg [8,9,10] calculated, using a "self consistent molecular method", that the adsorption energy of L-amino acids on one of the "chiral" forms of kaolinite is 0.03-0.19 kcal/mol greater than that for the D-enantiomer, and this differential adsorption can be the basis for spreading L-amino acids in nature. Moreover, he postulated [9] that addition of CN^- to ethyliminium cation on the surface of kaolinite can form a precursor of chiral L-Ala that is favored over the corresponding D-precursor by 0.36 kcal/mol. This mechanism can explain the preponderance of L-amino acids rather than D-amino acids. Furthermore, due to weak interactions assumed for the calculations [10], the enantiomeric forms of kaolinite leading to the formation of L-amino acids are more abundant than those leading to the D-enantiomer. But after Bonner [28,29], Julg's speculations [8,9,10] have no experimental evidences and are fully contradicted by recent experimental data. Thus at the present time, the main conclusion must be drawn that clays do not reveal any asymmetric adsorption and can not serve as sources of chirality in nature.

1.3. Asymmetric adsorption on quartz crystals

1.3.1. Properties and morphology of quartz crystals

There are seven crystal modifications of natural quartz: *alpha-* and *beta-*quartz; *alpha-*, *beta-*, and *gamma-*tridimite; and *alpha-* and *beta-*crystoballite. Below in Scheme 1.1. are shown the temperatures of transfers between the different quartz modifications.

At the conversion point (579 ^{0}C) optical rotation reveals a sharp jump of 4.5%. The value and sign of optical rotation of the quartz crystal depends on the direction of the beam of polarizing light passing through the crystal. If ϕ is the angle between the direction of the beam of light and the optical axis of the crystal (Fig. 1.1.) and $[\alpha]_{90}$ the specific optical rotation in the direction perpendicular to the optical axis, then the specific optical rotation at a given ϕ in the range 0° - 90° can be calculated by $[\alpha] = [\alpha]\ \mathrm{Cos}\ \phi + [\alpha]_{90}\ \mathrm{Sin}\ \phi$.

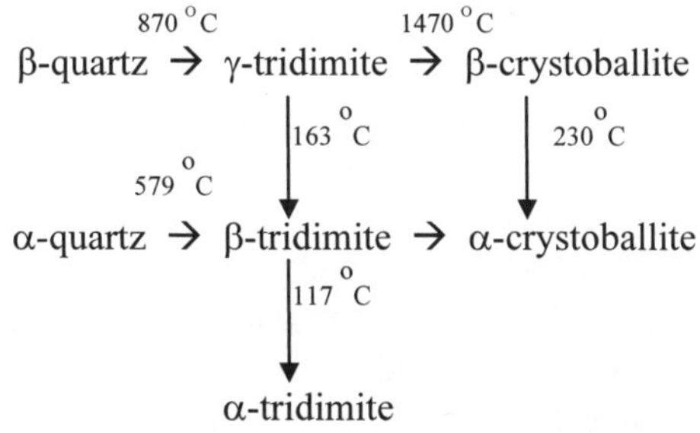

Scheme 1.1.

Measurement of the (-) rotation (α_{90}) in *d*-quartz (α_0) has definite difficulties because of double refraction. That is why the (+)-rotation only is usually accepted for *d*-quartz.

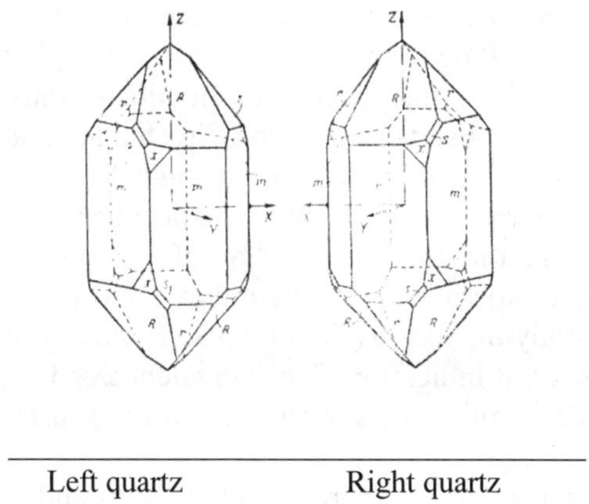

Figure 1.1. Morphology of quartz crystals (according to Cedy [33]).

The habits of an ideal quartz crystal (Fig.1.1.) has five simple forms: rhombohedrons, R, and trapezohedron, x, hexagonal, prism, and bipyramidal. The crystal has its main optical vertical axis of third-order and three horizontal axes of second-order directed to ribs of the prism. These axes are so called "electric ones". Examination of the positions of the faces of a crystal reveals the sign of its optical rotation. The crystal is a *d*-quartz if x and s faces are situated at the top of the face of the right side prism. The crystal is an *l*-quartz if the x and s faces are situated on the top of face at left (Fig. 1.1.). However, in nature, findings of optically pure crystals are rather rare. Mostly, crystals form twins. For example, inside individual quartz crystals may be areas of (+)- and (-)-rotating quartz, called "Brasilian twins", that differ in their sign of rotation. Other areas of quartz crystals may exhibit one sign of rotation, (+) or (-), but are oriented to each other in such a way that their electric axes are directed opposite to each other. These are called "Douphine twins". Both types of twins do not exhibit piezoelectric properties. On the other hand, pure, rather rare quartz crystals are optically and electrically active.

The optical activities of quartz crystals are caused by structural arrangements of Si and O atoms situated along helix axes in the crystal lattice. A right helix structure corresponds to a (+)-optical rotating quartz and a left helix to a (-)-rotating quartz [30,31,32].

Selection of optically pure crystals can be accomplished with the aid of a polariscope in which the thin plates of quartz are examined between crossed Nicol prisms in polarized light. These plates are sawed perpendicular to the optical axis of well formed natural crystals. Brasilian twins exhibiting bright interference patterns in crossed Nicol prisms must be broken out. Only optically pure areas are selected for further use, for example, for asymmetric adsorption and for asymmetric catalysis. Piezoquartz as plates of an oblique cut, called the "Cut of Curie", are used in radio electronics and can be identified by the sign of their optically rotation according to the method of Lemmlein [30,31,33]. According to this method, the etched plate of piezoquartz is preliminarily examined for the absence of Brasilian twins under a beam of light placed close to the observer's eye and directed at an angle of 3-5^0 to the observer's eye. As a result of reflections of light from the ribs of the three vicinal facets of the pyramids the shape of the peculiar light picture reveals *d*- or *l*-quartz. The correct method of selection of optically pure quartz crystals is very important for experiments in which they are used and also the method is important for studying the origin of optical activity in the nature. At present there are several indications for equivalence or for prevalence of one of the morphological forms of quartz crystals over the surface of Earth [34, 35, 36, 37, 38, 39,40,41,42,43,44].

Mason [34] and Tranter [35] performed calculations and alleged that a 1.4% excess of *l*- over *d*-quartz on the surface of Earth is the inevitable consequence of a parity-violating energy difference (PVED). They calculated that the *l*-quartz is stabilized by 10^{-17} kT per SiO_2 unit [35-37]. This allegation is based on a report [40] that there is 50.17% *l*-quartz and 49.3% *d*-quartz in 16,807 samples collected worldwide; that is an excess of 1.4% of *l*-quartz. But a more recent survey of 27,053 samples [38,39] yields 50.17 % *d*-quartz and 49.83 % *l*-quartz, which is a. 0.39 % excess of *d*-quartz. Therefore it is necessary to accept the fact that the terrestrial distribution of optically active quartz crystals is quite random and, consequently, no basis exists for calculation of PVED for quartz crystals [35-37]. Nevertheless, the quartz crystals and other minerals, for example, clays (see Chapter 1.2) are considered as possible sources of asymmetry in nature [1, 8-10,45,46].

1.3.2. Asymmetric adsorption on quartz crystals

Unlike the clays, owing to its chiral lattice, quartz must display certain selectivities for the adsorption of mainly neutral organic compounds possessing molecular or crystal-like chirality. Glucose, galactose and arabinose selectively adsorb on quartz crystals and quartz dissolves in solutions of these monosaccharides [47]. Racemic acids seem not to be resolved on quartz [48]; however, it has been shown that *d*- or *l*-quartz exert an orienting influence on the epitaxy of hemihedral crystals of Glu, Ala, and Gly. For example, the influence is seen for Glu, Ala, and Gly adsorbed on the surface of *l*-quartz crystals {1010}, and of Ala on *l*-quartz {1011}. These crystals are like the asymmetric epitaxy of crystals of (+) camphor on biotite and calcite [49].

In general, asymmetric adsorption on quartz crystals can not be effective. Selective adsorption of enantiomers or attempts for resolution of racemates have not shown positive results. Thus, attempts to resolve racemic *p*-phenylenebisiminocamphor, β-naphthylazomandelic acid, and the racemic free radical, phenyl-biphenyl-α-naphthylmethyl, have been unsuccessful [50]. Klabunovskii and Patrikeev [51] partially resolved racemic 2-butanol using a chromatographic technique based on counter current action. In a glass column vapors of racemic 2-butanol passed upward against a falling quartz powder. In this process, *l*-quartz preferentially adsorbed (+)-butanol. Then the butanol with a small excess of the (+)-enantiomer was removed from the quartz by heating in a desorbing zone. Both the (+)-enriched butanol and the quartz powder were recycled back into the process. After thirty cycles the (+)-2-butanol was recovered with an ee 0.22% [51].

Asymmetric adsorption on quartz powder was also shown by measurements of the adsorption capacity of *d*- and *l*-quartz in relation to (*S*)-(-)-2-methylbutan-1-ol (that is, the commercial (-)-amyl alcohol, AmOH, consisting of 75% (-)-alcohol in a mixture with 3-methylbutan-1-ol) adsorbed on samples of *d*- and *l*-quartz with equal surface areas, and the decrease in optical rotation was measured after adsorption [51]. It was found that *d*-quartz adsorbed more (-)-AmOH by 2.3% than *l*-quartz. Taking into account the 1 m^2 surface areas of the quartz samples, this 2.3% amounts to 0.06% (-)-AmOH more than *l*-quartz. In contrast, it is also necessary to note that Ohara et al.[52] tried in vain to resolve 2-butanol on optically active quartz.

The selective adsorption of enantiomers of transition metal complexes on quartz proved to be more successful. For the first time, in 1935, Tsuchida and colleagues [53] reported the partial resolution of several racemic Cr- and Co-complexes using chromatographic adsorption on columns composed of *d*- or *l*-quartz crystals originating from Kimbusen and Naegi deposits. The steric configuration of chlorobisdimethylglyoximoammine-Cobalt has not been determined. However, a solution of this complex over *l*-quartz gave a (+)-rotation and over *d*-quartz a (-)-rotation (0.02 ± 0.05° in a 4 cm polarimeter tube).

Resolution of a complex even to a very small extent could demonstrate that the original complex was a racemic mixture. Thus, complexes like [Co(dimethylglyoximo)$_2$(NH$_3$)Cl] were partly resolved on quartz powder. The spatial configuration of this complex was unknown. From the theoretical point of view it was expected that only the *cis*-form of the complex should exhibit optical activity. In agreement with this expectation, resolution of the tris(dimethylglyoximo)cobaltate(III) complex on quartz exhibited a poor but definite optical activity revealing the configuration of the complex to be *cis* rather than *trans* [53].

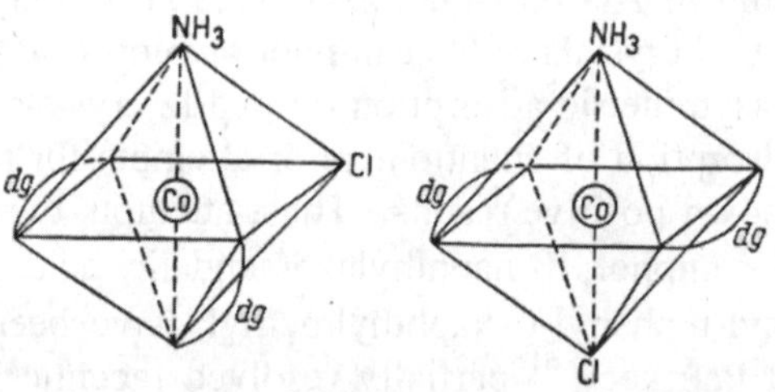

cis-complex *trans*-complex

Scheme 1.2. Structures of *cis*- and *trans*-Co(dimethylglyoximo)$_2$(NH$_3$)]Cl complexes (according to Klabunovskii [32b]).

Therefore it was sufficient to resolve this complex into enantiomers to determine its spatial structure. This complex does not form diastereomers and the only method of its resolution was asymmetric adsorption on crystals of quartz – a very inert material. In this manner, a solution of samples of the complex was vigorously shaken with quartz powder, the solution was filtered and its optical rotation measured. On *l*-quartz the optical rotation of the remaining solutions reached from + 0.05^0 to + 0.075^0 and on *d*-quartz they reached from -0.075^0 to -0.13^0. Thus the configuration of the complex proved to correspond to the *cis*-structure. The *d*-quartz preferentially adsorbs the (+)-complex and the *l*-quartz the (-)-complex. Brazilian twin quartz gave negative results.

Using the same method, Tsuchida et al. [53,54] found that the *d*-quartz preferentially adsorbs enantiomers of the following complexes:

(-)-[Co(en)$_2$]Br$_2$
(+)-[Co(en)$_2$ (NH$_3$)Cl]Br$_2$
(+)-[Co(en)$_2$Cl$_2$]Cl

(+)-$K_3[Co(ox)_3]$
(+)-$[Co\{Co(NH_3)_4(PhH)_8\}_3]Cl$

In 1938 Karagunis and Coumoulos [55] suggested such an effect on the resolution of racemic Cr-, Co-, and Pt-complexes and noted that "...resolution on optically asymmetric surfaces of minerals might have been the genesis of the first optically active organic molecules in Nature.".

Ikeda [56] found that *d*-quartz can partially resolve (1,2,3,4-tetrahydronaphthyl)hydroperoxide and cyclohexylhydroperoxide with specific rotations of $+20.0^0$ and -18.5^0. As the author indicated "...method of course is not quantitative one but can be used for decision of question is given compound racemic one or not…".

Kuebler and Bailar [57] studied the resolution on quartz powder (140-200 mesh) of the racemic complex: potassium dinitro(*N*-methyl-*N*-ethylglycinate)platinum. Resolution proved to be somewhat better than on crystals of quinine bisulfate. For example, a solution of the complex was shaken with quartz powder and after filtration was examined polarimetrically. Typically, optical rotations diminished with time and in 5-7 hrs fell to zero. On *l*-quartz rotations were $+0.028^0$, and on *d*-quartz $-0.024\,^0$.

Also complexes $[NaCo(en)(Ac)_4]$ (**I**) and $[BaCo(en)(Ac)_4]Br$ (**II**) were partially resolved on *d*-quartz powder [58]. Optical rotations for **I** on *d*-quartz were between -0.063^0 and $-0.012^0 \pm 0.003$-0.007^0, and for **II** on *d*-quartz, -0.018^0. The (+)-rotating forms of both complexes preferentially adsorbed on *d*-quartz.

The complex *N,N'*-bis(salicylidene-trimethylene-tetrammino)cobalt was partially resolved on *d*-quartz with a rotation of -0.03^0 (specific rotation – $300\,^0$) [59].

Complex	Maximum optical rotation	
	(-)-rotation	(+)-rotation
$K_3[Cr(ox)_3(H_2O)_3]$	0.70	0.60
$K_3[Co(ox)_3(H_2O)_3]$	0.50	0.30
$K[Cr(ox)_2(en)]$	0.17	0.20
$[Co(ox)(en)_2(H_2O)]Cl$	0.35	0.20
$[Co(en)_3]Cl_3$	1.0	0.15
$[Co(dg)_2(NH_3)_2]Cl$	0.80	1.00

ox = oxalate; en = ethylenediamine; dg = dimethylglyoxime

Table 1.2. The maximum optical rotation values of fractions of chromatographic resolution of racemic complexes with a quartz powder column (modified data adapted from Schweizer et al. [60]).

Karagunis and Coumoulos [55] tried to resolve the racemic $[Cr(en)_3(H_2O)_{(3\,½)}]Cl_3$ complex on quartz by the chromatographic method. Through a vertical column packed with activated (heating in vacuum) quartz powder was passed a solution of the complex in 85% ethanol. In the case of *d*-quartz the first fractions of eluent had a rotation of $+0.023^0$ followed by fractions with a rotation of -0.05^0. On *l*-quartz the opposite picture was observed; the first fractions had a rotation of -0.058^0 followed by fractions with rotations of $+0.054^0$.

Using the same method Schweizer and Talbott [60] found selective adsorption on *l*-quartz with a specific area of 4 m^2/g in resolution of several racemic Co and Cr complexes, such as $K_2[Cr(ox)_2]$ and $K_2[Co(ox)_2]$. The 0.01M solutions of the complexes were passed through a column of quartz powder (300 mesh). The eluate was divided into 20 fractions, among which fractions 6 through 10 exhibited progressive changes of sign of rotation. In Table 1.2. is shown the maximum optical rotations of the fractions. More effective resolution was observed when the isolated fractions were passed through a second and third column. Results received in resolution of the racemic complex $K_3[Cr(ox)_3(H_2O)]$ are shown on Table 1.3.

No. of fractions	optical rotation	No. of fractions	optical rotation
1	- 0.90	6	- 1.30
2	- 1.00	7	- 1.20
3	- 1.20	8	- 0.45
4	-1.30	9	+ 0.05
5	- 1.10		

Table 1.3. Optical rotations of fractions in resolution of racemic Cr-complex on quartz (Klabunovskii [1b]).

From the thin-layer technique, Kamai and colleagues [61] found partial resolution of racemic [(Bz)(Et)(pentyl)(Ph)As]Br on vibro-crushed *d*-quartz powder: the (-)-enantiomer adsorbed preferably. Chernyaev and coworkers [62] obtained partial resolution of the racemic complex $[Pt(Hal)(NH_3)NO_2]$. On *l*-quartz the (+)-complex was adsorbed preferentially and exhibited only (+) 0.02-0.05° at 579 nm, even so, this rotation proved to be enough for identification of the spatial structure of the complex.

Using *d*-quartz partial resolution was observed for the acetylacetonate chelates $[Cr(acac)_3]$ and $[Co(acac)_3]$, in which the (-)-complex was preferentially adsorbed [63]. Sievers and coworkers [64] found the partial gas chromatographic resolution of racemic $[Cr(F_6acac)]$ on *d*-quartz at 56°C, in which the first fraction had a specific rotation of 260°.

By shaking *d*-quartz powder with solutions of the racemic complex $[Be(PhCOCH_2)_2C{=}O]_2$ Bush and Bailar [65] obtained very small rotations of +0.011 and +0.026° ± 0.003°, but these were enough for evidence of tetrahedral coordination of Be in the complex.

It is well known that silica plays an important role in living matter. This was shown by the selective adsorption of some sugars on quartz crystals[66]. Consequently, Schwab and Wahl[67] tried to determine the asymmetric adsorption of silicic acid specimens recovered from the mushroom *Euspongia* and from the Japanese sponge. However, they did not find any asymmetric adsorption. It turned out that the structure of this silicic acid is similar to the structure of silica. It possesses no helical structure and cannot play any role in the formation of optically active compounds in organisms.

Amariglio et al. [68,69,70,71] failed to reproduce the results of Tsuchida and coworkers (1936) and Klabunovskii and coworkers (1951) on adsorption (and catalysis) over quartz crystals. She concluded that all former results were erroneous because of the small optical rotations of the substrates to be resolved. Such small rotations were of the size which might be obtained from the slight solubility of optical active quartz, and when combined with the experimental errors in the polarimetric measurements, these could account for the very small rotations observed from the adsorbates. But analysis of her reports [68-71] showed that critical examination was incorrectly carried out. For example, the samples of quartz in her experiments [68-71] were not well identified, and suspension of *d*- and *l*-quartz and even of glass pyrex powder gave the small optical rotation with the same sign (-0.05°) independent of the sign of the quartz used. Therefore, the negative results [68-71] cannot be consider valid. The same conclusion is reached with regards to catalysis using "quartz-catalysts " [66-69].

Based on the tentative and shallow negative results received by Amariglio and coworkers [66-69], Bonner [72,73,74,75] at first came to the tentative conclusion that "...the phenomena of asymmetric adsorption on quartz crystals as asymmetric adsorbate is open to question...". Later, Bonner and colleagues [72-75] tried to verify the asymmetric adsorption on quartz using a radiochemical method with labeled amino acids as substrates. This is a more precise technique then the polarimetric technique and revealed an excess of adsorption of one enantiomer of one of the amino acids as high as 20 %.

In the radioactive tracer experiments ^{3}H- or ^{14}C-labeled *D*- or *L*-Alanine hydrochloride (Ala HCl) in DMF solutions were equilibrated with finely powdered *d*- or *l*-quartz [72,73]. Differences in rates of radioactive disintegrations of solutions before and after equilibration gave a measure of the amount of Ala HCl adsorbed on the quartz samples. In general, *d*-quartz preferentially adsorbed *D*-Ala HCl and the *l*-quartz *L*-Ala HCl to the extents of 1-1.8%. These experiments used finely powdered samples of *d*- and *l*-quartz which had been prepared from known morphological handedness and optical rotation, but the authors failed to mention how the Brasilian twins (if any) were removed from quartz samples [72,73]. They did note [73] that samples of quartz were prepared by hydrothermal growth and that crystals were checked for uniform optical handedness as well as for enantiomerism by measurements of optical rotations at 589 nm. The crystals prepared had specific rotations of +21.2° and -26.724° for 1 mm plates (somewhat more than found for natural quartz). Table 1.4. summarizes results derived from the radioactive labeling experiments.

Adsorbate	% adsorbed by *l*-quartz	% adsorbed by *d*-quartz	Differential adsorption, %[a]
L-^{3}H-Ala HCl	29.35	28.33	1.02
L-^{3}H-Ala HCl	28.35	26.68	1.75
L-^{3}H-Ala HCl	22.21	21.03	1.18
D-^{14}C-Ala HCl	29.00	39.38	-1.30
D-^{14}C-Ala HCl	25.92	26.88	-0.96

[a] percent adsorbed by *l*-quartz minus percent adsorbed by *d*-quartz.

Table 1.4. Adsorption of ^{3}H- and ^{14}C-labeled *D*- and *L*-alanine hydrochloride on *d*- and *l*-quartz crystals (modified data adapted from Bonner et al. [72,73]).

Table 1.4. shows that the differential adsorption of *L*-Ala on *l*-quartz and of *D*-Ala on *d*-quartz was as much as 1.0-1.8%. Interestingly, results received by the same author on the asymmetric adsorption of racemic *DL*-Ala HCl [74] in which only one of the enantiomers (*D*- or *L*-Ala) was labeled with ^{3}H or ^{14}C showed differential adsorptions as high as 20% (Table 1.5).

	Labeled Enantiomer of DL-Alanine		
	L	*L*	*D*
% adsorbed by *l*-quartz	60.0	60.1	44.1
% adsorbed by *d*-quartz	40.0	39.9	55.9
Differential adsorption, %	20.0	20.2	11.8

Table 1.5. Differential adsorption of labeled *D*-enantiomer in *DL*-alanine on *d*-quartz (modified data adapted from Bonner et al.[74]).

These data in Table 1.5. show that the adsorption of one of the enantiomers from a racemate greatly exceeds its adsorption when alone. Of course one must neglect the possibility that during adsorption the two enantiomers exchange the label, which can distort the results.

To study the effect of the functional groups in Ala on its adsorption on the quartz surface, several derivatives were prepared. For example Ala HCl was esterified with isopropyl alcohol (Ala-iPr ester) and acetylated with F_3CCOCl. The results of such studies indicated that adsorption of Ala HCl is mainly attributable to the interaction of the NH_3^+ group with the quartz surface.

Adsorption of ^{3}H-labeled Ala-iPr ester on *d*- and *l*-quartz was studied using gas chromatography (GLC). Adsorption of this ester proceeds from chloroform solution followed by its desorption by chloroform or methanol. It was found that *l*-quartz preferentially adsorbs the *D*-enantiomer. Essential symmetric data were obtained when the adsorption-desorption process was repeated with *d*-quartz. The main results are summarized in Table 1.6.

Adsorbate	Percent Desorption from *l*-Quartz	Percent Desorption from *d*-Quartz
D-Ala-i-Pr	43.8	52.4
L-Ala-i-Pr	56.2	47.6
%ee (%*D*-%*L*)[a]	-12.4	+4.8

[a] ee = enantiomeric excess (*D*-*L*/*D*+*L*)

Table 1.6. Asymmetric adsorption-desorption of ^{3}H-*DL*-Ala-iPr ester on *l*- and *d*-quartz, measured by GLC (modified data adapted from Kavasmaneck et al. [75]).

Using labeled *DL*-Ala HCl these results were confirmed [74] by obtaining a differential adsorption of 12-20% (percent adsorption on *d*-quartz minus percent adsorption on *l*-quartz). The adsorption of *D*- and *L*-Ala-iPrHCl in chloroform solution was studied using GLC [75]. In this case *d*-quartz preferentially adsorbed the *L*-Ala ester and *l*-quartz the D-Ala ester with enantiomeric enrichments of up to 1.5-12.4%. Such large differential enrichments are very doubtful and can indicate possible artifacts, although these data were confirmed later by Furoyama and coworkers [76] by stereoselective adsorption on quartz [76,77]. According to Furoyama and colleagues [76] racemic ^{14}C-alanine from ethanol at 8°C was partly resolved on *d*- and *l*-quartz powder with a specific area of 0.08 sq.m./g (purchased from Toyo Communication Co.) with preferential adsorption of *L*-Ala on *l*-quartz.

Thus, these data also confirmed the earlier findings of Tsuchida [53] and Karagunis [55]. The positive results in asymmetric adsorption of amino acids, as shown in the work of Bonner and coworkers [72-75], "...required carefully contrived and rigorously maintained anhydrous conditions, with mere traces of moisture nullifying the effect..."

Attempts were unsuccessful in resolving several organic and inorganic racemates {Cr and Co acetylacetonates, lipoic acid (1,2-ditolan-3-valeric acid) and penicillamine (3-mercaptovaline) and tartaric acid} on chiral inorganic enantiomorphic crystals [78] other than quartz, such as cinnabar [79] $NaClO_3$, $NaBrO_3$, $NaIO_3$, $NiSO_4{\bullet}6H_2O$, sodium uranyl acetate, and benzil [80].

Sodium chlorate, $NaClO_3$ gives well formed crystals possessing optical activity. Its absolute configuration was investigated by means of optical rotatory dispersion (ORD). The dependence of specific rotation, [α], of a crystal of $NaClO_3$ with wave length, L, in nm in the range 237.6-875 nm can be given by [81]:

$$[\alpha] = 1.2387\ L^2/(L^2\text{-}0.0081)^2$$

Ferroni and Cini [82] tried to resolve the racemic chelate complex $Be[(PhCOCH)_2CO]_2$ using optically active crystals of $NaClO_3$ as chiral adsorbents, but results were uncertain (see Table 1.7.).

Sample of $NaClO_3$	Observed rotation of resolved complex [a]
d- (right rotation)	+ 0.067°
l- (left rotation)	0.00°
Racemate crystals with an excess of *d*-crystals (from Merck)	+ 0.071°
Racemate (50% *d*-crystals + 50% *l*-crystals)	+ 0.017°

[a] α_{436} of Be-complex

Table 1.7. Resolution of racemic beryllium complex on sodium chlorate crystals (modified data adapted from Ferroni et al. [82]).

In addition by Gillard (1979) it was noted that all resolution data on crystals which owe their chirality to crystal packing rather than to molecular chirality are very doubtful based on his own results showing the optical rotation of suspension of *l*-$NaBrO_3$ in benzene exhibit rotations comparable with those found previously [80].

Thus, optically active quartz crystals remain the only relatively effective stereoselective inorganic adsorbent and this fact must be used in discussions of the role of asymmetric adsorption on quartz crystals in the origin (or amplification) of optical activity on the surface of the Earth (see Bernal [1] and [28,29,83,84,85]).

1.4. Imprinting chiral adsorbents: specifically modified silica gels

Molecular recognition is one of the main phenomena of living matter. In enzymes recognition is the basis of the well-known "Lock-and-key" mechanism first discussed by E. Fischer. The generation of sites in inorganic materials with the ability of chiral recognition is of particular interest because such solids can be used for molecular separation including enantiomeric enrichments by adsorption and by chiral catalysis. The target to create complementary cavities in inorganic materials using molecular imprinting with a removable template is of theoretical and practical interest.

Polyakov was first to create complementary cavities on the surface of silica gel [86,87]. He removed water from silica gel in atmospheres of benzene, toluene, or xylene vapors and found that the pore structure of the resulting silica gel is influenced by the size and shape of the imprinting molecules in the gaseous atmosphere. Such silica gel dried in a benzene atmosphere adsorbed benzene well but toluene and xylene poorly. In such an environment, the adsorbate produced a matrix with the size and shape suitable for its adsorption. Objections that the structure of the silica gel still retains traces of the template compounds were removed by Dickey [88,89], who achieved the idea of Pauling [90,91] of imprinting of molecules or "molecular memory " using the adsorption of alkyl-oranges on silica gels formed in the presence of the same adsorbates. As shown below, adsorbents revealed strong specificity to that adsorbate which was originally formed with the same alkyl-orange: HSO_3-C_6H_4-N=N-C_6H_4-NR_2.

These data give very strong evidence of imprinting but they have poor reproducibility and were criticized [92]. Morrison et al. (1959) seemed to view the evidence as favoring an association mechanism in which the unextractable dye particles act as centers to attract their own kind more strongly than one of their homologues. Several positive examples were described later [93,94,95,96]

Erlenmeyer [97,98,99] and Haldeman and Emmett [100] confirmed the results of Dickey [88,89] and developed further possibilities of this method of preparation of specifically formed silicas.

Dye added to silica gel	Silica Gel modified with: Methyl orange	Ethyl orange	Propyl orange	Butyl orange
Methyl-orange (R = Me)	17(3.5)*	8(1.6)	3(1.1)	2(1.1)
Ethyl-orange (R = Et)	9(2.5)	30(9.0)	10(2.1)	3(2.2)
Propyl-orange (R = Pr)	7(2.3)	22(5.0)	30(20)	25(6.0)
Butyl-orange (R = Bu)	5(1.5)	13(2.8)	30(5.0)	30(15)

*Numbers in parentheses are the ratio of the adsorption of the dye on the imprinted silica gel to its adsorption on the non-imprinted silica gel.

Table 1.8. Specific adsorption of dyes (HSO_3-C_6H_4-N=N-C_6H_4-NR_2) on silica gels modified with these dyes; adsorption of Dye $5x10^{-5}$ mmol/kg adsorbent (modified data adapted from Dickey [88,89] and Davis [97]).

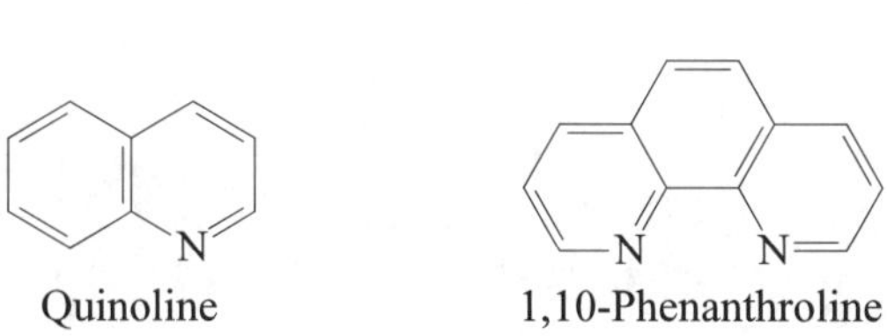

Quinoline

1,10-Phenanthroline

Scheme 1.3.

Erlenmeyer (1964) found that specifically prepared silica can preferentially adsorb compounds with analogous structures, for example, silica-Quinoline preferentially adsorbs 1,10-phenanthroline (Scheme 1.3.).

Waksmundzki and coworkers [101,102] and Bernhard [103] also received positive results. Patrikeev, Balandin, and colleagues [104] found that silica gel prepared in the presence of diketopiperazine catalysed the polycondensation of alanine esters into diketopiperazine faster then into tripeptide. Other examples of specific adsorption on silicas especially formed were given by Patrikeev and coworkers [105,106,107].

Later, various aspects of the catalytic actions of specifically formed adsorbents as catalysts were discussed by Jencks [108], who theorized that antibodies prepared in the presence of stable analogues that mimic transition states may be able to catalyze the corresponding reactions. Recently Lerner [109], and Mosbach [110] used transition states analogues to imprint the formation of antibodies that were able to function as catalysts for reactions from which the transition-state analogues were prepared. Confirmation of these concepts was obtained from studies of the Claisen rearrangement of (-)-chorismate to form prephenate [92,111].

Positive results were reported also by Morihara and colleagues [112,113,114,115] in a number of publications. Based on their experiments, the concept of "Footprint-catalysis" was developed. A silica gel surface was doped with Al^{3+} and the chiral template molecule *N*-benzoyl (*N'*-benzyloxycarbonyl-L-Alanine-amide). This embedded imprint molecule was removed by methanol extraction after aging and drying the silica gel [114]. The cavities thus formed displayed enantioselective catalysis for 2,4-dinitrophenolysis at 30°C in MeCN solution of the substrate benzoic *N*-benzyloxycarbonyl-L-Alanine anhydride. The observed maximal rates (10^{-2} L mol^{-1}) for "L-Silica", "DL-Silica", and control Silica (without imprinting) with L-substrate were 2.35, 1.31, and 0.81, respectively [116], and indicated a definite acceleration for the system: L-Substrate-L-Silica.

Also, they imprinted an aluminated silica with the transition state analogue for the butanolysis of benzoic anhydride, removed the imprinting molecule by extraction to form cavities in the structure of the silica, and were able to catalyse the butanolysis (Scheme 1.4.).

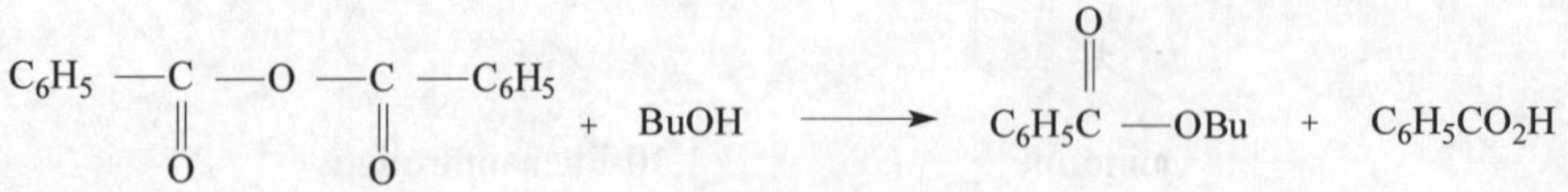

Scheme 1.4.

The authors [116] noted: "...This enantioselective catalysis should extend the scope of our strategy for tailored specific catalysts by the molecular imprinting method...". A review of these works exists [116]. Indeed, these results are of great interest, but it is necessary to note that Kaiser and Anderson [117] were unable to repeat the results with the benzamide imprinted silicas. Their failure probably was the result of a short lifetime for the specific properties of such Al-doped silica gels.

Heilman and Maier [118] also attempted to imprint silicas but were unable to obtain sufficient evidence for the connection of catalytic activity with imprinted sites.

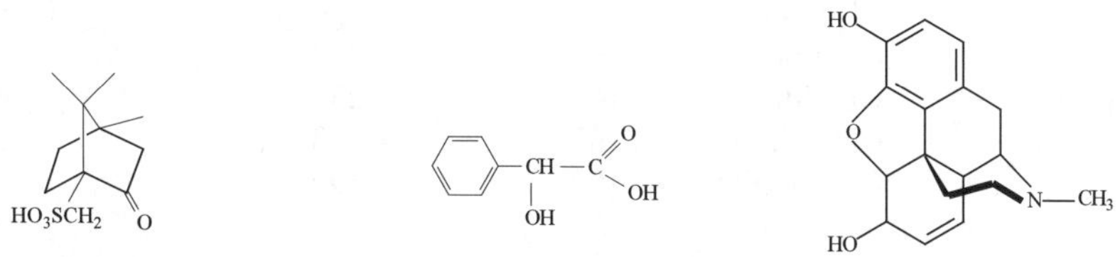

(1*S*)-(+)-10-camphorsulfonic acid   Mandelic acid   Morphine

Scheme 1.5.

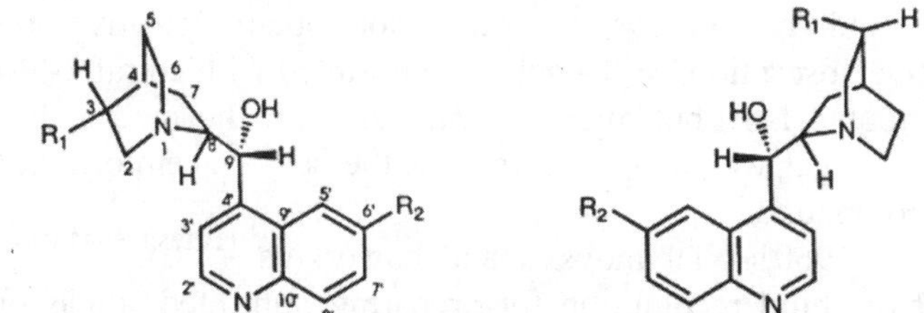

	R_1	R_2	
quinine	C_2H_3	OCH_3	quinidine
cinchonidine (CD)	C_2H_3	H	cinchonine (CN)
dihydroquinone	C_2H_5	OCH_3	dihydroquinidine
dihydrocinchonidine (HCD)	C_2H_5	H	dihydrocinchonine

Scheme 1.6. Structures of cinchona alkaloids.

Another promising development is the preparation of adsorbents which are capable of selective adsorption of enantiomers and which possess the ability to resolve racemates. Along these lines, Curti and Colombo [119,120] prepared a silica gel adsorbent formed in the presence of (1*S*)-(+)-10-camphorsulfonic acid (Scheme 1.5.) and resolved on this adsorbent the same racemic acid with an *ee* of 30%, and also they resolved the racemic mandelic acid with an *ee* of 10%. Beckett and coworkers [121,122] found selective adsorption of cinchona alkaloids and morphine on specific formed silica gels.

For elucidation of results the authors accepted the theory of "Fingerprints". Indeed, silica formed with alkaloids reveals a greater capacity for the same alkaloid in comparison to a control adsorbent without the imprinting molecules, and silica formed with quinine adsorbs more quinine and cinchonidine, more quinidine.

Adsorbate adsorbent	Control	Silica formed with Quinine	Silica formed with Quinidine
Quinine	146	394	196
Quinidine	142	232	220
Cinchonine	167	218	
Cinchonidine	200	190	

Table 1.9. Effect of structure of adsorbate molecule on the stereospecific adsorption on silica gels formed in the presence of quinine and quinidine (values at an equilibrium concentrations $20 \cdot 10^{-5}$ mol/kg adsorbent) (modified data adapted from Beckett et al. [122,123]).

But Maier and coworkers [123] could not obtain positive results for stereoselective adsorption. Silica gel was imprinted with chiral 2-(4-MeOPh)-3,3-dimethylbutan-1-ol, but after its removal from the matrix the resulting imprinted silica gel was unable to resolve the same racemic mixture of the imprinting compound.

In the 1960's Klabunovskii and coworkers [124,125,126,127,128] expanded the concept of chiral recognition for preparing imprinted silicas originating from chiral alkoxysilanes followed by hydrolysis. They used tetra(2-methylbutoxy)silane as a templating agent to induce chirality at the inner and outer surfaces of silica gel. By this method the start of action of forming the specific chiral center more closely resembles the formation of the polymeric net of silica. During the process of formation of the active surface of silica the following factors can effect the results:

1. the chiral structure of the original silane
2. the optically active intermediates formed during partial hydrolysis
3. the optically active alcohol formed in full hydrolysis [126].

A number of optically active menthoxybornyloxy-2-methylbutoxysilanes [125] and polymenthoxyalkyl- and polycarbmenthoxysiloxanes were obtained by hydrolysis of silanes [127]. Polymers incorporating the menthoxy group were thermostabile up to 200 °C.

The hydrolysis of (+)-(2-methylbutyl)trichlorosilane results in an optically active polymer, of which optical rotatory dispersion showed the absence of any contribution of the general polymer chain to the general optical rotation [129]. By hydrolysis of tetra(2-methylbutoxy)silane with full removal of the chiral group a specific silica gel was prepared [128]. The use of a MacBain quartz spring balance was used to measure adsorption isotherms. The adsorption of the racemic- and (+)-2-butanol vapors gave 23.8% excess more for the racemic than for the (+)-isomer, both at saturation and in a monolayer. This increase in adsorption of the racemic 2-butanol in comparison with the (+)-isomer was explained by its smaller molecular volume. Silicas formed in the presence of (+)-tartaric acid, (+)-mandelic acid, or (+)-Na tartrate proved to be less selective. Similar results were obtained in selective adsorption of *sec*-butylamine on specific formed silicas [129].

Patrikeev, Balandin, Klabunovskii and coworkers [130] used silica gel formed in the presence of colonies of the soil bacteria, *Bacillus mycoides*, that revealed "right" and "left" spiral forms. These "chiral" silica forms relate differently towards the optical isomers of some poisons, for example, (+)- and (-)-acrichine. On silica formed in the presence of "right" and "left" forms of *Bacillus mycoides* the adsorption of (+)- and (-)-linalool was measured in vacuum on MacBain quartz spring balance. Silica formed in the presence of the "left" *Bacillus mycoides* colony adsorbed 8 times more (-)-linalool than (+)-linalool [131].

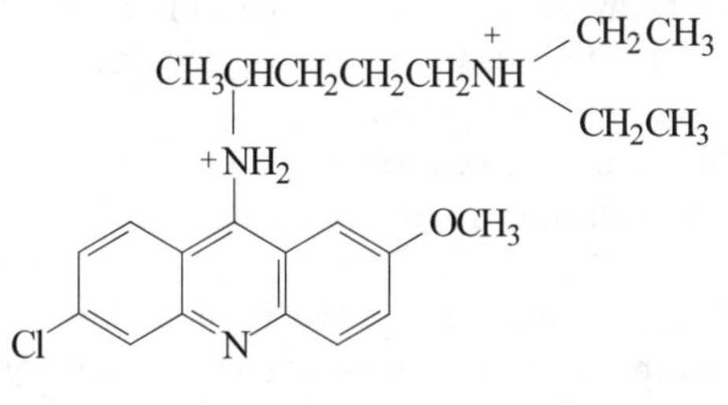

Acrichine

Linalool

Scheme 1.7.

Another concept of chiral molecular imprinting was developed by Wulff and colleagues [131,132,133] who used an organic polymer network as a matrix instead of silica. The main difference from all former methods consists of a template copolymerized and then removed to give a cavity in the polymer with highly specific binding sites rather than some sort of shape, such as a "fingerprint". This area of research is very promising, has grown rapidly [94-97,132], is used in chromatography, and is used for selective adsorption and for selective catalysts [133,134]. Thus, the imprinting in a polymer was made with the chiral complex [Rh-{(*R*)-PhMeCHOH}]. On that polymer the hydrogenation of

PhCOEt yields 1-phenylpropan-1-ol with an *ee* of 8% and 10% conversion [134,135].

References

[1] Bernal J.D. (1951) The Physical Bases of Life, Routlege & Regan Paul, London, p. 34.

[2] Cairns-Smith G. (1982) Genetic takeover and the mineral origin of life, Cambridge, NY.

[3] Grim R.E. (1968) Clay Mineralogy, McGraw Hill, NY. pp. 57-69.

[4] Wellner D. (1979) (Comments) *Sci.*, **206**, 484.

[5] Kahlen W., Wagner H.H. and Holderich W.F. (1998) Zeolite effect in the enantioselective *trans*-hydrogenation over a Co-Salen "Ship-in-the-botttle" complex, *Catal. Lett.*, **54**, 85-89.

[6] Degens E.T., Matheja J. and Jackson T.A. (1970) Template catalysis: asymmetric polymerization of amino acids on clay minerals, *Nat.* **227**, 492 -493.

[7] Bondy S.C. and Harrington M.E. (1979) L-Amino acids and D-glucose bind stereospecifically to a colloidal clay, *Science*, **203**, 1243-1244.

[8] Julg A. (1986) Adsorption sur la kaolinite et homochiralite L des acides amines des proteines, *Compt. rend.,* Ser.2, **303**, 1773-1776.

[9] Julg A. (1987) Synthese asymmetrique sur la kaolinite de l'amino-propionitrile, precurseur de l'alanine et homochiralite des acides amines des proteines, *Compt. rend.*, Ser.2, **305**,563 -565.

[10] Julg A. (1988) Stabilite relative des deux formes inverses optiques de la kaolinite. Application a l'homochiralite L des acides amines des proteines, *Compt. rend.* Ser.2, **306**, 1153 -1156.

[11] Field B.O. and Spencer J.E.D. (1990) The synthesis of amino acids and sugars on an inorganic template from constituents of the prebiotic atmosphere, *Origin of Life and Evolution of Biosphere*, **20**, 233-248.

[12] Gabel N.W. and Ponnamperuma C. (1967) Model for origin of monosaccharides, *Nat.* **216**, 453 -455.

[13] Goldanskii V.I. (1996) Formaldehyde as hypothetical primer of biohomochirality, in *Physical Origin of Homochirality in Life,* Amer. Inst. Phys. Conferences, Santa Monica CA, *Proceedings* 379, 1995 (Cline D.B. (ed.), AIP Press, Woodbury, N.Y. pp. 50-54.

[14] Osinovskii A.G. and Erofeev B.V. (1982) On opportunity of asymmetric condensation of formaldehyde into sacharides, in "*Coordinating Meeting on asymmetric catalysis*", 1982, Sci. Council Acad. Sci. on Catal. Inst. Org. Chem. Acad. Sci. USSR and Dept Chem. & Chem. Technol. Acad. Sci. Georgian SSR, Tbilisi State Univ., Abstracts, Borjomi.Georgia, p. 43-44.

---

[15] Osinovskii A.G. and Erofeev B.V. (1986) Asymmetric synthesis of carbohydrates based on formaldehyde on catalysts modified with L-amino acid esters in " *First Republic Meeting on Asymmetric Reactions*", Batumi, Gruz. SSR (Georgia), Abstracts, p.41.

[16] Jackson T.A. (1971) Evidence of selective adsorption and polymerization of the L-optical isomers of amino acids relative to the D-optical isomers on the edge of facets of kaolinite, *Experientia*, **27**, 242-243.

[17] Kavasmaneck P.R. (1978) Asymmetric adsorption by quartz: a model for the origin of optical activity, *Diss.Thesis*, Stanford Univ., 246 pp., Diss. Abstr. Int. B. 36 (9) 4508-4509; Chem. Abstr. 1976, **84**, 146534.

[18] Bonner W.A. (1979) Experiments on the abiotic origin and amplification of optical activity, Stud.Phys.Theor.Chem. (Origin Opt. Act.) *Nature* **7**, 5-20.

[19] Flores J.J. and Bonner W.A. (1974) Asymmetric polymerization of aspartic acid enantiomers by kaolin, *J. Mol. Evol.*, **3**, 49 -56.

[20] McCullough J.J. and Lemmon R.M. (1974) Question of the possible asymmetric polymerization of aspartic acid on kaolinite, *J. Mol. Evol.* **3**, 57 -67.

[21] Jackson T.A. (1979) Stereospecific sorption of L - amino acids by colloidal clay, *Sci.*, **296**, 483-484.

[22] Thomson T.D (1979) *Clays, Clay Minerals*, **21**, 242.

[23] Bondy S.C. and Harrington M.E. (1979) The preparation high-affinity binding of the biological forms of amino acids and hexose stereoisomers to bentonite, *Stud. Phys. Theor. Chem.* 7, (Origin of Optical Activity in Nature), p.141-149.

[24] a) Friebele E., Shimoyama A., Hare P.E. and Ponnamperuma C. (1981) Adsorption of amino acid enantiomers by Sodium montmorillonite, *Origin Life*, **11**, 173 -184.

b) Friebele E., Shimoyama A., Hare P.E. and Ponnamperuma C (1980) Adsorption of protein and nonprotein amino acids on a clay minerals; a possible role of selection on chemical evolution, *J. Mol. Evol.* **16**, 269-278.

[25] Bonner W.A., Van Dort M.A. and Flores J.J. (1974) Quantitative gas chromatographic analysis of leucine enantiomers. A comparative study, *Anal. Chem.* **46**, 2104-2106.

[26] Navarro-Gonzalez B., Khanna P.K. and Ponnamperuma C. (1993) in Ponnamperuma C. and Cella-Flores J. (Eds.) Chemical Evolólution: Origin of Life Proc. Trieste Conf. Chem. Evolution and Origin of Life, 1992, A Deepak Publ., Hampton, VA.

[27] Youatt J.B. and Brown R.D. (1981) Origin of chirality in Nature. A reassessment of the postilated role of bentonite, *Sci.,* **212**, 1145-1146.

[28] Bonner W.A. (1991) The origin and amplification of biomolecular chirality, *Origin Life, Evol. Biosphere*, **21**, 59-111.

[29] Bonner W.A. (1996) The Quest for Chirality, *Physical Origin of Homochirality in Life,* Amer. Inst. Phys. Conferences, Santa Monica CA, 1995, *Proceedings* 379, Cline D.B. (ed.) AIP Press, Woodbury, N.Y. pp.17-49.

[30] Lemmlein G.G. (1944) The relative number of left- and right-handed quartz crystals, *Coll. Mineralog. Soc.* (russ.) **73**, N 2-3, 94-100.

[31] Lemmlein G.G. (1936) On twisted quartz, *Dokl. Acad. Nauk SSSR* **4**, 279-282.

[32] Hylleraas E. (1927) Equilibrium position of the atom, double refraction and optical rotation power of *beta*-quartz, *Z. Phys.*, **44**, 871.

[33] Cedy W. (1946) Piezoelectricity. An Introduction to the Theory and Application of Electromechanical Phenomena in Crystals, 1st Ed. N.Y., London.

[34] Mason S.F. (1984) Origin of biomolecular handedness, *Nat.*, **311**, 19-23.

[35] Tranter G.E. (1985) Parity-violation energy differences of chiral minerals and the origin of biomolecular homochirality, *Nat.*, **318**, 172 -173.

[36] MacDermott A.J. (1993) The weak force and the origin of life, in *Chemical evolution. Origin of life,* (Ponnamperuma C. and Chela-Flores J. (eds.), Deepak Hampton, pp. 85-99.

[37] MacDermott A.J. (1996) The weak force and SETH: The Search for Extra-Terrestrial Homochirality, in *Physical Origin of Homochirality in Life,* Amer. Inst. Phys. Conf., Santa Monica CA, Proceedings 379, 1995, Cline D.B. (ed.), AIP Press. Woodbury, N.Y. pp. 241-254.

[38] Vistelius A.B. (1950), The frequency of enantiomorphic quartz types, *Zapiski Vsesoyusnogo Mineralog. Obshch. (Collections of All-Union Mineralogical Soc. USSR, russ.*), V.79, p. 191-195.

[39] Frondel C. (1978) Characters of quartz fibers, *Amer. Mineralogist* (*J. Mineral. Soc. Amer.*) **63**, N 1-2,17-27 .

[40] Frondel C. (1962) Silica minerals, in *Dana's System of Mineralogy. Silica minerals*, Wiley & Sons, NY. VII Auflage. Bd. 3. S.16.

[41] Heritsch H. (1952) Die Verteilung von Rechts-und Links Quarzen in Schriftgranaten, *Tschermaks Mineral*, Petrogr. Mitt. **3**, 115-125.

[42] Kozlovski M.I. (1962) Spiral growth and dissolution of crystals, in *Growth of Crystals*, Shubnikov A.V. and Sheftel N.N. (eds.).V.3, Consultants Bureau, N.Y. p.101-106.

[a] Klabunovskii E.I. and Thiemann W. (2000) The role of quartz in the origin of optical activity on Earth, *Origins Life Evol. Biosphere* **30**, 431-434.

[b] Klabunovskii E.I.(2001) Can enantiomorphic crystals like quartz play role in the origin of homochirality on Earth, *Astrobiology* **1**, 127-131.

[c] Klabunovskii E.I. (2002) The role of enantiomorphic in the origin of chirality on Earth and other planets, In *Second European Workshop on Exo/Astrobiology*, Graz, Austria, Abstr., p. 9.

[44] Gross G. (1972) Zerstoerungsfreie Statisch-morphologische Untersuchung eines Quartzfundes, *Schweiz. Mineral. Petrogr. Mitt.* **52**, 523-535.

[45] Rein D. (1992) *Die wunderbare Haendigkeit der Molekuele vom Ursprung der Leben aus der Asymmetrie der Natur*, Birkhuser Verlag, Basel, Boston, Berlin.

[46] Thompson M.R. (1937) How to tell right-handed and left-handed quartz crystals, *Rocks and Minerals* N 12.

[47] Holzapfel L. (1951) Silizium Verbindungen in biologischen Systemen. Organische Kieselsaeure - Verbindungen. XX Mitt., *Z. Electrochem.* **55**, 577.

[48] Seifert H. (1955) Ueber orientierende Abscheidungen von Aminosaeure auf Quarz, *Naturwiss.* **42**, 13.

[49] Kleber W. and Meyer K (1957) Asymmetrische effecte bei der Epitaxie von d-Campher auf Biotit und Calcit, *Naturwiss.* **44,** 373 -374.

[50] Henderson G.M. and Rule H.G. (1939) A new method of resolving a racemic compounds. *J. Chem. Soc.,* 1568 -1573.

[51] [a]) Klabunovskii E.I. and Patrikeev V.V. (1951) On mechanism of asymmetrizing action of metal catalysts deposited on right and left quartz, *Dokl. Acad. Sci. USSR.* **78,** N3, 485 –487, *Chem. Abstr.* 1951, **45**, 7860a.
[b]) Harada K. (1977) The recent studies on the origin of optical activity. *Kagaku no Ryoki*, **31**, 1-11.

[52] Ohara M., Fujita I. and Kwan T. (1962) The selective adsorption of optical antipodes as revealed by the chromatographic technics, *Bull. Chem. Soc. Jpn*, **35**, 2049 -2051.

[53] Tsuchida R., Kobayashi M. and Nakamura A. (1935) The configuration of chlorobisdi-methylglyoximoammine-Cobalt, *Bull. Chem. Soc. Jpn.* **11**, 38-40.

[54] Nakamura A. and Tsuchida R. (1954) Synthesis of the tris-(dimethylglyoximo)-cobaltate (III). *J. Amer. Chem. Soc.* **76**, 3103.

[55] Karagunis G. and Coumoulos G. (1938) A new method of resolving a racemic compounds, *Nat.* 142,162 -163.

[56] Ikeda Y. (1954) Partial resolution of racemic hydroperoxydes by asymmetric adsorption on rock quartz, *Bull. Liberal Arts*, Coll. Wakayama Univ. (Nat. Sci.) **4**, 27-28; *Chem. Abstr.* **49**, 9583 (1955).

[57] Kuebler J.R. and Bailar J.C. (1952) The stereoisomerism of complex inorganic compounds. XIV. Studies upon the stereochemistry of saturated tervalent nitrogen compounds, *J. Amer. Chem. Soc.,* **74**, 3535 -3538.

[58] Bush D.M. and Bailar J.C. (1953) Stereochemistry of complex inorganic compounds. XVII. The stereochemistry of hexadentate ethylene-diaminetetraacetic acid complexes, *J. Amer. Chem. Soc.* **76**, 4574-4575.

[59] Das Sarma B. and Bailar J.C. (1955) The stereochemistry of metal chelates with polydentate ligands P.1, *J. Amer. Chem. Soc.* **77**, 5476-5480.

[60] Schweizer G.K. and Talbott C.K (1950) Resolution of optically active inorganic stereoisomers by adsorption on quartz, *J. Tennessi Acad. Sci.* **25**, 143-147 (1950); *Chem. Abstr.* **46**, 11004.

[61] Kamai G.Kh., Klabunovskii E.I., Gatilov Yu.F. and Khodakov G.S. (1961) Resolution of tertiary arsenic compounds into optical isomers using asymmetric adsorption on natural dissymmetric adsorbents, *Dokl. Acad. Sci. USSR*, **139**, 1112-1113, *Chem. Abstr.* 1962, **56**, 41f.

[62] Chernyaev I.I., Korablina L.S. and Muraveiskaya G.S. (1965) On optical resolution and photochemical isomerization of asymmetric *cis*-diammines, *Zh. neorg. khim*. (J. Inorg. Chem. (russ.), **10**, 1045-1052.

[63] Moeller T. and Gulyas E. (1958), The partial resolution of certain inner complexes by means of a chromatographic technique, *J. Inorg. Nucl. Chem*., **5**, 245 -248.

[64] Sievers R.E., Moshier R.W. and Morris M.L. (1962) Resolution of Chromium (III) -hexafluoroacetylacetonate by gas chromatography, *Inorg. Chem.* **1**, 960-967.

[65] Bush D.H. and Bailar J.C. (1954) The optical stability of Berillium complexes, *J. Amer. Chem. Soc.* **76**, 5352-5353.

[66] McLaren A.D. and Estermann E.F. (1954) The adsorption and reactions of enzymes and proteins on kaolinite, *Acta Biochem. Biophys.* **61**, 158-173.

[67] Schwab G.M. and Wahl B. (1956) Beitraege zur Struktur der organischen Kieselsaeure, *Naturwiss.* **43**, 513 -514.

[68] Amariglio-Simon A. (1967) Theses, Etude critique des reactions asymetriques sur quartz et des conditions de stereospecificite dans les interactions moleculaires, Univ. Nancy.

[69] Amariglio A., Amariglio H., and Duval X. (1968) La synthese asymetrique, *Ann. chim*. **3**, 5-25.

[70] Amariglio A., Amariglio H. and Duval X. (1968) Essais de reactions dissymetrique sur quartz optiquement actif, *Helv. Chim. Acta*. **51**, 2110-2132.

[71] Amariglio A. and Amariglio H. (1971) in *Chemical Evolution and Origin of Life*, Buvet E.R. and Ponnamperuma C. (eds.), North-Holland Publ. Amsterdam, p. 63 .

[72] Bonner W.A., Kavasmaneck P.R. and Martin F.S. (1974) Asymmetric adsorption of alanine by quartz, *Sci*. **186,** 143-144.

[73] Bonner W.A., Kavasmaneck P.R., Martin F.S. and Flores J. (1975) *Origin Life*, **6**, 367.

[74] Bonner W.A. and Kavasmaneck P.R. (1976), Asymmetric adsorption of DL-alanine hydrochloride by quartz., *J. Org. Chem*., **41**, 2225-2226.

[75] Kavasmaneck P.R. and Bonner W.A. (1977) Adsorption of amino acid derivatives by d- and l-quartz, *J. Amer. Chem. Soc*., **99**, 44-50.

[76] Furoyama S., Sawada M., Hashiya K. and Morimoto T. (1982) Asymmetric adsorption of alanine by quartz powder from ethanol solution, *Bull. Chem. Soc. Jpn.* **55**, 3394-3397.

[77] Kawashiro K., Morimoto S., Yoshida H. and Sugiura K. (1977) The synthesis and chromatography of peptide nitriles, *Origin Life*, **8**, 347-353.

[78] Gillard R.D. and da Luz de Jesus J.D.P. (1979) Studies on the origin of asymmetry. Part I. Optically active crystals as selective adsorbents, *J. Chem. Soc., Dalton Trans.* 1779 -1782.

[79] Chandrasekhar S.(1953) The optical rotatory dispersion of cinnabar, *Proc. Indian Acad. Sci.,* **37A**, 697-703.

[80] Chandrasekhar S. (1954) The rotatory dispersion of benzil, *Proc. Indian Acad. Sci.*, **39A**, 243-253.

[81] Chandrasekhar S. (1953) A redetermination of the rotatory dispersion of Sodium chlorate, *Proc. Indian Acad. Sci.,* **37A**, 458- 67.

[82] Ferroni E. and Cini R. (1960) The resolution of complex antipodes by optically active solids, *J. Amer. Chem. Soc.,* **82**, 2427-2428.

[83] Kondepudi D.K., Kaufman R.J. and Singh N. (1990) Chiral symmetry breaking in sodium chlorate crystallization, *Sci.* **250**, 975-976.

[84] Bonner W.A., Blaire N.E. and Dirbas F.M. (1981) Experiments on the abiotic amplification of optical activity, *Origin Life*, **11**, 119-134; 1975, **6**, 187-194.

[85] a) see also "Generation and Amplification of Chirality in Chemical Systems, Thiemann W. (ed.). *Proc. Second International Symposium,* Univ. Bremen, July 16 ,1980, Origin of Life, 1981, V.**11**, 194 pp., D.Reidel Publ.Co., Dordrecht.

b) Bonner W.A (1979) in *Origin of Optical Activity in Nature*, Stud. Phys. & Theoret. Chem., N 7, Walker D.C. (ed.), Chem. Dept. Univ. British Columbia, Canada, Elsevier Sci. Publ. Co.. Amsterdam, Oxford, NY, pp. 5-20.

[86] Polyakov M.V. (1931) Adsorption properties and structure of silica gel, *Zh. Fiz. Khim.* (J. Phys. Chem., russ.) **2**, 799-805.

[87] Vysotskii Z.Z. and Polyakov M.V. (1956) The preparation of specific adsorbents, *Zhurn. Fiz. Khim.* (J. Phys. Chem., russ.) **30**, 1901-1902.

[88] Dickey F.H. (1949) The preparation of specific adsorbents, *Proc. Natl. Acad. Sci. USA*, **35**, 227-229.

[89] Dickey F.H. (1955) Specific adsorption, *J. Phys. Chem.* 59, 695-707.

[90] Pauling L. and Campbell D.H. (1942) The production of antibodies in vitro, *Science*, **95**, N 2469, 440-441.

[91] Pauling L. (1940) A theory of the structure and process of formation of antibodies, *J. Amer. Chem. Soc.,* **62**, 2643-2657.

[92] Morrison J.L., Worsley M., Shaw D.R. and Hodgson G.W. (1959) The nature of the specificity of adsorption of alkyl orange dyes on silica gels, *Can. J. Chem.* **37**, 1986-1995.

[93] Kempe M., Fischer L. and Mosbach K. (1993) Chiral separation using molecularly imprinted heteroaromatic polymers, *J. Mol. Recogn.* **6**, 25-29.

[94] Kempe M. and Mosbach K.(1994) Chiral recognition of *N-alpha*-protected amino acids and derivatives in non-covalently imprinted polymers, *Int. J. Peptide Protein Research*, **44**, 603-606.

[95] Kempe M. and Mosbach K. (1994) Direct resolution of naproxen on a non-covalently molecularly imprinted chiral stationary phase, *J. Chromatogr.* **664**, 276-279.

[96] Ramstrom O., Yu C. and Mosbach K. (1996) Chiral recognition in adrenergic receptor binding mimics prepared by molecularly imprinting polymers, *J. Mol. Recogn.* **9**, 691-696.

[97] Bartels H., Prijs B. and Erlenmeyer H. (1966) Ueber spezifisch adsorbierende Silicagele, *Helv. Chim. Acta* **49**, 1621-1625.

[98] Erlenmeyer H. and Bartels H.(1964) Ueber das Problem der Aehnlichkeit in der Chemie. Ueber spezifisch adsorbierende Silicagele. II., *Helv. Chim. Acta,* **47**, 1285-1288.

[99] Erlenmeyer H. and Bartels H. (1965) Ueber spezifisch adsorbierende Silicagele. IV., *Helv. Chim. Acta* **48,** 301-303.

[100] Haldeman R.G. and Emmett P.H. (1955) Specific adsorption of alkyl orange dyes on silica gel, *J. Phys. Chem.*, **59**, 1039-1043.

[101] Waksmundzki A.J., Oscik J., Rozylo J. and Matusewicz J. (1962) Structure and adsorption properties of specific adsorbents, *Folia Soc. Sci. Lublin* **2**, 149-154.

[102] Waksmundzki A., Wawrzynowicz T. and Wolski T. (1962) Adsorption properties of silica gels precipiated in the presence of some alkaloids, *Ann. Univ. Mariae Curie-Sklodowska, Liblin-Polonia.* Sect. AA, 17, 27-38.

[103] Bernhard S.A. (1952) The preparation of specific adsorbents, *J. Amer. Chem. Soc.,* **74,** 4946-4947.

[104] Patrikeev V.V., Kozarenko T.D. and Balandin A.A. (1962) On specific polycondensation of amino acid esters, *Izv. AN SSSR. Otdel. Khim. Nauk.* pp. 170-171.

[105] Patrikeev V.V., Smirnova Z.S. and Maksimova G.I. (1962) Some biological properties of specifically formed silica, *Dokl. Acad. Nauk SSSR* **146**, 707-709.

[106] Balandin A.A., Patrikeev V.V., Sholin A.F., Bychenkova L.Ya. and Taber A.M. (1966) The giving of stereospecificity to hydrogenation catalysts by using specific formed carrier-silica gel, *Dokl. Acad. Sci. USSR.*, **169**, 1362-1365.

[107] Sholin A.F., Patrikeev V.V. and Balandin A.A. (1967) Selective adsorbing silica gels, the possibility of their applications in the catalysis. Nature of

their specificity, *Proc. Int. Symp. on Mechanism of Heterogeneous Catalysts and Properties of Catalysts*, Sofia, Abstracts, Publ. Bulgar Acad. Sci., 1967, p. 72.

[108] Jencks W.P., *Catalysis in Chemistry and Enzymology*, McGraw Hill, N.Y., 1969.

[109] Tramontano A., Janda K.O. and Lerner R.A. (1986) Catalytic antibodies, *Science*, **234**, 1566-1570.

[110] Ekberg B. and Mosbach K. (1989) Molecular imprinting: a technique for producing specific separation materials, *Trends Biotechnol.* **7**, 92-96.

[111] Wulff G. (1988) Selective binding to polymers via covalent bonds, The construction of chiral cavities as specific receptor sites, *Pure Appl. Chem.*, **54**, 2093-2102.

[112] Matsuishi T., Shimada T. and Morihara K. (1992) Definite evidence for enantioselective catalysis over "Molecular Footprint" catalytic cavities chirally imprinted on a Silica (alumina), *Chem. Lett.*, 1921-1924.

[113] a) Morihara K., Kurihara S. and Suzuki J. (1988) Footprint Catalysis. I. A new method for designing "Taylor-Made" catalysts with substrate-specificity: silica (alumina) catalysts for butanolysis of benzoic anhydride, *Bull. Chem. Soc. Jpn.* **61**, 3991-3998.

b) Morihara K., Nishihata E., Kojima M. and Miyake S. (1988) Footprint Catalysis. II., Molecular recognition of footprint catalytic sites, *Bull. Chem. Soc. Jpn.* **61**, 3999-4003.

[114] Morihara K., Takiguchi M. and Shimada T. (1994) Footprint catalysis. XI., Molecular footprint cavities with chiral amines and their chiral molecular recognition, *Bull. Chem. Soc. Jpn.* **67**, 1078-1084.

[115] Morihara K., Kurokawa M., Kamata Y. and Shimada T. (1992) Enzyme-like enantioselective catalysis over chiral 'Molecular Footprint' cavities on a silica (alumina) gel surface, *J. Chem. Soc., Chem. Comm.*, 358-360.

[116] Davis M.E., Katz A. and Ahmad W.R. (1996) Rational catalysts design via imprinted nanostructured materials, *Chem. Mater.*, **8**, 1820-1839.

[117] Kaiser G.G. and Anderson J.T. (1992) Sorbents for liquid chromatography based on the footprint principle, *Fresenius J. Analyt. Chem.*, **342**, 834-838.

[118] Heilman J. and Maier W.F. (1994) Selective catalysis on silicon dioxide with substarte-specific cavities, *Angew. Chem., Int. Ed. Engl.,* **33**, 471-473.

[119] Curti R. and Colombo U. (1951) Separatione degli antipodi ottici dell'acido camfosolfonico mediante cromatografia su assorbenti specifici, *Chim. Ind.* **23**, 103.

[120] Curti R., Colombo U. and Clerici F (1952) Cromatografia con adsorbenti specifici, *Gazz. Chim. Ital.* **82**, 491-502.

[121] Beckett A.H. and Anderson P. (1959) "Footprints" in adsorbents, *J. Pharm. Pharmacol. Suppl.* **11**, 258 T- 260 T.

[122] Beckett A.H. and Youssef H.Z. (1963) Active sites in stereoselective adsorbents as models of drug receptors and enzyme active sites, *J. Pharm. Pharmacol. Suppl.* **15**, 253 T-266 T.

[123] [a)] Hunnis, M., Rufinska A. and Maier W. F. (1999) Selective surface adsorption versus imprinting in amorphous microporous silicas, *Micropor. Mesopor. Mater.* **29**, 389-403.

[b)] Maier W.F. (1998) Amorphous microporous mixed oxides, new selective catalysts with chemo- and shape-selective properties, *Prep. Symp. Amer. Chem. Soc., Div. Fuel Chem.* **43**, 534-537.

[124] Klabunovskii E.I., Volkova L.M. and Agronomov A.E. (1961) On new method of preparation of stereospecific silicas, *Izv. Akad Nauk. SSSR. Otdel. Khim. Nauk*, p. 2101, *Chem. Abstr.* 1968, **56**, 9928a.

[125] Klabunovskii E.I. and Volkova L.M. (1967) On optically active alkyloxy- and alkylchloro-silanes as precursors for preparation of stereospecific silicas and organosilicium polymers, *Izv. Akad. Nauk. SSSR. Ser. Khim.* pp. 863-865, *Chem. Abstr.*1967, **67**, 43354u.

[126] Andrianov K.A., Mamedov A.A., Volkova L.M. and Klabunovskii E.I. (1969) Optically active menthoxy-, menthoxymethylethoxy(menthoxy)-silane, *Izv. Akad. Nauk. Ser. Khim.* pp 2305-2307, *Chem. Abstr.* 1970, **72**, 43888z.

[127] Klabunovskii E.I., Agronomov A.E., Volkova L.M. and Balandin A.A. (1963) On adsorption of racemic and (+)-isomer of *sec*-butanol on specific silicas, *Izv. Akad. nauk. SSSR. Otdel. Khim. Nauk,* pp. 228-234, *Chem. Abstr.* 1963, **58**,13170h.

[128] Klabunovskii E.I., Godunova L.F. and Balandin A.A. (1966) On optically active organosilicium polymers, *Vysokomol. soed.* (russ.) **8**, 441-443, *Chem. Abstr.* 1966, **64**, 19815f.

[129] Sholin A.F., Patrikeev V.V., Agronomov A.E. and A.A. Balandin (1968) Adsorption of d- and *dl-sec*-butylamine and benzene on titanogels, *Kolloid. Zh.* (*Colloid J.*, russ.) **30**, 450-454.

[130] Patrikeev V.V., Balandin A.A., Klabunovskii E.I., Mardashev Yu.S. and Maksimova G.I. (1960) Selectivity of action in relation of optical isomers of adsorbents formed in the presence of bacteria, *Dokl. Akad. Nauk SSSR.*, **132**, 850-852, *Chem. Abstr.*1961, **55,** 5649c.

[131] Wulff G. and Sarhan A. (1972) Ueber die Anwendung von enzymanalog gebauten Polymeren zur Racemattrennung, *Angew. Chem.* **84**, 364.

[132] Wulf G., Lauer M. and Disse B. (1979) Ueber enzymanalog gebaute Polymere X. Ueber die Synthese von Monomeren zur Einfuerung von Aminogruppen in Polymere in definiertem Abstand, *Chem. Ber.,* **112**, 2854-2865.

[133] Wulff G., Gross T. and Schoenfeld R. (1997) Enzyme models based on molecularly imprinted polymers with strong esterase activity, *Angew. Chem.* (Engl.) **36**, 1962-1964.

---

134 a) Locatelli P., Gamez P. and Lemaire M. (1998) Molecular imprinting of polymerized catalytic complexes in asymmetric catalysis, *J. Mol. Catal.*, 135, 89-98.

b) Locatelli P., Gamez P. and Lemaire M. (1998) Molecular imprinting. Polymerised catalytic complexes in asymmetric catalysis, *Stud. Surf. Sci. Catal.* **108**, 517-522.

135 Gamez P.B., Dunjic B., Pinel C. and Lemaire M. (1995) "Molecular imprinting effect" in the synthesis of immobilized Rhodium complex catalyst, *Tetrahedron Lett.* **36**, 8779-8782.

Chapter 2

HETEROGENEOUS HYDROGENATION CATALYSTS BASED ON QUARTZ

Abstract

This chapter summarizes data about the application of chiral metal catalysts supported on optically active quartz crystals in hydrogenation and other reactions. Despite the low enantioselective efficiency of these catalysts, recent result show that almost 100% enantioselectivity results when they are involved in autocatalytic processes.

2.1. General

Adsorption of molecules on a catalytic surface is fundamental to any heterogeneous catalytic process. Therefore, it is necessary to distinguish between cases in which the catalyst itself is chiral, e.g. contains on its surface chiral complexes directing asymmetric reactions, and cases in which the catalyst surface is achiral but the catalyst is supported on a chiral carrier. In both cases the mechanism of action seems to be the same and reactions proceed through intermediate chiral complexes, including complexes on the surface of the catalyst, or through chiral fragments on the chiral carrier. It might be expected that both cases lead to asymmetric catalysis.

There were several attempts to prepare asymmetric catalysts without a chiral carrier. It was shown [1] that irradiation of thin films of AgCl by right or left circularly polarized light (R-CPL and L-CPL) resulted in their ability to rotate a plane of linear polarized light. It is, of course, not a true optical activation of the material, because the grains of remaining AgCl do not reveal any chiral structure. Indeed, under the action of polarized light the definite orientation of grains of AgCl occurs as a result of the so called Weigert effect. However, repetition of these experiments met many difficulties in reproducing the results [2,3]. But Ghosh [4] succeeded in the preparation of layers of 30 nm thick AgCl by irradiating a thin film of AgCl with R-CPL or L-CPL. These thin films of AgCl produced an observed optical rotation of -0.06°, or a specific rotation about -10000°. Prepared in such a way optically active films can be covered with Pt or Pd and after removing the AgCl (and Ag) by treatment with nitric acid the resulting thin films of Pt or Pd also exhibit optical activity with an observed rotation of -0.10°.

These films were suggested to be used as chiral catalysts for absolute asymmetric synthesis. But further communications in this field were absent; thus it is assumed that such experiments failed. Similar catalytic systems are of interest because in this manner it will be possible to turn common catalysts without chiral carrier into chiral catalysts like the "Bredig ferment" [5,6]. When the Et_2N group is introduced into the cellulose of cotton fibres, they are transformed into active catalyst which can reproduce the asymmetric activity of carboxylase, the enzyme for the asymmetric decomposition of enantiomers of camphorcarboxylic acid, and the activity of oxynitrilase for catalyzing the synthesis of (-)-mandelonitrile. A catalyst of this type was obtained in 1932 by Bredig and Gerstner [6]. These authors regarded this example as the first case of heterogeneous asymmetric catalysts. Indeed, the mechanism of the chiral action of a catalyst by this sort depends on asymmetric adsorption on the optically active cellulose carrier. Such asymmetric adsorption on natural materials is well documented by the numerous reports of stereoselective adsorption of enantiomers and of chromatographic techniques resolving racemates of amino acids on cellulose columns. Thus, Yuasa [7] succeeded in resolving without derivatization a number of racemic amino acids on native cellulose columns.

A similar mechanism of action was found in the case of a Pd catalyst supported on silk fibroin [8,9] in the asymmetric hydrogenation of C=N bonds in prochiral compounds. The mechanism of action consists of the formation of the chiral complexes on the silk fibres of the Pd-fibroin which act as chiral catalysts. This principle was later developed for a number chirally modified catalysts that are very effective in the asymmetric hydrogenation of *beta*-keto esters [9,10].

2.2 Asymmetric hydrogenation on metal-quartz catalysts

The asymmetric adsorption of some organic and complex compounds on quartz crystals was described in Chapter 1.2. In 1932 Schwab and coworkers were first to show that chiral quartz crystals can be used as chiral supports for metal catalysts [11-14]. Seven years later Stankiewicz, in his dissertation [15], confirmed these results, and later Terentyev and Klabunovskii extended the scope of asymmetric reactions using "Metal-Quartz Catalysts" [16-24].

Schwab and Rudolph [11] prepared chiral catalysts by supporting the metals on the surfaces of ferreted fine powdered optically active quartz crystals, which proved to be active during asymmetric dehydrogenation and dehydration of racemic butan-2-ol. The dehydration-dehydrogenation reactions of butan-2-ol (Scheme 2.1.) were carried out in the vapor phase at

400-500°C and atmospheric pressure with catalysts composed of Cu, Ni, or Pt deposited in thin layers on *d*- or *l*-quartz.

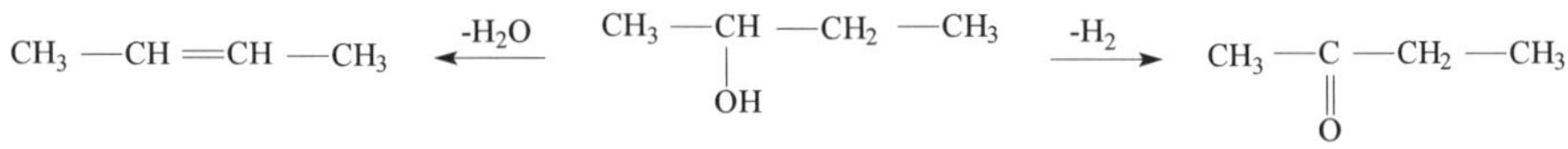

Scheme 2.1. Dehydration and dehydrogenation of butan-2-ol.

The reactions resulted in very small preferential decompositions of one of the enantiomers of racemic butan-2-ol. The remaining unreacted butan-2-ols exhibited small optical rotations. Catalysts were prepared by immersion of fine powdered crystals of Dophinee quartz into solution of salts of metals in concentrations needed for formation of monoatomic layers of metals on the surfaces of the quartz crystals. These impregnated crystals were ignited, the salts decomposed, and the resulting catalysts reduced in flowing hydrogen. The catalysts were placed in a circulating catalytic reactor system in which racemic butan-2-ol vapors were entrained in nitrogen or air and passed over the catalyst bed. The resulting liquid reaction products were investigated polarimetrically. Optical rotations were extremely small and did not exceed 0.2-0.3°. Polarimetric measurements of liquid products are considered valid if rotations exceed ± 0.03°. Likely, as a result of the asymmetric action of these chiral catalysts, the decomposition of the enantiomers of racemic butan-2-ol proceeded with different rates and resulted in unreacted butan-2-ol with an optical rotation of approximately 0.10°. For a Ni-quartz catalyst a maximal optical rotation of +0.13° was observed for the product. If that product was passed again through a new bed of the same catalyst, the optical rotation could be increased up to +0.20°, as shown on Figure 2.1.

Catalysts based on *l*-quartz predomonantly decomposed the (+)-enantiomer of butan-2-ol, so the product mixture had a (-)-optical rotation. Considering the mechanism of action of that chiral catalyst the authors [12] concluded that a maximal effect may be attained at small coverings of metal (close to a monoatomic layer) on the chiral surface of the optically active quartz. Increasing the content of metal increased the rate of reaction but decreased the asymmetric effect and as well as the resulting optical rotation of the product (Figure 2.2.).

At high reaction temperatures (400-500°C) the rate of racemization of the optically active butan-2-ol becomes fast enough that the optical rotation of the product goes through a maximum as a function of reaction time. Results received with a 0.10% Ni-*d*-quartz catalyst at 550-560°C are shown below in Table 2.1.

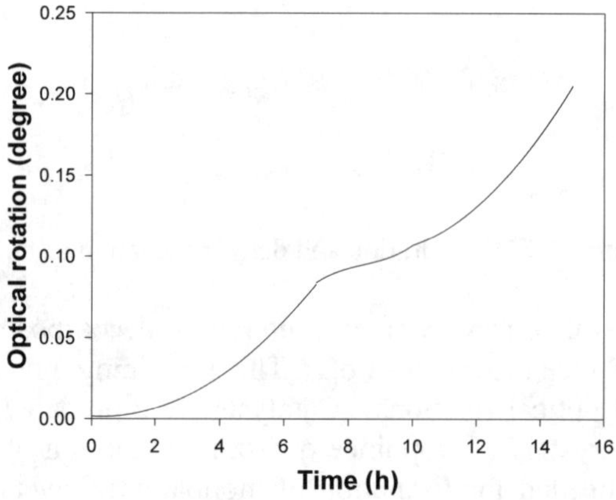

Figure 2.1. The increase in optical rotation during the decomposition of racemic butan-2-ol over Cu-*d*-quartz catalyst at 400°C, after addition of a fresh portion of the catalyst at 8h (Schwab et al. [11,12]).

Figure 2.2. Dependence of optical rotation of butan-2-ol on reaction time and on extent of surface coverage by Pt on the *d*-quartz.
(1) 1 atomic layer of Pt, (2) 1.5 atomic layers of Pt, (3) 2 atomic layers of Pt (adapted from Schwab et al. [12] and Klabunovskii [32]).

Reaction time (h)	Conversion (%)	α (°)	*ee* (%)
5.2	3.6	+ 0.016	0.07
8.0	5.8	+ 0.073	0.32
Reaction mixture was passed through a new bed of the catalyst.			
5.2	3.0	+ 0.03	0.15
9.2	5.4	+ 0.12	0.53
13.7	7.8	+ 0.05	0.21
18.7	11.0	+ 0.01	0.03

Table 2.1. Asymmetric decomposition of butan-2-ol over a 0.10% Ni-*d*-quartz catalyst at 550-560^{0}C; length of tube, 20 cm (adapted modified data from Schwab et al. [12]).

In 1938 Stankiewicz [15] reproduced the asymmetric 2-butanol decomposition experiments of Schwab (1932) at a higher temperature and atmospheric pressure and extended studies to the asymmetric decomposition of racemic menthol and 3-methylheptan-3-ol. Over a Cu-*d*-quartz catalyst the latter substrate produced a maximal optical rotation of -0.26^0 and the decomposition of 2-butanol gave a reaction mixture with a rotation of +0.25^0.

For the preparation of catalysts the crystals of *d*-quartz from the Idar deposit were used (quartz crystals from the mineral-collection of Crantz (Bonn) did not give good results). Crystals were sawed into plates of 12-15mm thick perpendicular to the optical axes and plates were carefully studied in a polariscope for absence of the Brasilian twinning area. An analogous procedure for obtaining *l*-quartz was applied. Crystals were then crushed in a mortar to 2-4 mm particles and washed with king's-water. Specific areas were estimated at 8 cm^2/g. Catalysts were prepared by two methods:

1) by high vacuum cathodic scattering of Pt on quartz powder with periodic shaking to obtain uniform layers
2) by immersion of quartz powder into water solutions of salts of Cu and Ni.

After calcinations and reductions in flowing hydrogen at 450^0 the resulting catalysts were used in the vapor phase decomposition of alcohols. It was calculated that the Cu-catalysts received in such a way contained 19 atomic layers and the Ni catalyst 160-240 atomic layers.

Carrying out vapor phase experiments in N_2 or N_2/air and attaining a polarimetric accuracy of $\pm 0.01^0$, Stankiewicz [15] found that such catalysts were more effective than those containing monoatomic layers, which contradicted the data of Schwab (1934).

For the case of the asymmetric decomposition of racemic butan-2-ol, the asymmetric effect depends on the conditions of the reaction and results in either dehydration or dehydrogenation (Scheme 2.1.). The asymmetric dehydrogenation of butan-2-ol on Cu-*d*-quartz in N_2-air mixture as flowing transport gas results in levorotatory products (Table 2.2.).

Temperature of reaction (°C)	α_D (°)
280	- 0.08
320	- 0.15
330	- 0.16
340	- 0.21
390	- 0.08

Table 2.2. Temperature dependence of the optical rotation of the product mixture in the asymmetric dehydrogenation of butan-2-ol over a Cu-*d*-quartz catalyst in transport flow (nitrogen + air) (modified adapted data from Stankiewicz [15]).

In flowing dinitrogen, however, mainly the dehydration of butan-2-ol occurred and the remaining 2-butanol was dextrorotatory (Table 2.3.).

Table 2.3. shows that 500-505°C is the optimal temperature range at which the (-)-enantiomer of butan-2-ol reacts somewhat faster than the (+)-enantiomer, and under this condition of dehydration the orientation of a molecule of butan-2-ol on the surface of the catalyst changes in accord with data of Schwab (1934).

In flowing CO_2 at 500°C mainly dehydrogenation proceeds to the (-)-product with a rotation of -0.13°. But these results disagree with the data of Schwab (1934), which can be explained by too fast of a rate (9 mmol/min).

Temperature of reaction (°C)	Conversion (%)	Rate (mmol/min)	α_D (°)
460	10.6	0.89	+ 0.07
470	17.6	1.18	+ 0.05
500	26.8	1.0	+ 0.10
505	22.6	0.50	+ 0.14
510	25.0	0.70	+ 0.06

Table 2.3. Temperature dependence of asymmetric dehydration of butan-2-ol over Cu-*d*-quartz catalyst in dinitrogen transport gas (modified adapted data from Stankiewicz [15]).

On a Cu-*l*-quartz catalyst in flowing N_2 dehydration gives a product with rotation up to -0.25° (Table 2.4).

Temperature of reaction (°C)	Conversion (%)	Rate (mmol/min)	α_D (°)
500	5.4	1.6	- 0.07
515	5.4	0.8	- 0.09
525	5.4	0.8	- 0.11
530	3.9		- 0.25
545			- 0.04

Table 2.4. Temperature dependence of asymmetric dehydration of butan-2-ol over Cu-*l*-quartz catalyst in transport gas flowing N_2 (modified adapted data from Stankiewicz [15].)

Decomposition studies of racemic menthol and 3-methylheptan-3-ol gave similar results (Scheme 2.2.). Thus dehydration of racemic menthol over Cu-*l*-quartz in a flowing N_2-air mixture at 430-460°C results in a product with a rotation of -0.16°, which indicated preferential decomposition of the (+)-enantiomer of menthol. The same picture was obtained, but to a somewhat greater extent, for the dehydrogenation of 3-methylheptan-3-ol on Cu-*l*-quartz in which the (+)-enantiomer reacted and the product exhibited a rotation of -0.12°.

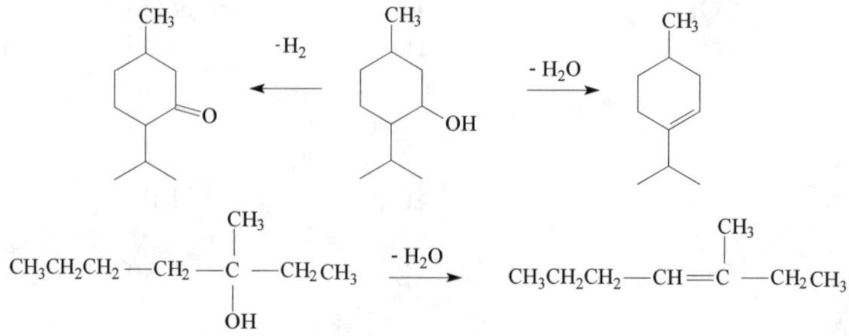

Scheme 2.2. Dehydration and dehydrogenation of menthol and dehydration of 3-methylheptan-3-ol.

Dehydration of racemic menthol on Ni-*l*-quartz at 560°C gave products containing 62.4% menthenes and a rotation of -0.15°. This catalyst was strongly carbonized during reaction and lost its activity, but after oxidation and reduction with hydrogen it gave a product with a rotation of -0.08^0.

Increasing the amount of Ni on quartz somewhat diminished the rotation from -0.15 to -0.13°. On Pt-quartz catalysts 2-butanol produced a product with -0.14° at 353°C. Increasing the reaction temperature resulted in larger amounts of butenes and lower rotations.

In the dehydration of 3-methylheptan-3-ol at 170°C Pt-*l*-quartz proved to be the most effective catalyst with an observed rotation reaching -0.26°. Thus it was shown that Pt-quartz catalysts give particularly large effects, presumably owing to uniform deposits of Pt-particles on the surfaces of the quartz. Table 2.5. below summarizes the main results received by Stankiewicz [15].

In the fifties Terentyev and Klabunovskii [16-24] described details of the asymmetric decomposition of racemates and asymmetric synthesis using catalysts supported on optically active quartz crystals. Asymmetric resolution of racemic 2-butanol was extended by a number of new reactions of absolute asymmetric syntheses: hydrogenation of 2-phenylcinnamic acid and cyano-ethylation of cyclohexanone. Reactions were studied in vapor and liquid phases at ambient and elevated temperatures. Catalysts consisting of the metals Cu, Ni, Pt, Pd, andAg, with NaOH and aluminum alkoxides supported on Dophinee quartz or piezoquartz were used. Asymmetric decomposition of butan-2-ol was carried out in the liquid phase in flowing or circulating systems in the vapor phase at atmospheric pressure. Optical rotations of products were measured on a Schmidt & Haentsch polarimeter with an accuracy of 0.01°.

Alcohol	Catalyst	Optimal reaction temperature (°C)	Transport gas	α_D*	Correlation**
Butan-2-ol					
	Cu-*d*-quartz	340	$N_2 + O_2$	- 0.21	*d* and +
	Cu-*d*-quartz	505	N_2	+ 0.14	*d* and -
	Cu-*d*-quartz	500	CO_2	- 0.13	*d* and +
	Cu-*l*-quartz	530	N_2	- 0.25	*l* and +
	Pt-*l*-quartz	355	N_2	- 0.14	*l* and +
3-Methylheptan-3-ol					
	Cu-*d*-quartz	160-210	N_2	+ 0.10	*d* and -
	Cu-*l*-quartz	275	$N_2 + O_2$	- 0.12	*l* and +
	Pt-*l*-quartz	170	$N_2 + O_2$	- 0.26	*l* and +
Menthol					
	Cu-*l*-quartz	450	$N_2 + O_2$	- 0.16	*l* and +
	Ni-*l*-quartz	560	$N_2 + O_2$	- 0.15	*l* and +
	Pt-*l*-quartz	460	$N_2 + O_2$	- 0.15	*l* and +

Table 2.5. Asymmetric Dehydrogenation-Dehydration of Racemic Alcohols over Metal-Quartz Catalysts (summarized adapted data from Stankiewicz[15]).

* Optical rotation of the mixture after the reaction

** Correlation between the quartz used (*d* or *l*) and the sign of the optical rotation.

Crystals of natural quartz obtained from the Institute of Crystallography of the Academy of Sciences of the USSR in Moscow were sawed into plates 10-15 mm thick perpendicular to the crystals' optical axes and were selected by examination in a polariscope. The areas of Brasilian twinnings were removed and *d*- and *l*-Quartz crystals in piezoquartz plates were identified by methods of Lemmlein (1939) (see Chapter 1.3.1.). The selected *d*- and *l*-quartz crystals were carefully crushed in a mortar under water to avoid twinning and amorphisation occurring as in the cases of strong action on the crystals, e.g. crushing in ball-mill or vibro-mill (see Khodakov and Rebinder [25]). Then the powder was purified in hot king's water, washed, and sieved. The fraction of powder with a specific area of 180 cm^2/g was used for preparation of the catalysts. Metals were deposited on the quartz by two methods: by precipitation from salts, drying, ignition, reduction in flowing hydrogen; and by cathodic scattering in high vacuum with periodically shaking of the quartz sample to achieve uniform surface covering by the metal. In some cases, when reproducibility of selectivity was low owing to sintering or carbonization and segregation of the metal particles on the surface of the carrier, the catalysts were repeatedly crushed in a mortar and used again. By this operation the border zones between the metal and quartz were renewed and reproducibility was recovered. The extent of surface coverage of the catalysts composed of Cu on quartz could be divided into groups with broad distribution of metal layers: 0.4-73 atomic layers for the "Precipitation" catalysts and close to a monoatomic layer, 0.4-0.8 atomic layers for the "cathodic scattering" catalysts. Results received with these two groups of catalysts are shown on Table 2.6. [23].

It has been shown that enantioselectivity is increased at low conversions particularly with the catalysts using cathodic scattering of metals (0.4-0.8 atomic layers on the surface of quartz). The temperature dependence of asymmetric decomposition of butan-2-ol revealed two maxima of optical rotation of products: at temperatures between 320 - 400° butan-2-ol mainly dehydrogenates and at higher temperatures, above 400°, dehydration takes place. Therefore, in the reaction on Cu-*d*-quartz catalysts two maxima of optical rotation were found: at 340°C $\alpha = 0.21°$ and at 530°C $\alpha = 0.25°$ [15].

Cu content (%)	n	Temperature (°C)	Conversion (%)	Rate (%/min x 10^2)	$[\alpha]_D$*	*ee** (%)
0.0075	0.78	320	1.2	3	+ 0.08	0.36
		370	1.9	3	+ 0.16	0.34
		423	7.6	12	+ 0.09	0.41
		459	14.5	30	+ 0.18	0.81
0.003	0.38					
		400	4.4	4	+ 0.14	0.63
		420	10.3	4	+ 0.10	0.45
		458	13.6	14	+ 0.16	0.72

Table 2.6. Asymmetric decomposition of racemic butan-2-ol on Cu-*d*-quartz catalysts with low metal contents.
n = Cu atomic layers on quartz; * = values of the unreacted butan-2-ol (adapted modified data from Terent'ev et al. [23]).

It was found that enantioselectivity of the catalyst diminished with reaction time due to poisoning of metal-quartz border zones on the surface. This caused the optical rotations of products to pass through maxima. The time of reaching maximal rotation depends on the extent of coverage of the surface of the quartz with metal. To obtain good results reaction time must be restricted to 30-50 min and the metal content must not exceed 0.001%. Increasing the Cu content on the surface of quartz accelerated the rate of the decomposition of butan-2-ol, but simultaneously increased the rate of its racemisation too. In summary, Table 2.7. gives results obtained from the asymmetric dehydrogenation-dehydration of racemic 2-butanol with various metal-quartz catalysts.

Type of quartz	Metal and mode of preaparation	Reaction temperature (°C)	α_D of products
l-quartz			
	Cu*	360	- 0.12
	Ag**	350	- 0.08
	Ni*	65	- 0.12
	Pt** #	280	+ 0.08
d-quartz			
	Cu*	360	+ 0.16
	Cu**	460	+ 0.18
	Ag**	470	+ 0.10
	Pd***	360	+ 0.06
	Ni*	388	+ 0.11

Table 2.7. Optical rotation after asymmetric dehydration and dehydrogenation of racemic mixtures of 2-butanol over metal-quartz catalysts (length of tube: 20 cm).
* nitrate solution and reduction at 300°C;
** cathodic scattering;
*** chloride solution;
\# reaction in flowing N_2+O_2 (modified adapted data from Terent'ev et al. [16]).

It was found that Cu-*d*-quartz catalysts are more effective than the others in the asymmetric decomposition of 2-butanol and that the sign of optical rotation of the product after the dehydrogenation reaction always corresponded to the sign of rotation of the quartz. Ag-quartz catalysts prepared by cathodic scattering contained 0.03-0.08% Ag and gave an α_D of +0.106° on Ag-*d*-quartz and -0.07° on Ag-*l*-quartz at the optimal temperature of 370°C. Two maximal values of optical rotation were observed on Ni-quartz catalysts, too: -0.012° on Ni-*l*-quartz and +0.16° on Ni-*d*-quartz.

Cu-, Ag-, and Pt-quartz catalysts were used for the asymmetric isomerization of the racemic mixture of methyloxirane (Scheme 2.3.).

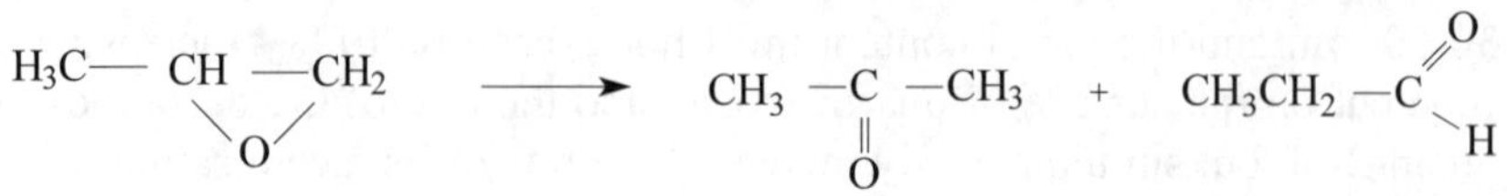

Scheme 2.3. Isomerization of methyloxirane.

Maximal rotations of the partially isomerized products at 180°C on Cu-*l*-quartz catalyst was -0.055° and on Ag-*l*-quartz catalyst was -0.052°. Another example of asymmetric decomposition is the asymmetric hydrogenation of racemic 2-methylcyclohexanone (Scheme 2.4.).

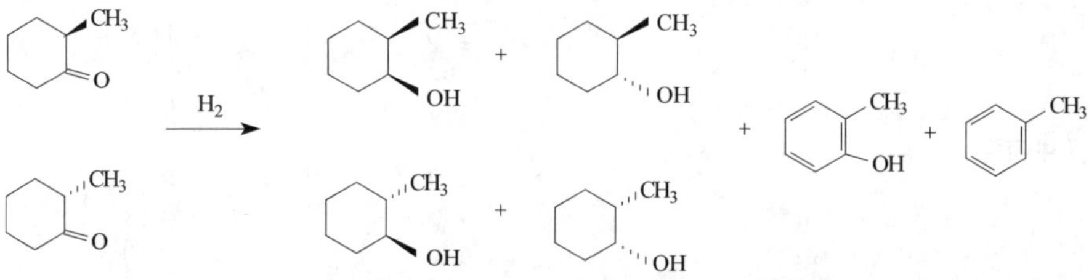

Scheme 2.4. Kinetic resolution and diastereomeric hydrogenation of 2-methylcyclo-hexanone (adapted from Terent'ev, Klabunovskii et al. [19]).

After reaction at 244°C the rotations of the reaction products were found to be -0.042° on Pt-*l*-quartz and +0.039° on Pd-*d*-quartz. Optical activity in the product could arise as a result of both kinetic resolution of the racemic starting material and creation of a new chiral center in the 2-methylcyclohexanol by enantioselective hydrogenation of the C=O group. The same transformation was realized in the liquid phase hydrogenation of racemic *alpha*-pinene in decalin solution (Scheme 2.5.).

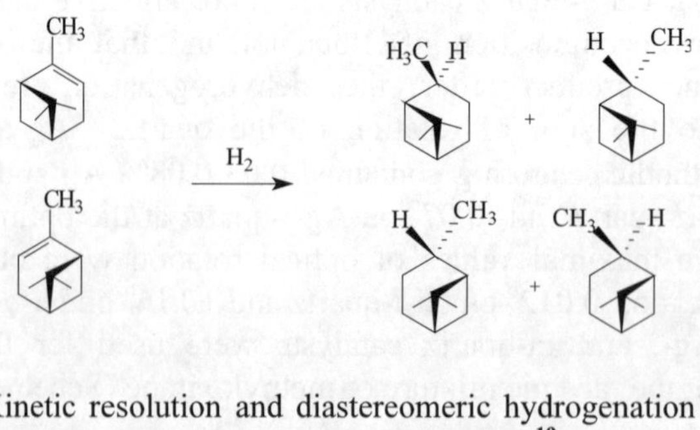

Scheme 2.5. Kinetic resolution and diastereomeric hydrogenation of *alpha*-pinene into pinane (adapted from Terent'ev, Klabunovskii et al. [19]).

Rotation was +0.046° on Ni-*l*-quartz and -0.046° on Ni-*d*-quartz. Besides the asymmetric resolution of racemates, metal-quartz catalysts were used to accomplish pure asymmetric syntheses. The first attempt was realized by Schwab (1934) in the hydrogenation of tiglic acid [(*E*)-2-methylbut-2-enoic acid] over Ni-quartz catalysts.

At low surface coverages of Ni on quartz no reaction occurred, probably as a result of poisoning of the small amount of metal. At high Ni coverages hydrogenation took place but at 200°C where no asymmetric effects would be expected.

Terent'yev and Klabunovskii [19,23] hydrogenated the ethyl-(2-phenylcinnamate) (Scheme 2.6.) in decalin solution at 135°C on a Ni-*d*-piezo-quartz catalyst and received a product with α_D = -0.09° ± 0.007°. On Ni-*l*-quartz the product had an optical rotation of +0.04°.

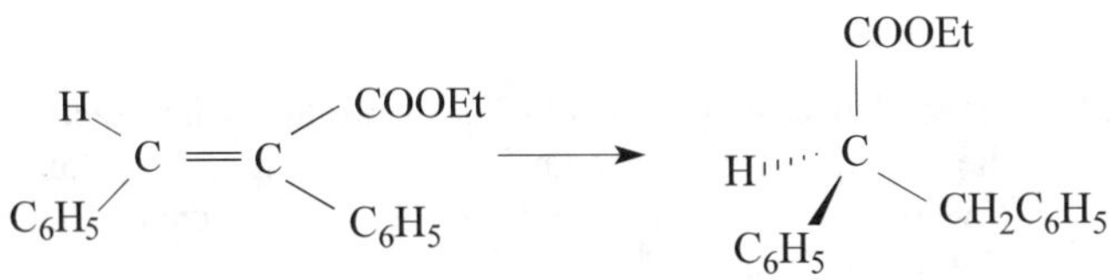

Scheme 2.6. Hydrogenation of ethyl 2-phenylcinnamic acid.

A more complicated case occurred in the addition of acrylonitrile to 2-methyl-cyclohexanone (Scheme 2.7.). This cyanoethylation reaction was studied at ambient conditions on a new chiral catalyst: NaOH on *d*- or *l*-quartz [19].

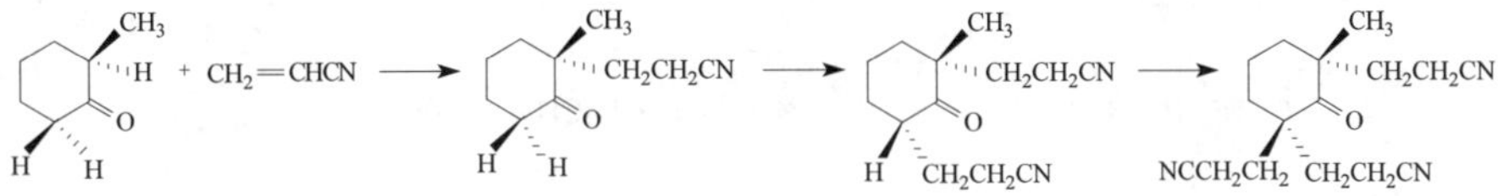

Scheme 2.7. Asymmetric cyanoethylation of 2-methylcyclohexanone.

The original ketone exists as a racemic mixture. During the reaction a new chiral center appears in the intermediate, but it disappears in the final product. In accordance with the reaction path in Scheme 2.7., the optical rotation proceeds through a maximum at α_{589} = 0.157° as Fig. 2.3. shows.

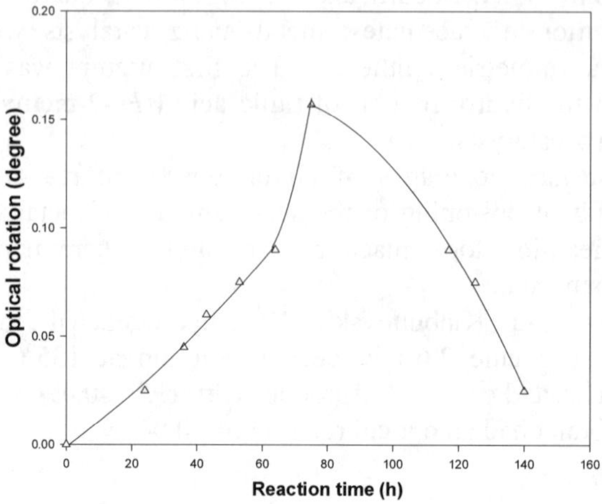

Figure 2.3. Dependence of optical rotation of product with reaction time during the asymmetric cyanoethylation of 2-methylcyclohexanone with acrylonitrile on a NaOH-quartz catalyst (adapted from Terent'ev, Klabunovskii et al. [20]).

In 1953, Ponomarev and Zelenkova [26] applied the Ni-quartz catalysts for the asymmetric hydrogenation of furan derivatives (120 °C, 135-150 bar, liquid phase) in order to elucidate the mechanism of reaction. It was well known that in the hydrogenation of 4-(2-furyl)butan-2-ol, **1**, 4-(tetrahydrofuran-2-yl)butan-2-ol, **2**, and 2-methyl-1,6-dioxaspiro[4.4]nonane, **3**, were formed (Scheme 2.8.).

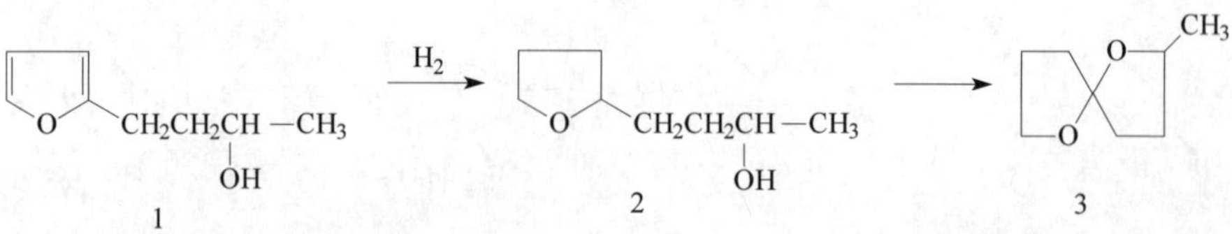

Scheme 2.8. Asymmetric hydrogenation of 4-(2-furyl)butan-2-ol.

In this reaction the spirane can be formed but its stereostructure was unknown. The spirane molecule can be chiral only in that case when the two five-membered rings are perpendicular to each other. Optical rotation of the spirane received in this reaction confirmed the structure of the supposed spirane [26]. Thus, over Ni-*l*-quartz catalyst, spirane **3** with α_{589} = -0.066°, while over Ni-*d*-quartz catalyst, spirane **3** with $\alpha_{589} = +0.06^{\circ}$ and **2** with $\alpha_{589} = +0.04^{\circ}$ were obtained. Results showed that the spirane can exist in two enantiomeric forms.

Owing to the small specific area of quartz powder as carriers, in metal-quartz catalysts and their low ability for adsorption, these catalysts in general proved to be of interest only in connection with the theory of the origin of chirality in Nature. Moreover as was noted above, the use of quartz as a carrier demands a very careful special selection procedure for preparation of optically pure samples. Therefore, it is not surprising that several attempts to realize asymmetric catalysis with the use of quartz crystals were not always successful. For example, Tsuchida et al. [27] in 1936 attempted to repeat the results of works of Schwab (1932) but failed. Likewise, Amariglio [28-31] in 1967 attempted to reinvestigate studies on adsorption and metal-catalysts based on quartz crystals but also failed. The hydrogenation of butan-2-one and the asymmetric dehydrogenation-dehydration of racemic butan-2-ol were studied on Ni- and Pt-quartz catalysts. Catalysts were prepared by immersion of quartz powder in solutions of Ni or Pt salts, followed by liofilic drying, calcination and reduction in flowing hydrogen. Decompositions of butan-2-ol was carried out in flowing systems at atmospheric pressure. Products were condensed and examined polarimetrically with an accuracy of 0.01°. From the decomposition of racemic butan-2-ol on 0.03-0.04% Ni-*l*-quartz and 0.16-0.30% Pt-*l*-quartz catalysts at 200-300°C optical rotations of products did not exceed 0.001°.

Hydrogenations of butan-2-one were carried out at 140-150°C. The optical rotation of the product was observed to be 0.10°, which fell over time to 0.02°. Other generally disappointing irregularities also occurred. Independent of the sign of the rotation of quartz, all products were levorotatory, and the optical activity disappeared in several hours. Quartz crystals without metal layers also catalyzed the dehydration of butan-2-ol at 350° and gave rotations of -0.075° and -0.085°, which after filtration of the products fell to -0.005°. Thus the products obtained with the Pt-*l*-quartz catalyst at 307°C had an initial rotation of -0.045° that fell to zero in 3 hrs.

Based on their negative results in attempts to repeat the works of Schwab (1934) and Klabunovskii (1950) and others authors Amariglio et al. [28-31] criticized these works and claimed that the results were artifacts. Also using gas chromatography with quartz powder packing they could not resolve racemic propylene oxide, butan-2-ol, 1-phenylethanol, and other compounds[31].

But careful examination of the published works [28-31] leads to the conclusion [32] that their criticism is flawed and their own experiments are insufficient as follows:

1) Quartz plates were purchased from Centre National d'Etude des Telecommunications and as indicated by Amariglio-Simon (1967) made of synthetic quartz rather then of natural quartz crystals. In their publications [34,35] they did not indicate the method of identification and selection of *d*- and *l*-quartz crystals.

2) Samples of quartz crystals were crushed into micron size particles using vibro-set with W-carbid. They did not indicate the condition of the pounding of the crystals although it is well known that crushing of the optically active quartz particularly with vibro-mil can cause twinnings formation, amorphisation of quartz, and its hydrophilisation [25].

3) Moreover, the amorphisation of quartz powder was confirmed by Amariglio-Simon (1967) because treatment with HF of the crushed quartz crystal powders with specific areas of 2 m^2/g diminished this value to 0.2 m^2/g.

4) The method of purification of the quartz powder after that treatment also was not indicated [28] and the catalysts of high catalytic activity contained rather thick layers of metal on the surface of the quartz (extent of coverage was 15-26%) [28-31], which exceeded significantly the characteristics of the previous catalysts used [11-24].

5) Polarimetric measurements were carried out on a polarimeter with very large half-shadow angle (5°) that could not give sufficient accuracy in the estimation of the optical rotation of products.

6) Products of dehydrogenation-dehydration of butan-2-ol were examined in the polarimeter without distillation, after only filtration.

7) All products gave only levorotatory compounds independently of the optical sign of the quartz used. Thus Pt-*l*-quartz catalyst, based on quartz washed with HF, gave in dehydrogenation of butan-2-ol rotation of catalysate with optical rotations of -0.085° and -0.09° that went to zero in several hours after sedimentation of the suspension of quartz. In another experiment a rotation of -0.058°was obtained, but that disappeared in 4-5 hrs. The same (-)-rotation in solution was received even from the suspension of crushed quartz or pyrex glass and silica.

All these facts placed into question the experiments and general criticism of Amariglio [28-31]. Moreover it should be noted that at 306°C during conversion of α-quartz to β-quartz the twinning of quartz proceeds [33]; therefore catalysis by quartz crystals is not recommended at high temperatures.

Unfortunately, explanations [32] of the problems associated with the results of Amariglio et al.[28-31] have not been incorporated into considerations of quartz as a chiral adsorbent [34].

2.3 General comments on adsorption and catalysis using quartz crystals

Published results on asymmetric catalysis and asymmetric adsorption using quartz crystals have a number of problems, the main one being a very small effectiveness of the catalysts or adsorbents in producing high optical rotations in products: observed angles of rotation of products did not exceeded 0.2°. Optical rotations were measured by polarimeters at one wavelength, 589 nm. At that time, 40-60 years ago, other methods of examining enantiomers (enantiomeric chromatography, spectropolarimetry, radiochemical analysis), were not yet available. The structures of the catalysts (surface of metal in metal-quartz catalysts, the state and dispersion of the metal, and others) were not employed. In a number of experiments, catalysts contained increasing amounts of metal on the surface of quartz, up to 100-200 atomic layers, that likely excluded the action of the chiral quartz surface.

Nevertheless, despites these experimental problems, consistent regularities were observed; for example, in the decomposition of butan-2-ol the sign of optical rotation of the product, caused by unreacted butanol partially enriched with one enantiomer, corresponds to the sign of rotation of the quartz. That result means that the other enantiomer of butanol was preferentially decomposed on the catalyst. However, this correlation can change depending on the nature of the catalyzing metal and on the character of the reaction: dehydration or dehydrogenation. Thus, with Cu-*d*-quartz a (+)-rotation was observed, but with Pd-*d*-quartz a (-)-rotation resulted. Moreover, the direction of reaction, dehydrogenation or dehydration, can change the correlation. And in some experiments the rotations did not give the "mirror picture" of order values of rotations; for example, on Pt-*d*-quartz the rotation of the reaction mixture was +0.13° but on Pt-*l*-quartz it was only 0.032°, and on Ni-*d*-quartz α_D = 0.048° but on Ni-*l*-quartz only 0.02-0.03°. Such results may have been caused by a lack of good uniformity in the quartz samples as well as differences in specific areas of the carrier.

Another complicating factor is that the surface of quartz is the only asymmetric factor in the metal-quartz catalysis and its specific areas in most cases were very small (only 44 cm^2/g [11-14]). This coupled with the fact that the amount of metal deposited on quartz was rather high, so the extent of racemization of butanol during reaction would be high, which detracts from the effectiveness of the catalyst. Thus, quartz appeared not to be an effective chiral carrier for catalysis or adsorption in asymmetric experiments. Nevertheless, in general the data using quartz crystals are of interest and received positive evaluations in several publications [34,35].

2.4 On the mechanism of "Quartz-Catalysis"

For the mechanism of asymmetric catalysis on metal-quartz catalysts there is no unifying point of view. There are the concepts of "contact activation", "specific adsorption", and "formation of specific chemical bonds" [12-14]. It has been proposed that on the surface of the catalyst intermediate surface diastereomers are formed which are decomposed at different rates. The data on the action of quartz-catalysts have been rather thoroughly discussed [36-42]; in addition positive estimations of data were given by Bonner [34] and Harada [35].

Schwab, Rost, and Rudolph [12] in 1934 proposed that the critical stage in asymmetric catalysis on metal-quartz catalysts is contact activation of the substrate adsorbed on metal-aggregates, which transfers asymmetry from the carrier-quartz to the reacting molecule of the substrate, rather than adsorption on the asymmetric adsorbent-quartz crystal. Schwab and Rost [13] in 1941 considered the catalyst to be an inorganic model of an enzyme in which the working group is represented by both the catalyzing metal and the asymmetrizing carrier-quartz. It was noted that "...only the borders between the metal and the free quartz surface are responsible for asymmetric action...".

Indeed, during the asymmetric decomposition of racemic butan-2-ol on Cu-*d*-quartz catalyst, the change of the sign of optical rotation of the product from (-) to (+) during dehydrogenation indicated no adsorption of the substrate on the asymmetric surface of quartz; however, “contact activation” of butan-2-ol takes place and asymmetry of the quartz spread from the border-zones of the Cu-quartz where the asymmetric reaction was carried out. This point of view also agreed with the fact that a significant asymmetric effect occurred in the cases of thin, close to monoatomic, layers of metals on quartz surfaces, which allowed for development of extensive networks of metal-quartz border-zones.

Such a point of view was first put forward by Bredig and Gerstner [6,36]. They studied the asymmetric induction action of a catalyst composed of an amine immobilized on fibers of native cellulose (Et_2NH-Cellulose) and found that this catalyst is active in the asymmetric formation of (-)-mandelonitrile from benzaldehyde and HCN. After saponification of the product, the (+)-mandelic acid was obtained with an ee of 22%. Considering the mechanism of this catalytic transformation, the authors [36] suggested that the optically active cellulose transfers its asymmetry to the substrate through the catalytically active amine. Kuhn [39,40,42] considering the action of "Bredig's catalysts" noted that these catalysts are very similar to enzymes. Adsorption spectra of these catalysts indicated that the amino group became part of the dissymmetric center as evidenced by the reaction results. Therefore, according to Kuhn’s experiments, it is impossible to distinguish the asymmetric and catalytic functions in such catalysts.

The mechanism of conversion of an achiral molecule into a chiral one was found by Kuhn in circular dichroism spectra. The type of natural quartz used, amethyst crystals, contained colloidal inclusions of Fe. The circular dichroism spectra revealed bonds involving Fe. It would be of importance from that point of view if the films of metals removed from surfaces of quartz crystal would show circular dichroism spectra, too.

The experimental data indicates that metal-quartz catalysts are more selective if they contain a thin, near monoatomic layer of metal on their surfaces. In accordance with this requirement the complicated net of border zones between metal and quartz must be created. Only on these borders can effective asymmetric reactions take place. Guided by this point of view Klabunovskii and Patrikeev [22,32] considered the role of quartz as an asymmetric adsorbent that resolved the enantiomers and preferentially accumulated one of them on the border zones between metal and quartz where the reaction proceeds. In the case of the reaction of racemic butan-2-ol on Cu-*l*-quartz catalyst, (*S*)-(+)-butan-2-ol adsorbed and decomposed preferentially resulting in (-)-optical activity for the product mixture (Scheme 2.9.).

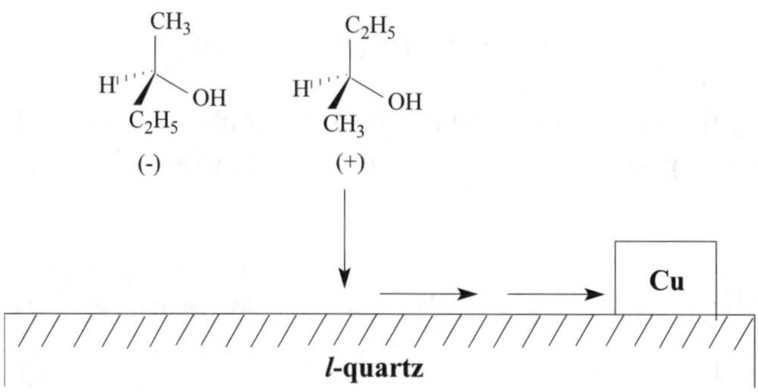

Scheme 2.9. Asymmetric adsorption and decomposition of the racemic mixture of (*R*)-(-)- and (*S*)-(+)-butan-2-ol over Cu-*l*-quartz catalyst (adapted from Klabunovskii et al. [22,32]).

2.5. Evaluation of absolute configuration of molecules using data from "Quartz-catalysis"

Several authors have attempted to use data from quartz catalysis to determine the absolute configuration of molecules. Although the question of determination of absolute configuration of organic compounds or complexes can be considered as solved owing to the use of spectral structural analysis methods, interest continues in the search of other ways of determination of absolute configuration. This search is stimulated by the opportunities to examine changes of configurations of molecules during reactions and thus to determine the final spatial mechanism of the reactions [32]. As suggested earlier, a method was used for calculation of the configuration of the simplest organic compounds, but it proved to be insufficient. This was demonstrated by calculating the optical rotation of (*R*)-butan-2-ol (Scheme 2.10.) using the method given by Boys [37] and the theory of optical activity by Kirkwood [38]. This calculation gave erroneously a (+)-rotation for this compound.

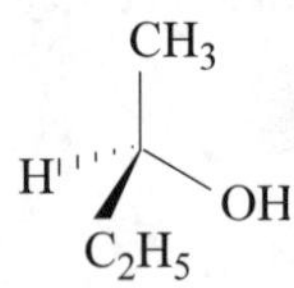

Scheme 2.10. (*R*)-butan-2-ol.

In contrast, the calculation method given by Kuhn [39,40] resulted in the (+)-butan-2-ol being genetically connected with the (*S*)-(+)-lactic acid (Scheme 2.11.).

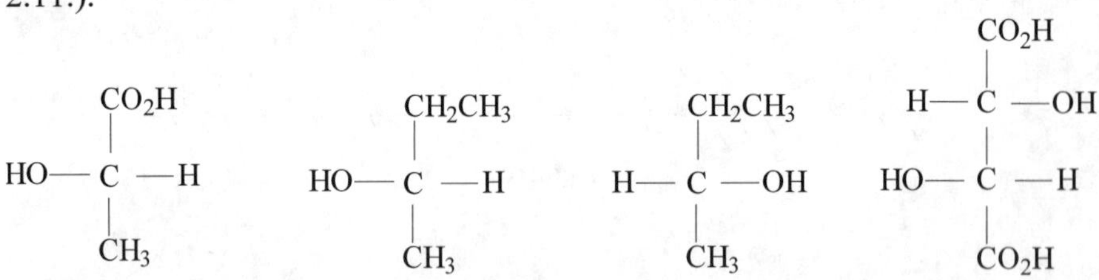

(*S*)-(+)-lactic acid (*S*)-(+)-butan-2-ol (*R*)-(-)-butan-2-ol (2*R*,3*R*)-tartaric acid

Scheme 2.11.

In a critical examination of methods for determining absolute configuration based on calculations Volkenshtein [41] came to conclusions based on the better grounded calculations of Kuhn [42] and the one-electron model of Condon,

Altar, and Eyring [43], whose method of calculation allows one to derive configuration from a genetic spatial D-series of organic compounds originating from D-glyceraldehyde.

In 1884 Fischer noted that it is necessary to distinguish the spatial configuration of a molecule and its optical rotation. Thus, both (+)- and (-)-rotating amino acids and other compounds belong to the L-configurational series. As the original structure for the D-series Fischer suggested the structure of D-glucose. To this D-series belongs the (2*R*,3*R*)-(+)-tartaric acid, the absolute configuration of which was later determined by X-ray diffraction [44-46] and confirmed the earlier predictions of Fischer.

As (2*R*,3*R*)-(+)-tartaric acid can be genetically connected by chemical interactions with the D-series of organic compounds originating from D-glyceraldehyde, also (+)-tartaric acid can be considered a more reliable structure than D-glyceraldehyde. It turned out that the spatial formula adopted arbitrarily by Fischer proved to be the correct absolute configuration. On the other hand, Waser [47] ascribed the opposite configuration for (+)-tartaric acid, but that was not confirmed.

As found by Schwab et al. [11-14], metal-*d*-quartz catalysts preferentially adsorb and decompose (*R*)-(-)-butan-2-ol and the remaining unreacted butanol has the (+)-optical rotation. The structure of quartz consists of endless networks of SiO_2 units forming spatial right- or left-handed helices [38]. Using data on adsorption and catalysis on quartz crystals the absolute configuration of butan-2-ol can be determined:

i) if the correlation of the spatial helix structure of quartz to its sign of optical rotation is known

ii) if it is assumed that the left handed-helix structure of butan-2-ol makes better contact with the "right-helix" structure of quartz. The (+)-*beta*-quartz has the "right-helix" spatial structure [48].

Condon, Altar, and Eyring [43] came to the conclusion that the absolute configuration of (-)-butan-2-ol has a “right-helix" structure if the dehydration of butanol over Cu-quartz catalysts results mainly in (*E*)-but-2-ene and has a "left-helix" if dehydration results in but-1-ene. This conclusion followed from the data of Schwab (1934) and is based on the idea that the chiral structure of the surface of quartz complements the reacting molecule of butan-2-ol.

Results received in the studies of “quartz-catalysis” allow one to suggest that the mechanism of action of quartz catalysts is a result of asymmetric adsorption on the surface of optically active quartz. The effectiveness of asymmetric catalysis is determined by the ability of the catalytic system to form the labile diastereomeric complex intermediate, which will be determined by the extent of spatial correspondence of the configuration of the reacting molecule to the structure of the surface of the catalyst (or adsorbent); in this regard, the left-spiral structure of the surface of the catalyst (adsorbent) will correspond to the right-spiral model of the reacting (forming) molecule. But this conclusion is preliminary because the absolute configuration was unknown; the correspondence of sign of optical rotation of the quartz to the spatial arrangement of helix crystal lattice was unknown. The authors [43] attempted to use the results of calculations of Hilleraas [48] who assigned to (+)-rotating quartz the right-spiral structure. Besides the mechanism was unknown for the decomposition of butan-2-ol on the surface of metal-quartz. Schwab (1934) noted that in the dehydration of butan-2-ol over Cu-quartz catalyst the butene is formed but the stereochemistry of the products was unknown. Having in mind the preliminary character of their conclusions about the configuration of the butan-2-ol in Scheme 2.9., the (-)-rotation was attached in accord with the theory of Kuhn [39,40,42]. Such a conclusion was valid for a model of an intermediate state of butanol on the surface of the catalyst on which the reaction results in formation of but-1-ene. This can be realized only by orientation of butan-2-ol as a left-spiral model (the arrangement of groups at the chiral center corresponds to a left-spiral). However, in 1958 De Vries [49] by X-ray diffraction showed that (+)-rotating *alpha*-quartz has a left-spiral configuration rather than right-spiral and the (-)-rotating quartz belongs to the space group P3,2 and has a right-spiral lattice. These relationships are quite opposite to those accepted earlier [43] and based on calculations of Hilleraas [48]. Therefore the concepts of Condon, Altar, and Eyring [43] become very doubtful. However, from the model of a reacting molecule of butan-2-ol on the surface of a Cu-quartz catalyst it follows that if dehydration of butan-2-ol preferentially forms but-1-ene then a molecule of butan-2-ol must react in the left-spiral conformation, and if (*E*)-but-2-ene is formed, then it must react in the right-spiral conformation. The works of Schwab (1934), Terent'yev, and Klabunovskii (1953) did not give any data on structure of butenes formed in the reaction over Cu-catalyst at high temperature.

To attempt to solve this question Balandin, Klabunovskii et al. [50,51] have shown that the dehydration of butan-2-ol directly forms a considerable portion of (*E*)-but-2-ene on the catalyst surface rather than forming it as a result of the isomerisation of but-1-ene or (*Z*)-but-2-ene.

Labeling but-1-ene with ^{14}C and utilizing GLC-analysis allowed for the determination [51] that quite a definite part of (*E*)-but-2-ene is formed directly from the catalytic dehydration of butan-2-ol, rather than through but-1-ene. Thus, as in the case of the dehydration of butan-2-ol, (*E*)-but-2-ene was formed preferentially, so butan-2-ol in its surface reacting state adopts the right-spiral conformation (having (-)-rotation) and, therefore, is more easily adsorbed on the (+)-rotating quartz which has the left-spiral structure. Indeed, (+)-rotating quartz adsorbs (and then decomposes) preferentially the (-)-butan-2-ol. According to De Vries [49] (+)-quartz has left-spiral structure and hence (-)-butan-2-ol has (*R*)-configuration, in accordance with the calculations of Kuhn [39,40,42] and of Volkenshtein [41]. Thus, taking into consideration the results of the works of Klabunovskii and Balandin et al. [50,51], the uncertainty of determining the absolute configuration of butan-2-ol using the method of "Quartz-catalysis" was removed. These data confirm the selection of the absolute configuration of D- and L-glyceraldehydes and related compounds and demonstrate the important role of phenomena of complementary structures in forming the chiral spiral structure.

It is of interest that a positive optical rotation resulted from passing linear polarized microwave irradiation of wave lengths 3-20 nm through randomly oriented right-handed helices made from "chiral" models consisting either of right spirals of copper wire or of models made of similar chiral tetrahedral [52].

In 1953 Frank [53] proposed an autocatalysis model in which the molecule of the product can catalyze its own formation. For a long time this theoretical model had no experimental confirmation until the works of Soai's group [54-59], who were able to realize this idea in an asymmetric alkylation reaction: addition of diisopropyl-zinc to 2-alkinylpyrimidine-5-carbaldehyde leading to the corresponding alkanol ((*R*)-1-(2-*tert*-butylethinyl-5-pyrimidyl)-2-methylpropan-1-ol) (Scheme 2.12.) with an enantiomeric excess of 51%; in contrast, in the presence of an asymmetric inductor, L-Valine, only of 1% enantiomeric purity was realized.

Recently it has been shown that optically active quartz crystals as asymmetric inductors become very effective in autocatalytic enantioselective reactions. Soai et al.[54-59] have shown that in asymmetric autocatalysis, the action of small amounts of chiral reaction products (involved in the reaction cycle) may enhance the enantioselective excess by a factor of 94 after introduction of an intermediate into the reaction. Optically active synthetic quartz crystals were used in this reaction with ratios of 1:1.9 quartz to aldehyde and 1:2.2 quartz to diisopropyl-zinc.

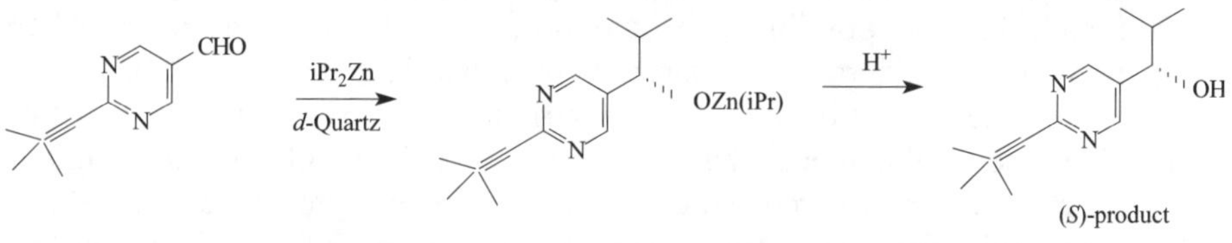

Scheme 2.12. Quartz promoted asymmetric synthesis with iPr_2Zn (Soai et al. [57]).

The most important results were the following. Using *d*-quartz as an asymmetric inductor gave the alkanol with an (*S*)-configuration and 93%

optical purity, while *l*-quartz produced the alkanol with a (*R*)-configuration and 97% enantiomeric purity [57].

For these experiments synthetic hemihedral *d*- and *l*-quartz crystals from Toyo Communication Equipment Co. were received. The crystals were identified only by shapes and then powdered by a ball-mill and by wet- and dry-type methods; therefore, the twinnigs was excluded for average particle sizes 2.9-7.6 μm. (according to a private communication of Prof. K. Soai to one of authors of this book, E.K.). The surface of quartz powder was cleaned by HCl, MeOH, acetone, water, HF and water, and dried in vacuo at 100°C. Reactions were carried out in toluene at 20°C. The ratio of quartz to 2-alkinylpyrimidine-5-carbaldehyde (A) and diisopropyl-zinc (B) was 120:9:20. The reaction mixture was stirred for 12h, the mixture of A and B was added, the mixture was stirred for 5h, the mixture of A + B was added again for amplification of asymmetric effect, and finally the mixture was stirred for 5h. The alkanol product was obtained with an *ee* of 96% with an *S*-configuration using *d*-quartz and with an *R*-configuratio and an *ee* of 97% using *l*-quartz.

The same very high optical yields of this alkanol, *ee* 97% *S* and 98 % *R*, were obtained in the presence of enantiomorphic crystals *d*- or *l*-sodium chlorates, $NaClO_3$, respectively (Sato et al. [58]). Similar high effects probably can be obtained in the presence of other natural crystals, like epsomite, potassium bichromate, and cinnabar in future investigations.

The mechanism of these processes can be explained in terms of amplification of very small chiral effects created by the formation of the intermediate, and its transformation into the alkanol (Scheme 2.11.) because the chiral crystal itself induces a very small enantiomeric enrichment in the formed pyrimidinic alkanol and the effect is a result of the involvement of a chiral crystal in the intermediate complex.

These results have a threefold significance: first, as a practical method of amplifying small chiral effects in organic reactions, second, as a possible way to discover and assign the chirality of substances, such as natural minerals, that can be used as chiral catalysts (or carriers), and third, as a way to search for the possibility of the origin of chirality on Earth, where enantiomorphic minerals like quartz can exist (see Klabunovskii [60,61]).

In the case of quartz as asymmetric inductor, Soai [57] noted: "...results provides chemical evidence that chiral quartz may be involved in the origin of the chirality of organic molecules through catalytic asymmetric synthesis...". Of course this statement is true in the relation of origin of chirality rather than homochirality. Earlier, Tranter [62] and McDermott [63] gave a calculation of parity violation of energy difference (PVED) for *l*-quartz stabilized by 10^{-17} kT per SiO_2 unit of quartz structure based on the allegation that a possible source of homochirality of organic materials is generated by excesses of *l*-quartz on the surface of Earth. But analyses of literature by Klabunovskii and

Thiemann [64] showed that this calculation of PVED for quartz crystal seems to lack a sound physical basis because in all examined quartz locations the amounts of *d*- and *l*-quartz crystals are equal. Hence the quartz crystals are erroneously considered as a possible source of one handedness in nature, although local formation of optically active molecules, for example, via autocatalytic processes, appears quite possible in revealing chirality rather than homochirality (Klabunovskii [60,61]).

In the works of Soai [55] a very important result was found: leucine, which was prepared by irradiation with circular polarized light of racemic leucine as a result of asymmetric decomposition of racemate (see Bonner [65]), could be used in the above mentioned addition reaction as achiral inductor of very low enantiomeric purity (2%) with autocatalytic amplification occurring. As a result, optically active alkanol in the reaction in Scheme 2.12. was obtained with 26% optical purity (see Avalos et al. [66], Szabo-Nagy and Keszthelyi [67], and Mislow [68]).

In similar ways circular polarized light irradiation of very low efficacy combined with enantiomorphic crystals and autocatalytic amplification reactions, could be the origin of homochirality of organic molecules (Bonner [65]).

For another way, based on PVED of enantiomers [67] Mislow [68] recently indicated: "...if an enantiomeric excess as tiny as 10^{-5}% is sufficient to initiate the amplification [55] then an *ee* value even as small as 10^{-9}% is surely capable in principle of initiating a similar cascade of processes leading to an optically active product. There is no reason to consider that for chiral initiators a critical *ee* value exists, somewhere between 10^{-9}% and 10^{-5}%, below which automultiplication of chirality can no longer take place..."

Indeed, even without any chiral inductor it was possible in the above mentioned autocatalytic reaction of chiral alkanol synthesis, Scheme 2.12., to obtain high optical yields (up to 91%) of both *R*- and *S*-products. These results can be compared with separation of racemates during crystallization [70].

This means that an absolute asymmetric synthesis from optically inactive prochiral molecules, free of chiral contaminations and run under chiral conditions can be realized, even without the action of circular polarized light or other chiral auxiliaries.

Indeed, Singleton and Vo [69] adopted the enantioselective reaction used in Soai's reaction [54], Scheme 2.12., to examine the replicative growth in enantomeric excess using 2-methylpyrimidine-5-carbaldehyde and diisopropyl-zinc without either an asymmetric inductor or a chiral intermediate as catalyst. In the first experiment an almost racemic product with an *ee* of 0.00003% and *R*-configuration was prepared. Then using 2.5-10% of each product solution containing the intermediate and without addition of any origin asymmetric inductor, 48 successive reactions were carried out preparaing the final product with an *ee* of 22% with the same *R*-configuration.

Of course, these reactions are not true examples of absolute asymmetric synthesis [69], because the first excess of the *R*-enantiomer formed on the base of stochastic distribution served as the trigger for the next repeated reactions similar to the iniating crystal in "spontaneous" crystallization.

Soai et al [59] repeated these reactions without adding an asymmetric inductor carrying out 37 separate runs rather then successive reactions as in Singleton's experiments, and they obtained both *R*- and *S*-products, showing almost stochastic distribution: *S*-products were formed 19 times and *R*-products 18 times, with *ee* values of up to 91 %. The small deviation from stochastic distribution was "…due to unknown chiral factors…"[59]

It needs to be noted that this case is not a true asymmetric synthesis as Soai claimed, because during the 37 runs approximately an equal number of *R*- and *S*-alkanols were obtained with a stochastic distribution that was similar to racemic separations in "spontaneous" crystallization (Kondepudi et al. [70]).

References

1. Zocher H. and Coper K. (1928) Ueber die Erzeugung der Anisotropie von Oberflaechen, *Z. Phys. Chem.* **132**, 295-302.
2. Cotton A. (1929) Action de la lumiere polarisee sur certaines plaques photographiques preparees a partir de solution d'argent colloidal, *Compt. rend.* **189**, 599-603.
3. Ghosh J.C. and Banerjee T. (1939) Der Mechanismus photochemischer Reaktionen an einigen anorganischen Kolloiden. Der Einfluss des Lichtes in verschiedenen zirkularpolarisierten Zustaenden, *Koll.Z.* **86**, 372-387.
4. Ghosh J.C. (1939) The production of optically active substances and metal films of Ag, Pt, and Pd by means of circularly polarized light, *J. Indian Chem. Soc.* **16,** 51-62.
5. Nikolaev L.A. (1964) Biocatalysts and their models (*russ.*), High school Publisher, Moscow.
6. Bredig G., Gerstner F. and Lang H. (1935) Katalyse mit organischer Faser. II., *Biochem. Z.* **282**, 88-98.

[7] Yuasa S., Flory D., Basile B. and Oro J. (1984) Abiotic synthesis of purines and other heterocyclic compounds by the action of electric discharges, *J. Mol. Evol.* **21**, 76-80.

[8] Akabori S., Sakurai S., Izumi Y. and Fujii Y. (1956) An asymmetric catalyst, *Nature* **178**, 323-324.

[9] Izumi Y. and Tai A. (1975) Stereo-differentiating Reactions. The Nature of Asymmetric Reactions. Kodansha, Tokyo and Academic Press, N.Y, San Francisco, London (see p.102).

[10] Klabunovskii E.I. and Vedenyapin A.A. (1980) Asymmetric Catalysis. Hydrogenation on Metals, Nauka, Moscow (see p.148).

[11] Schwab G.M. and Rudolph L. (1932) Katalytische Spaltung von Racematen durch Rechts-und Linksquarz, *Naturwiss.* **20,** 363-364.

[12] Schwab G.M., Rost F. and Rudolph L. (1934) Optisch-asymmetrische Katalyse an Quartzkrystallen, *Koll.Z.* 68, 157-165.

[13] Schwab G.M. and Rost F. (1941) Fermentmodelle, in *Handbuch der Katalyse,* Schwab G.M. (ed.), B.3., Biokatalyse, Springer Verlag, Wien, S. 545-582.

[14] Rost F. (1935) Die Entstehung der optischen Asymmetrie, *Angew. Chem.* **48**, 73-88.

[15] Stankiewicz A. (1938) Inaugural Diss. Thesis. Aufspaltung von Racematen durch heterogene Katalyse. Albertus-Universitaet zu Koenigsberg (Tag der muendlichen Pruefung 24.02.1936, Leitung F.Eisenlohr), 43 pp.

[16] Terent'yev A.P., Klabunovskii E.I. and Patrikeev V.V. (1950) Asymmetric synthesis by means of catalysts deposited on right and left quartz, *Dokl. Acad. Nauk. SSSR*, **74**, 947-950; *Chem. Abstr.* 1951, **45**, 3798c.

[17] Terent'yev A.P. and Klabunovskii E.I. (1953) Catalytic asymmetric synthesis. I. Asymmertric decomposition of racemic 2-butanol over Cu-quartz catalysts. Comm. I., in *Collection of papers on General Chemistry (russ.),* V.2, 1521-1529; *Chem. Abstr.* 1955, **49**, 5262f.

[18] Terent'yev A.P. and Klabunovskii E.I. (1953) Catalytic asymmetric synthesis. Comm. II. Asymmetric decomposition of 2-butanol over quartz-catalysts, in *Collection of papers on General Chemistry* (*russ.*),V.2, 1598-1604; *Chem. Abstr.* 1955, **49**, 5262g.

[19] Terent'yev A.P. and Klabunovskii E.I. (1953) Catalytic absolute asymmetric synthesis. Comm. III. Reactions of isomerization, dismutation and hydrogenation, in *Collection of papers on General Chemistry*, (*russ.*), V.2, 1605-1611; *Chem. Abstr.* 1955, **49**, 5263a.

[20] Terent'yev A.P. and Klabunovskii E.I. (1953) Catalytic absolute asymmetric synthesis. Comm. IV., in *Collection of papers on General Chemistry* (*russ.*) V.2, 1612-1616; *Chem. Abstr.* 1955, **49**, 5263c.

[21] Klabunovskii E.I. (1951) Asymmetric synthesis by means of catalysts based on optically active quartz, *Thesis*, M.V. Lomonosov Moscow State Univ, Moscow, USSR.

[22] Klabunovskii E.I. and Patrikeev V.V. (1951) On mechanism of asymmetrizing action of metal catalysts deposited on right and left quartz, *Dokl. Akad. Nauk, SSSR*, **78**, 485-487; *Chem. Abstr*. 1951, **45**, 7860a.

[23] Terent'jev A.P. and Klabunovskii E.I. (1951) For problem of absolute asymmetric synthesis, Sci. Papers of Moscow State Univ. **151**, 145-180.

[24] [a] Klabunovskii E.I. (1960) Absolute asymmetric synthesis and asymmetric catalysis, in Florkin M. (ed.), *Aspects of the Origin of Life*, Pergamon Press, N.Y., Oxford, 105-115.
[b] Klabunovskii E.I. (1959) Absolute asymmetric synthesis and asymmetric catalysis, in Clark F. and Singe R.L.M. (eds.) *The Origin of Life on the Earth,* Moscow, 1957, Pergamon Press, London, N.Y., 158-168.
[c] Levi G. (1987) Optical activity and catalysis, *Science in USSR*, N1, 101-107.

[25] Khodakov G.S. and Rebinder P.A. (1960) Influence of the medium on the formation of an amorphous phase in quartz during mechanical dispersion, *Dokl. Acad. Nauk SSSR*. (*russ.*) **131**, 1316-1318.

[26] Ponomarev A.A. and Zelenkova V.V. (1953) For the study of furane compounds. IV. Experiments in the asymmetric synthesis of spiranes and tetrahydrofuryl-alcohols, *Zh. Obtsh. Khim.* (*J. General Chem., russ.*) 23, 1543-1547.

[27] Tsuchida R., Kobayashi M. and Nakamura A. (1936) The configuration of chlorobisdimethylglyoximoammine-Cobalt, *Bull. Chem. Soc. Jpn*. **11**, 38-40.

[28] Amariglio-Simon A. (1967) Etude critique des reactions asymetriques sur quartz et des conditions de stereospecificite dans les interactions moleculaires, *Dissertation Thesis*, Univ. Nancy.

[29] Amariglio A., Amariglio H. and Duval X. (1968) La synthese asymetrique, *Ann. Chim*. (*Paris*), **3**, 5-25.

[30] Amariglio A., Amariglio H. and Duval X. (1968) Essais de reactions dissymetrique sur quartz optiquement actif, *Helv. Chim. Acta* **51**, 2110-2132.

[31] Amariglio A., Amariglio H. and Duval X. (1969) Resolution of various racemates by gas chromatography on optically active quartz column, *Comptes rendus* (*Ser.C*) **268**, 1981-1984.

[32] [a] Klabunovskii E.I. (1968) *Stereospecific catalysis* (russ.), Nauka, Moscow, (see, p. 258), *Chem. Abstr*. 1969, **70**, B109520r.
[b] Klabunovskii E.I. (1963) *Asymmetrische Synthese*, VEB Deutscher Verlag der Wissenschaften, Berlin.

[33] Tsinzerlink E.V. (1941) Quartz twinning control under *alpha-beta-*conver-sion, *Compt. rend. Acad. Sci. USSR*, **33**, 365-367; *Chem. Abstr.* 1941, **38**, 4848.

[34] Bonner W. (1996) The quest for chirality, in *Physical origin of homochirality in life,* AIP Conference, Santa Monica, CA, 1995, *Proceedings* 379, Cline D.B. (ed.), AIP Press, Woodbury, N.Y., p. 17-49.

[35] [a] Harada K. (1978) First asymmetric syntheses: Origin of optical activity, *Kagaku Kogyo* **29**, 569-573.
[b] Harada K. (1980) The origin of optical active compounds and their evolution, *Kagaku Sosetsu*, **30**, 145-153.

[36] Bredig G and Gerstner F. (1932) Asymmetrisch Katalyse mit organischer Faser (Ein neues Fermentmodell), *Biochem. Z.* **250**, 414-429.

[37] Boys S.F. (1934) Optical rotatory power. II. The calculation of the rotation power of a molecule containing four refractive radicals at the corners of an irregular tetrahedron, *Proc. Roy .Soc. (A)* **144**, 675–692.

[38] Kirkwood J.G. (1937) The theory of optical rotatory power, *J. Chem. Phys.* **5**, 479-491.

[39] Kuhn W. (1935) Absolute Konfiguration der Milchsaeure (Mit Bemerkungen ueber Voraussetzung Annahmen bei der Berechnung der optischen Drehung, *Z. Phys. Chem. (B)* **12**, 23-67.

[40] Kuhn W. (1962) Rotatory dispersion and vibrating momentum of optical active absorption bonds, *Proc. 4th Int. Meet. Molec. Spectr.*, Bologna, 1959, p. 35-48, (see also *Z. Elektrochem.* **68**, 28-38).

[41] Volkenshtein M.V. (1940) Quantum mechanics of optical activity, *Uspekhi Khimii (Russ. Chem. Rev., russ.)* **9**, 1252-1279 (especially see p. 1272).

[42] Kuhn W. (1952) Das Problem der absoluten Konfiguration des tetraedrischen Kohlenstoffatoms, *Z. Elektrochem.* **56**, 506-524.

[43] Condon E.U., Altar W. and Eyring H.(1937) One-electron rotatory power, *J. Chem. Phys.* **5**, 753-775.

[44] Bijovoet J.M., Peerdeman A.F. and Van Bommel A.J. (1951) Determination of the absolute configuration of optically active compounds by means of X-rays, *Nature* **168**, 271-272.

[45] Burwell R.L.,Jr. (1960) Stereochemistry in Heterogeneous Catalysis, in *Techniques Nouvelles en Catalyse Heterogene*, Selwood P.W., Burwell, R.L.,Jr., Eischens R.P. (eds.), Soc. Editions Technip, Paris, pp. 38-78.

[46] Speakman J.S. (1954-1958) Crystallography and stereochemistry, in *Progress in stereochemistry*, Klyne W. and de la Mare P.B.D. (eds.), Vol. 1-2., Butterworth Sci. Publ., London.

[47] Waser J. (1949) The absolute configuration of *d*-tartaric acid, *J. Chem. Phys.* **17**, 498-499.

48 Hylleraas E.A. (1927) Equilibrium position of the atom, double refraction and optical rotation power of *beta*-quartz, *Z. Phys.* 44, 871-886.

49 De Vries A. (1958) Determination of the adsolute configuration of the *alpha*-quartz, *Nature* **181**, 1193.

50 Balandin A.A., Klabunovskii E.I. and Litvin E.F. (1961) On composition of mixture of butenes forming at catalytic dehydration of 2-butanol, *Izv. Akad. Nauk SSSR, Otdel. Khim. Nauk*, 1863-1870; *Chem. Abstr.* 1962, **56**, 7112c.

51 Isagulyants G.V., Derbentsev Yu.I., Klabunovskii E.I. and Balandin A.A. (1964) On mechanism of catalytic dehydration of 2-butanol on the surface of alumina, *Izv. Acad. Nauk SSSR, Ser. Khim.* 985-990; *Chem. Abstr.* 1964, **61**, 8158h.

52 Tinoko I. and Woody R.W. (1964) Optical activity of oriented helixes, *J. Chem. Phys.* **40**, 160-165.

53 Frank F.C. (1953) On spontaneous asymmetric synthesis, *Biochim. Biophys. Acta* **11**, 459-463.

54 Soai K., Shibata T., Morioka H. and Choji K (1995) Asymmetric autocatalysis and amplification of enantiomeric excess of a chiral molecule, *Nature* **378**, 767-768.

55 Shibata T., Yamamoto J., Matsumoto N., Jonekubo S., Osanai S. and Soai K. (1998) Amplification of a slight enantiomeric imbalance in molecules based on asymmetric autocatalysis. The first correlations between high enantiomeric enrichment in a chiral molecule and chirality polarizing light, *J. Amer. Chem. Soc.* **120**, 12157-12158.

56 Sato I., Omiya D., Igarashi H., Kato K. Ogi Y., Tsukiyama K. and Soai K. (2003) Relationship between the time, yield and enantiomeric excess of asymmetric autocatalysis of chiral 2-alkynyl-5-pyrimidyl alkanol with amplification of enantiomeric excess, *Tetrahedr. Asymm.* **14**, 975-979.

57 Soai K., Osanai S., Kadowaki K, Yonekubo S., Shibata T. and Sato I. (1999) *d*- and *l*-Quartz promoted highly enantioselective synthesis of a chiral organic compound, *J. Amer. Chem. Soc.* **121**, 11235-11236.

58 Sato I., Kadowaki K. and Soai K. (2000) Asymmetric synthesis of an organic compound with high enantiomeric excess induced by inorganic ionic sodium chlorate, *Angew. Chem. Int. Ed.* **39**, 1510-1512.

59 Soai K., Sato I., Shibata T., Komiya S., Hayashi M., Matsueda Y., Imamura H., Hayase T., Morioka H., Tabira H., Yamamoto J. and Kowata Y. (2003) Asymmetric synthesis of pyrimidyl alkanol without adding chiral substances by the addition of diisopropylzinc to pyrimidine-5-carbaldehyde in conjunct-tion with asymmetric autocatalysis, *Tetrahedr. Asymm.* **14**, 185-188.

60 Klabunovskii E.I. (2001) Can enantiomorphic crystals like quartz play the role in the origin of homochirality on Earth, *Astrobiology* **1**, 127-131.

[61] Klabunovskii E.I. (2002) The role of enantiomorphic crystals in origin of chirality on Earth and other planet, in *Programme and Abstracts of Second European Workshop on Exo/Astrobiology,* Graz, Austria, Sept.16-19, Lammer H. et al.(eds.) p. 66.

[62] Tranter G.E. (1985) Parity violation energy difference of chiral minerals and the origin of biomolecular homochirality, *Nature* **318**, 1712-1713.

[63] McDermott A.J. (1996) The weak force and SETI: the search for extra-terres-trial homochirality, *Physical origin of homochirality in life*, Amer. Inst. Phys. Conf., Santa Monica CA, Febr. 1995, *Proceedings*, Cline D.B. (ed.), Press Woodbury, N.Y., p. 241-254.

[64] Klabunovskii E.I. and Thiemann W. (2000) The role of quartz in the origin of optical activity on Earth, *Origin Life Evol. Biosphere* **30**, 431-434.

[65] Bonner W.A. (1996) The quest for chirality, *Physical origin of homochirality in life*, Amer. Inst. Phys. Conf., Santa Monica CA, Febr. 1995, *Proceedings*, Cline D.B. (ed.), Press Woodbury, N.Y., p. 17-49.

[66] Avalos M., Babiano R., Cintas P., Jimenez J.L. and Palacios J.C. (2000) Chiral autocatalysis: where stereochemistry meets the origin of life, *Chem. Commun.* 887-892.

[67] Szabo-Nagy A. and Keszthelyi L. (1999) Experimental evidences for parity violating energy differences between enantiomers, *Adv. Biochirality*, Chapter 28, p. 367-375.

[68] Mislow K. (2003) Absolute asymmetric synthesis: A commentary, *Coll. Czech. Chem. Commun.* **68**, 849-864.

[69] Singleton D.A. and Vo L.K. (2002) Enantioselective synthesis without discrete optical active additives, *J. Amer. Chem. Soc.*, **124**, 10010-10011.

[70] [a] Kondepudi D.K., Kaufman R.J. and Singh N. (1990) Chiral symmetry breaking in sodium chlorate crystallization, *Science* **250**, 975-976.

[b] Kondepudi D.K. Laudadio J. and Asakura K. (1999) Chiral symmetry breaking in stirred crystallization of 1,1'-binaphthyl melt, *J. Amer. Chem. Soc.* **121**, 1448-1451.

Chapter 3

HETEROGENEOUS CATALYSTS SUPPORTED ON CHIRAL CARRIERS

Abstract

Chapter 3 presents data about enantioselective hydrogenation reactions on metal catalysts supported on chiral carriers. Discussed are palladium supported on modified silica gels, effective chiral colloidal catalysts, template catalytic systems, palladium-silk, palladium-wool, and palladium polypeptide catalysts.

3.1. Pd-silica catalysts

Biemer et al. [1-4] reported the enantioselective behaviors of a Pd catalysts supported on specially prepared silica gels, which have been precipitated from Na silicate with HCl in the presence of optically active alkaloids: sulfates of quinine (Q), quinidine (Qd), cinchonine (Cn), or cinchonidine (Cnd). These catalysts proved to be active in the asymmetric hydrogenation of 2-methyl-cinnamic acid (Scheme 3.1.) [1].

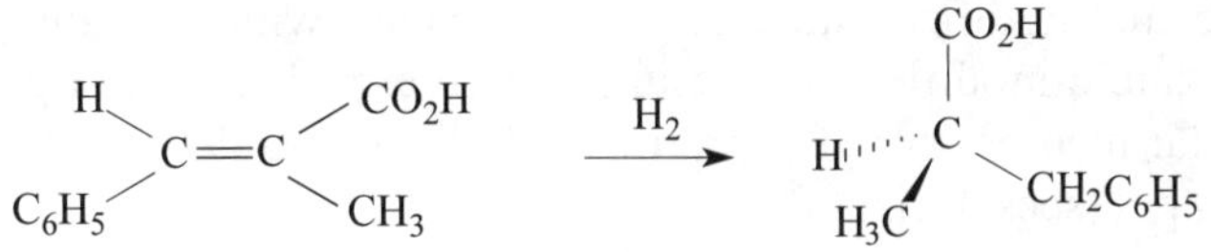

Scheme 3.1. Hydrogenation of 2-methylcinnamic acid.

The results of the asymmetric hydrogenations on the modified silica gels are given in Table 3.1.

The mechanism proposed by Padgett and Biemer in 1964 [1] was based on the idea that the catalysts have two adsorption sites which differ in their affinity for the two enantiofaces of the substrate molecules.

Catalysts	*ee* (%)	$[\alpha]_{589}$
Pd-Q-Sg	3.21	+0.87 *
Pd-Qd-Sg	1.66	+0.45
Pd-Cn-Sg	1.74	+0.47
Pd-Cnd-Sg	3.25	+0.88
Pd-Q-Sg (aged 6 months)	3.14	+0.85

* Observed α_{589} = +0.065 in benzene (Prof. R.L. Beamer's private communication to one of the authors, E.K.).

Table 3.1. Asymmetric hydrogenation of 2-methylcinnamic acid on Pd catalysts supported on specifically modified silica gels (Sg) (modified data adapted from Beamer [1]).

Enantioselectivity of the catalysts is due to the surface structure of the silica gel, which has been modified during precipitation in the presence of optically active alkaloids. The resulting modified silica gel acts like polyamino acids, as an optically active support for the metal (for opportunities of optical activation of the surface of silica as chiral adsorbents, see Chapter 1.3). It is necessary to note that the hydrogenations of 2-methylcinnamic acid on silica gels, modified with both Qn and Qnd and also with Cn and Cnd, result in products with only (+)-rotations. Qn and Qnd are not enantiomers but diastereomers; see formulas in Scheme 5.1.6. They possess opposing configuration about C8 and C9 but configurations at C3 and C4 are the same in both alkaloids. Configurationally Cnd and Cn correspond to Qn and Qnd. This indicates that the C3 and C4 centers in the alkaloid modifier molecules have no significant role in the formation of asymmetric centers. On the other hand, since the modified silica gel was extracted with methanol to remove traces of alkaloid-modifiers, the results can be explained only by the concept of "molecular memory" or "fingerprints" of alkaloid molecules on the surface of the silica gel (see Morihara et al. [5]).

3.2. Metal colloids as asymmetric catalysts

Colloid catalysts with chiral stabilizing agents can be used as asymmetric catalysts in the reactions of prochiral compounds. One such catalyst is the well-known "Skita-catalyst"(colloidal Pt or Pd with gum-arabicum as an optical active polysaccharide as a protective colloid). These catalysts have been used very often in a number of hydrogenations of unsaturated compounds, including prochiral molecules, but never were their asymmetrizing action noted. Nevertheless, including chiral components as protective colloids in such catalysts allowed for the discovery of asymmetric effects in their action. Indeed, Balandin, Klabunovskii et al. [6] found small enantioselectivity in the hydrogenation of 2-oxoacids over Skita-catalysts (Scheme 3.2.).

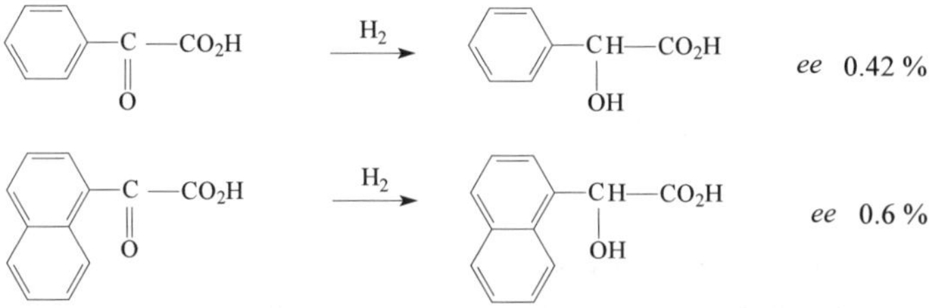

Scheme 3.2. Asymmetric hydrogenation of 2-oxoacids.

For a long time this sort of catalysts received no attention as an asymmetric catalyst and only recently were the above mentioned results repeated and significantly improved.

Boennemann and Braun [7] found a very effective enantioselective colloidal catalyst. A platinum sol stabilized by DHCnd revealed very high enantioselectivity in the hydrogenation of ethyl pyruvate (EtPy) into (*R*)-(+)-ethyl lactate (*R*-EtLa) with an *ee* of 78% at 12°C and atmospheric pressure in a *quasi*-homogeneous phase of a solution of AcOH + MeOH (Scheme 3.3.).

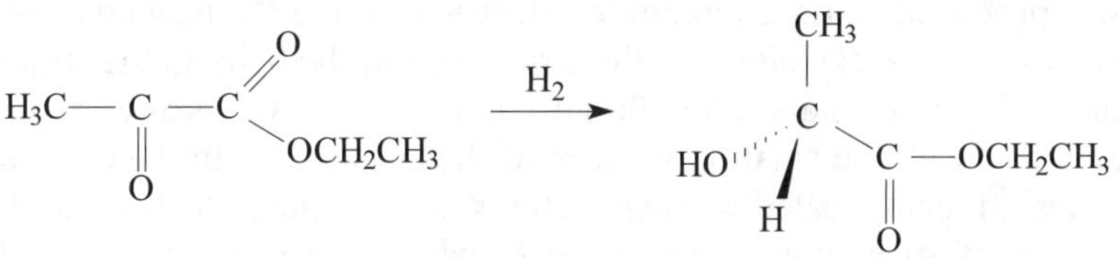

Scheme 3.3. Hydrogenation of ethyl pyruvate.

The particle size of Pt varied in the range of 1.5-4 nm. A decrease of DHCnd concentration in the protective colloid leads to reduced activity at constant enantioselectivity as a result of agglomeration of the colloid. Catalytic activity decreases with increasing particle size as a result of decreasing dispersion but has no effect on *ee*, contrary to results with heterogeneous catalysts, for example, Pt-Alumina-Cnd (Blaser et al. [8]). This Pt-colloid catalyst can be supported on charcoal or silica and hydrogenates EtPy with an *ee* of 80% at 100 bar hydrogen, which is comparable to heterogeneous Pt-alumina-Cnd catalysts.[8]. Blaser has also found that polyvinylpyrrolidon (PVP) (Scheme 3.4.) stabilizes Pt-colloid-Cnd, which hydrogenates EtPy into EtLa with an *ee* of 65% at 100 bar in propan-2-ol solution (Boennemann [7]).

The Pt-PVP-Cnd catalyst with a composition of 10:300:50, proved to be active in the hydrogenation of EtPy at 24°C and 4.6 bar hydrogen. Besides, Collier et al. [9] reported the hydrogenation of EtPy with an *ee* of 40% at 17% conversion using a Pt colloid prepared by metal evaporation and stabilized with MeEtCO + DHCnd [9,10].

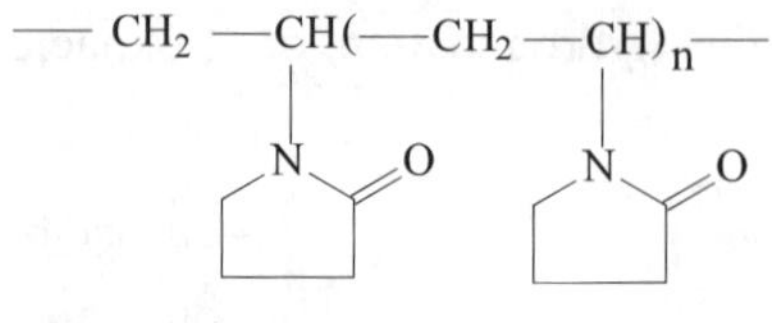

Scheme 3.4. Polyvinylpyrrolidon.

3.3. Asymmetric template catalysis

An asymmetric reaction can be catalyzed by the surface of a chiral crystal and on the surrounding chiral arrangements, such as on the surface of a cholesteric liquid crystal, in the temperature interval of the existence of the mesophase, where the helix structure is formed. However, the experimental data about optically active cholesteric liquid crystals as chiral catalysts are contradictory.

The stereospecific Claisen transformation [11] and the enantio-selective decarboxylation of ethylmalonic acid [12] were found in the presence of cholesterylbenzoate and the formation of (*S*)-(-)-[Cr(ox)$_3$] [13] and (-)-helicene in photoreactions were reported on the (-)-rotating liquid-crystalline Tween-80, which is a polyoxyethylene-sorbitane-monooleate (n = 20) [14,15].

On the other hand, it was reported that either several reactions could not be reproduced or no asymmetric effect was found [16]. Negative results in several works were explained by the large steps of the helix in the structure of cholesteric liquid crystals (300-400 nm) and by the absence of chirodiastaltic interaction [17] with the small molecules of the substrates. In the case of large molecules of complicated stereostructures it was possible to wait for the appearance of such interactions as revealed by the appearance of induced circular dichroism (ICD) in the absorption bands of chromophors of components and products of the reaction.

Pavlov et al. [18] studied the hydrogenation of 2-acetamidocinnamic acid **1** into *N*-acetylphenylalanine **2** at atmospheric pressure and a temperature range of 50-70°C (Scheme 3.5.) using a cholesteric liquid crystal (cholesteryl tridecanoate, Me(CH$_2$)$_{11}$COO-Cholesteryl).

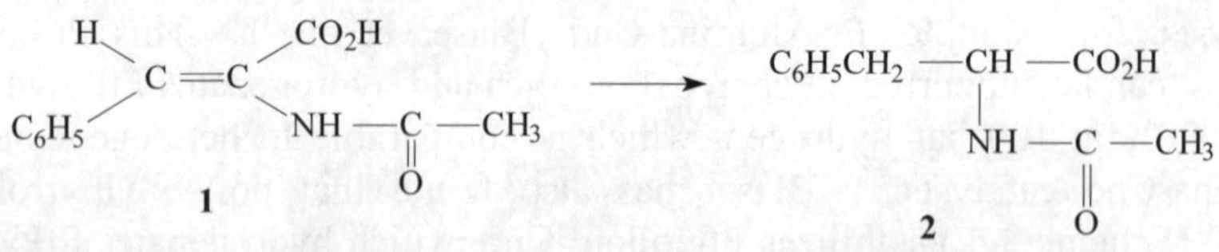

Scheme 3.5. Hydrogenation of 2-acetamidocinnamic acid.

They found that substrate **1** and the complex catalyst [RhCl(PPh$_3$)$_3$] in the presence of the cholesteric liquid crystal (CTD) in butanol solution should induce circular dichroism with maximal value in the temperature range of the cholesteric mesophase. Addition of hydrogen to **1** in this system results in optically active *N*-acetylphenylalanine **2**. In BuOH solution CTD adopts a helix structure therefore the mixture of CTD + **1** revealed the ICD spectrum in the band of absorption of **1** (Fig. 3.1.) and the solution of cholesterinstearate + catalyst also gave an ICD spectrum (Fig. 3.2.). In the latter case the cholesterinstearate was used instead of CTD because of its better solubility. The appearance of the ICD spectra for **1** and for the catalyst gave evidence of the existence of the helix structure of CTD as a chiral matrix. As the reaction progresses the maximal values of ICD increase indicating formation of a helix arrangement of the chiral matrix. It was found that the asymmetric hydrogenation of **1** results in optically active **2** with a maximal *ee* of 16% at 62°C just in the middle of the temperature interval of stability of the mesophase of the liquid crystal CTD (Fig. 3.3.).

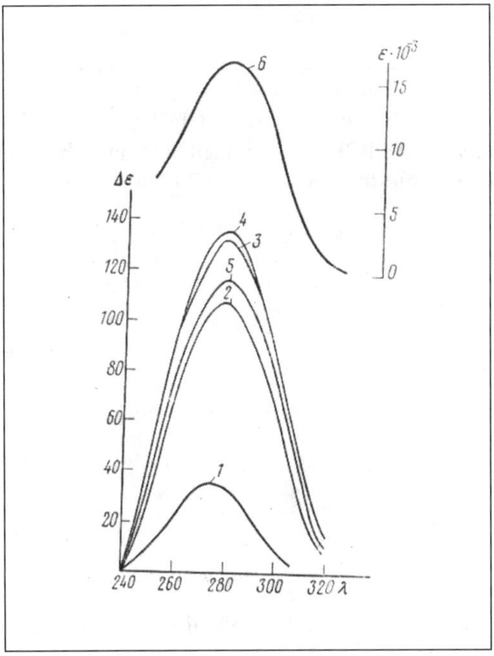

Figure 3.1. Induced CD (curves 1-5) and UV (curve 6) spectra of substrate *N*-acetylaminocinnamic acid included into chiral liquid crystal matrix of cholesteryl tridecanoate in 1-butanol at 54 °C.
(1 at 15 min, 2 at 30 min, 3 at 60 min, 4 at 90 min and 5 at 120 min), (adapted from Pavlov et al.[18]).

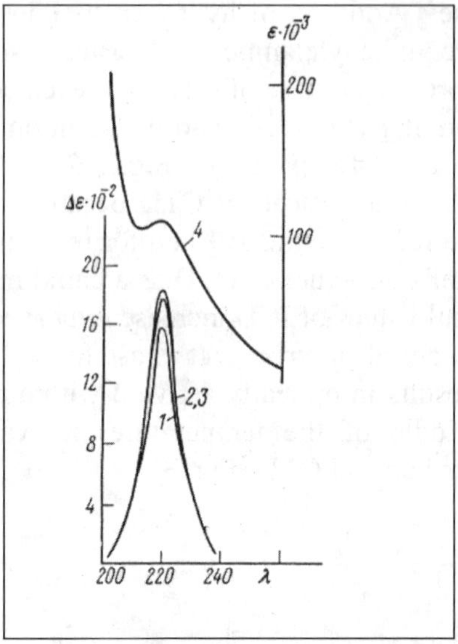

Figure 3.2. Induced CD (curves 1-3) and UV (curve 4) spectra of Wilkinson catalyst, $RhCl(PPh_3)_3$, included into the chiral matrix of the liquid crystal cholesteryl stearate at 67°C in 1-butanol (adapted from Pavlov et al. [18]

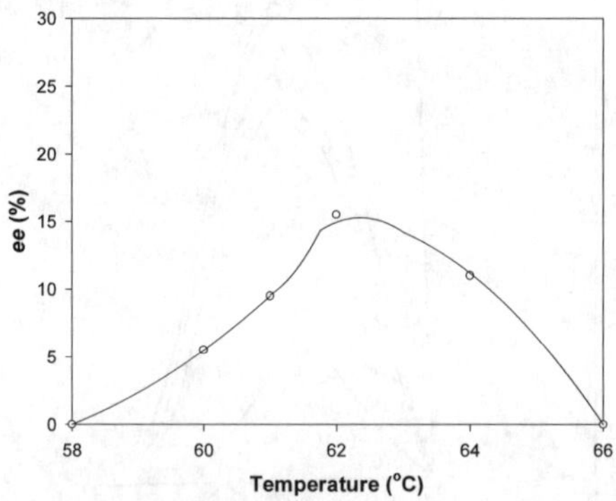

Figure 3.3. Dependence of the optical yield (ee %) of 2 on temperature in the hydrogenation of 1 (0.107g in butan-1-ol) included in the chiral matrix CTD (1.66g) under the action of catalyst $RhCl(PPh_3)_3$ (0.013g) (adapted from Pavlov et al. [18]).

3.4. Metal catalysts supported on polymers

Optically active natural or synthetic polymers are of significant interest as dissymmetric adsorbents and carriers for chiral catalysts because the structures of polymers have been studied in detail and can be changed in desirable directions that cannot be reached in the case of natural materials.

3.4.1. Pd-silk catalysts

In 1956 Akabori et al. [19-21] reported an entirely new type of asymmetric catalysts, Pd deposited on natural silk, which was active in the hydrogenation of C=C and C=N bonds. Silk fibroin has a rigid structure, and can coordinate metal cations. So, if silk fibroin is boiled with aqueous Pd-chloride and after the formation of the chelate complex, it is reduced, one obtains a Pd catalyst in which Pd atoms are embedded in an asymmetric environment.

Such a catalyst can exhibit enantioselectivity, such as in the hydrogenation of **1** (in Scheme 3.6.) into L-(+)-glutamic acid **2** with an *ee* of 39% at 50°C and 80 bar of hydrogen and in the reduction of **3** into D-(+)-phenyl-alanine, **4**, with an *ee* of 26% at 70°C and 90 bar of hydrogen.

More effective, with an *ee* of 35%, was the hydrogenation of the azlactone of the 2-acetamidocinnamic acid, **5**, into D-(+)-*N*-acetylphenyl-alanine, **6**.

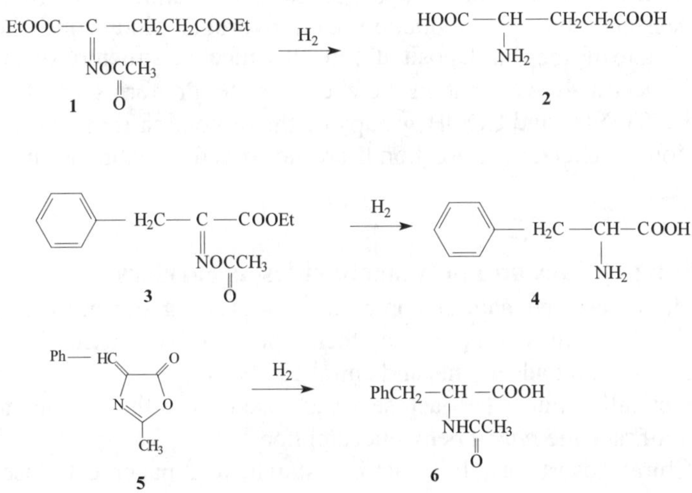

Scheme 3.6.

These Pd-silk catalysts are of interest but after their publications [19-22] there was no confirmation in the literature on the enantioselectivity of these catalysts, and moreover, the same authors [23-25] noted bad reproducibility of these data. The results varied with the type of silk.

Thus D-(+)-Phe was obtained over Pd catalyst supported on:

i) fibroin of cultural silkworm (*ee* 35.5%)
ii) fibroin of natural silkworm (*ee* 31.9%)
iii)acetylated fibroin of cultural silkworm (*ee* 65.8%).

Moreover, if the silk-fibroin was dissolved in copper-ammonia solution and reprecipitated, its chiral properties were lost. It seems therefore, that the asymmetric properties depends not only on the asymmetric structures of the fibroin but also on the asymmetric structure of the whole molecule [25c].

Although these catalysts were not very promising as asymmetric catalysts, they proved to be very perspective liquid phase hydrogenation catalysts. They have very low densities, may be spun or woven into sheets or cloth, and can be readily separated from reactants.

The stereoselectivity of the silk-catalysts was demonstrated by the production with almost 100% *ee* of *L*-proline on Pd-silk or *L*-ornithine on Pt-silk [22]. A catalyst of 6% Pd-silk was used also in the pilot scale for hydrogenation of nitrotoluenes into amines with 80% yield [20].

By an analogous method Pt-silk and Rh-silk catalysts were prepared and possessed activities for hydrogenation of nitro-aromatic compounds, ketones, and alkenes.

X-Ray studies of the catalysts revealed crystallinity of structure both in chelate state and after reduction with hydrogen, indicating the highly dispersed state of the Pd deposited onto the micellar structure of the fibroin fibre. IR-spectra showed that in the chelate state, Pd forms bonds with the NH_2, OH, CONH, and COOH groups of the protein carrier. During the hydrogenation of chelates, these bonds are destroyed revealing Pd in a close to atomic state.

3.4.2. Polypeptides and polysaccharides as carriers

Carbohydrates are the natural compounds possessing the most asymmetric centers in molecules that make them prospective auxiliaries for the preparation of chiral adsorbents and chiral catalysts.

Optically pure (+)-lactose was used for the chromatographic resolution of racemic *para*-phenylenecamphor.

Chiral adsorbents, like lactose, starch, and paper cellulose, gained applications in analytical and practical uses. Optically pure lactose was used by Henderson and Rule [26] for the resolution of racemic *para*-phenylene-bis-iminocamphor in chromatographic columns. The (+)-enantiomer was more strongly adsorbed than the (-)-enantiomer.

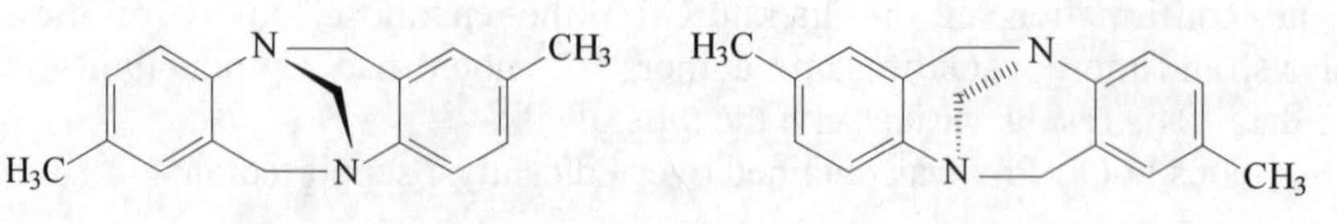

Scheme 3.7. Enantiomers of Troger's base.

Carrying out the process of resolution repeatedly they obtained the opticaly pure (-)-product with an $[\alpha]_{546}$ = -1992°. On lactose Prelog and Wieland [27] succeded in resolving racemic "Troger's base ", which contains chiral nitrogen atoms (Scheme 7).

Synthetic optically active polymers as carriers for chiral catalysts were used in several works. Petrov et al.[28] studied the diastereomeric liquid phase hydrogenation of (-)-menthyl 2-phenylcinnamate into (-)-menthyl *L*-2,3-diphenyl-propionate (menthyl 2-phenyl-2,3-dihydrocinnamate) over Pd-poly-(-)-(2-methylbutyl)metacrylate and Pd-poly-(-)menthylmetacrylate which proceeded according to Prelog's rule. Comparisons of the extent of conversion of hydrogenation over catalysts on different polymers have shown that optically active polymers have no asymmetrizing action on thr reaction because they do not form the stable complexes (metal-carrier-substrate) that must be responsible for the asymmetric effect. Thus Pd deposited on optically active poly-L-Leucine resulted in hydrogenation of 2-acetaminocinnamic acid to phenylalanine only with an *ee* of 5.4% [1].

Beamer et al. [29] used Pd-polymer catalysts in the hydrogenation of the C=C bonds in 2-acetamidocinnamic acid to *N*-acetylphenylalanine (Scheme 3.5.) and 2-methylcinnamic acid to 2-methyl-3-phenylpropanoic acid (Scheme 3.8.). The results are in Table 3.2.

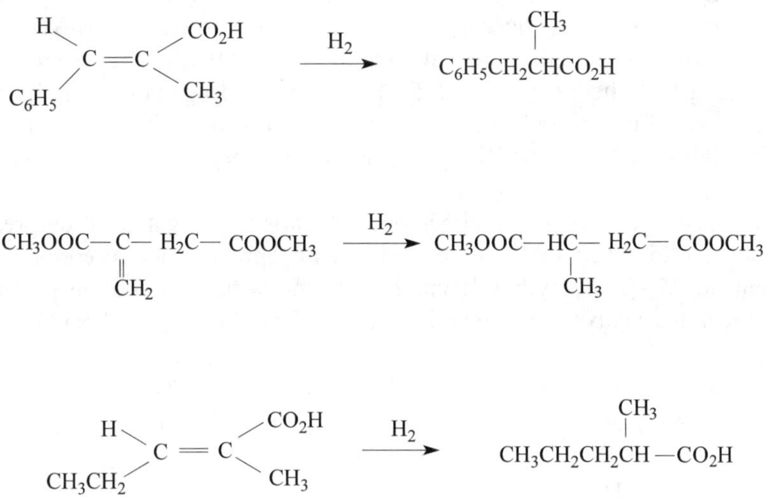

Scheme 3.8.

Catalyst	*ee* % of 2-methyl-3-phenylpropanoic acid	*ee* % of *N*-acetylphenylalanine
Pd-Poly-*S*-Val	+0.90	+4.25
Pd-Poly-*S*-Leu	-11.80	-5.4
Pd-Poly-Benzyl-*S*-AspA	+1.43	+0.95
Pd-Poly-benzyl-*L*-Glu	-4.15	-6.0

Table 3.2. Hydrogenations on Pd-polymer catalysts.

Smith et al. [30] have found a new type of chiral heterogeneous catalyst based on Pt, Pd, Rh or Ru dispersed onto a *beta*-cyclodextrine-epichlorohidrin copolymer (CDP). The ability of these catalysts to catalyze the enantioselective hydrogenation of C=C bonds was tested with the hydrogenation of dimethylitaconate into dimethyl methylsuccinate and (*E*)-2-methylpent-2-enoic acid into 2-methylvaleric acid (Scheme 3.8.).

The hydrogenation proceeds at 25°C and atmospheric pressure. The enantioselectivity of these catalysts vary depending on the structure of the substrates, the type and the loading of metals and the extent of the conversion. The best results are produced by the 2% Ru in the hydrogenation of (*E*)-2-methylpent-2-enoic acid, where the *ee* was 7.9%. Increasing the metal loading from 2% to 4% resulted in a dramatic decrease in *ee* (to 1.4%) probably as a result of the cange in metal particle sizes. However, in the hydrogenation of dimethylitaconate, the increase of the loading from 2 to 4% does not affect the *ee* (5.8 and 5.2%). The hydrogenation of dimethylitaconate at 25°C resulted in an increase of *ee* up to 15.7% on 2% Pt-CDP. Other catalysts (Ru, Pd or Rh) gave, however, worse results (*ee's* were 1.6 and 2.5%).

Hydrogenation of ethyl acetoacetate was investigated in the presence of Ru complex $[Ru(CO)_3CL_2]_2$ bounded with optically active copolymer of Na-maleate-(*S*)-(2-methylbutyl) vinyl ether and with poly-sodium glutamate. Only the latter catalyst revealed a low *ee* of 3% at 100°C and 80 bar [31].

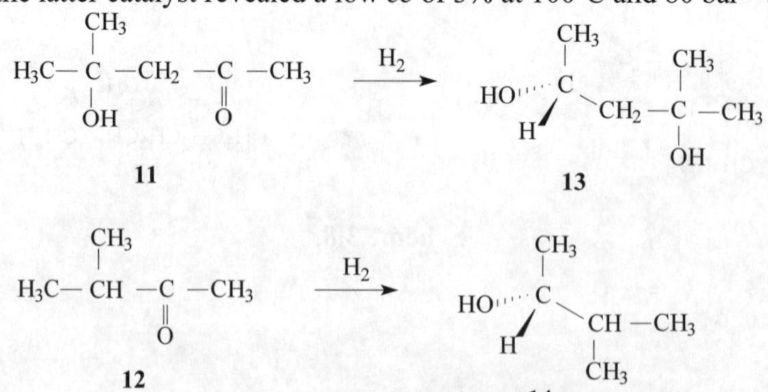

Scheme 3.9.

Recently, strikingly new results were obtained with a biopolymer, in which a natural wool carrier combined with $PdCl_2$ in EtOH to form catalytic complexes (Yin et al. [32]). These catalysts hydrogenated diacetone alcohol (4-hydroxy-4-methylpentan-2-one) into (*R*)-2-methylpentan-2,4-diol and 3-methylbutan-2-one into (*R*)-3-methylbutan-2-ol (Scheme 3.9.) with *ee*'s of 73% and 100%, respectively. These catalysts could be used repeatedly without decreasing enantioselectivity.

References

[1] Padgett R.E. and Beamer R.L. (1964) Stereospecific hydrogenation using palladium-on-silica gel catalysts, *J. Pharm. Sci.* **53**, 689-690.

[2] Beamer R.L., Fickling C.S. and Ewing J.H. (1967) Stereospecific hydrogenation using Pd-on-poly-*L*-leucine, *J. Pharm. Sci.* **56,** 1029-1032.

[3] [a)] Beamer R.L., Belding R.H. and Fickling C.S. (1969) Stereospecific hydrogenation. Comm. III. Pd-on-poly-(gamma-*S*-glutamate) and Pd-on-poly-(*S*-aspartate), *J. Pharm. Sci.* **58**, 1142 –1144.

[b)] Beamer R.L., Belding R.H. and Fickling C.S. (1969) Stereospecific hydrogenation. Comm. IV. Pd-on-poly-(*S*-valine) and Pd-on-poly-(*S*-leucine), *J. Pharm. Sci.* **58**, 1419-1421.

4. Beamer R.L and Brown W.D. (1971) Stereospecific hydrogenation. Comm.V. Hydrogenation rates using Pd-on-poly-(*S*-leucine) and Pd-(poly-*S*-valine), *J. Pharm. Sci.* **60**, 583-588.

[5] Morihara K., Kurokawa M., Kamata Y. and Shimada T. (1992) Enzyme-like enantioselective catalysis over chiral 'Molecular Footprint' cavities on a silica (alumina) gel surface, *J. Chem. Soc., Chem. Comm.* 358-360.

[6] Balandin A.A., Klabunovskii E.I. and Petrov Yu.I. (1959) Configurational relationship in stereospecific catalysis, *Dokl. Akad. Nauk SSSR (russ.)* **127**, 557–560; *Chem. Abstr.* 1960, **54**, 241a.

[7] Boennemann H.B. and Braun G.A. (1996) Enantioselective hydrogenations on platinum colloids, *Angew. Chem. Int. Ed. Engl.* **35**, 1992-1995.

[8] Blaser H.U., Jalett H.P., Monti D.M., Baiker A. and Wehrly J.T. (1991) Enantioselective hydrogenation of ethyl pyruvate: effect of catalyst and modifier structure, *Stud. Surf. Sci. Catal.* **67,** 147-156.

[9] Collier P.J., Goulding T., Iggo J.A. and Whyman R. (1995) Studies of the Pt-cinchona alkaloid catalyst for enantioselective *alpha*-keto esters hydrogenation, in Jannes G. and Dubois V. (eds.), *Chiral Reactions in Heterogeneous Catalysis*, Plenum Press, N.Y. p. 105-110.

[10] Koehler J.U. and Bradley J.S. (1998) A kinetic probe of the effect of a stabilizing polymer on a colloid catalyst: accelerated enantioselective hydrogenation of ethyl pyruvate catalysed by poly(vinylpyrrolidone) stabilized Pt, *Langmuir* **14**, 2730-2735.

[11] Gillard R.D. and da Luz de Jesus J.D.P. (1979) Studies on the origin of asymmetry. Part I. Optically active crystals as selective adsorbents, *J. Chem. Soc., Dalton Trans*. 1779-1782.

[12] Penzien K. and Schmidt G.M. (1969) Reactions in chiral crystals: an absolute asymmetric synthesis, *Angew. Chem. Int. Ed. Engl*. **8**, 608-609.

[13] Saeva F.D., Sharpe P.E. and Olin G.R. (1975) Asymmetric synthesis in a cholesteric liquid crystal solvent, *J. Amer. Chem. Soc*. **97**, 204-205.

[14] Thiemann W. and Teutsch H. (1990) Possible amplification of enantiomeric excess through structural properties of liquid crystals-model for origin of optical activity in the biosphere, *Origin of Life Evolution of Biosphere* **20**, 121-126; 1986, **16**, 420.

[15] Teutsch H. (1988) Dissertation, Thesis "Zur Entstehung optischer Aktivitaet in der Biosphere: Ausgewaelte Photoreactionen als Asymmetrische Synthesen", Bremen Univ.

[16] Eskenazi C., Nicoud J.F. and Kagan H.B. (1979) Asymmetric induction in cholesteric media revisited, *J. Org. Chem.* **44**, 995-999.

[17] Craig D.P. and Mellor D.P. (1978) Discriminating interaction between chiral molecules, in *Topics in Current Chemistry, Bonding and Structure,* Part 1, N 63, p. 1-48.

[18] Pavlov V.A., Spitsyna N.I. and Klabunovskii E.I. (1983) Enantioselective hydrogenation in the presence of cholesteryltridecanoate as chiral liquid crystal matrix, *Izv. Acad. Sci. SSSR. Ser. Khim. (russ.)* 1653-1656; *Chem. Abstr*. 1983, **99**,194048j.

[19] Akabori S., Sakurai S., Izumi Y. and Fujii Y. (1956) An asymmetric catalyst, *Nature* **178**, 323-324.

[20] Akabori S., Izumi Y., Fujii Y. and Sakurai S. (1956) An asymmetric catalyst, *J. Chem. Soc. Jpn. (Pure Chem. Sect.)* **77**, 1374-1378.

[21] Akabori S., Sakurai S., Izumi Y. and Fujii Y. (1957) An asymmetric catalyst - a hydrogenase model, *Biochimia* **22**, 154-156.

[22] Itoh T. (1963) A synthesis of *L*-proline from *L*-glutamine, *Bull. Chem. Soc. Jpn.* **36**, 25 -29.

[23] Izumi Y., Imaida S., Fukawa H. and Akabori S. (1963) Asymmetric hydrogenation with modified Raney nickel. I. Studies on modified heterogeneous catalysts. II., *Bull. Chem. Soc. Jpn.* **36**, 21-25.

[24] Izumi Y. (1971) Methods of asymmetric synthesis - enantioselective catalytic hydrogenation, *Angew. Chem. Int. Ed. Engl.* **10**, 871–881.

[25] [a] Izumi Y., Akabori S., Fukawa H., Tatumi S., Imaida M., Fukuda T. and Komatu S. (1965) Asymmetric hydrogenation with modified nickel, *Proc. 3rd Int. Congr. Catal.*, 1964, Amsterdam, North-Holland Publ., Amsterdam, p. 1364-1370.

[b] Izumi Y. (1959) Studies with silk-palladium catalyst. I-III. *Bull. Chem. Soc. Jpn.* **32,** 932 - 945.

[c)] Akabori S. (1959) A comment to the paper of prof. Klabunovskii, in *Proc. of the First Int. Symp. on the origin of life on the Earth,* Moscow 19–24 Aug. 1957, by Clark F. and Singe R.L.M. (eds.) Pergamon Press, London, N.Y. 1959, p.183.

[26] Henderson G.M. and Rule H.G. (1938) A new method of resolving a racemic compounds, *Nature* **141**, 917-918.

[27] Prelog V. and Wieland P. (1944) Ueber die Spaltung der Troeger'schen Base in optische Antipoden, ein Beitrag zur Stereochemie des dreiwertige Stickstoffs, *Helv. Chim. Acta* **27**, 1127-1131.

[28] Petrov Yu.I., Klabunovskii E.I. and Balandin A.A. (1967) Hydrogenation of menthyl ester of *alpha*-phenylcinnamic acid on palladium catalysts deposited on optically active polymers, *Kinet. Katal.(Kinetics and Catalysis, russ.)* **8**, 1384-1386; *Chem. Abstr.* 1968, **68**, 86557a.

[29] Beamer R.L., Smith J.D., Andrako J. and Hartung W.H. (1960) Palladium catalysts. X. Substrate-specific and stereospecific centers, *J. Org. Chem.* **25**, 798 801.

[30] Smith G.V., Cheng J. and Song R. (1996) Enantioselective hydrogenation of prochiral C=C bonds over noble metal catalysts supported by *beta*-cyclodextrin polymer, *Chemical Industries Series* (*Catalysis of Organic Reactions,* Malz R.E., Jr., ed.), (Dekker), **68,** 479-483.

[31] Braca G., Carlini C., Ciardelli F. and Sbrana G. (1976) Polymeric and monomeric Ru-carboxylate complexes: preparation and catalytic activity in heterogeneous and homogeneous phase, in Bond, G.C., Wells P.B. and Tomkins F.C. (eds.) *Proc. 6th Int. Congr. Catal.*, London, 1976, V.1, Chem.Soc. Burlington House, London, p.528-538.

[32] Yin M.Y., Yuan G.L., Huang M.Y. and Jiang Y.Y. (1999) Catalytical behavior of a wool-Pd complex in asymmetric hydrogenation of diacetone alcohol and 3-methyl-2-butanone, *J. Mol. Catal., A. Chem.* **147,** 89-92.

Chapter 4

HYDROGENATION ON CATALYSTS BASED ON Ni AND BIMETALS

Abstract

This Chapter summarizes more than 250 works of new types of metal catalysts, based on nickel and other metals including bimetals, which are active in the enantioselective heterogeneous hydrogenation of *beta*-keto carboxylic acid esters to *beta*-hydroxy carboxylic acid esters, and also are important in the production of monomers for chiral biodegradable polyesters. Cited are works mainly by groups working since the 1950's, which were published from the initial period of discovery up to the present time. Over 30 monographs and reviews on the elaboration of Ni-dissymmetric systems characterize the steps of development of this field from the first modest attempts at asymmetric hydrogenations up to the current excellent effective catalytic systems which produce optical yields of hydroxyesters up to 98% and are comparable to the effectiveness of homogeneous chiral metal-complex catalysts. But unlike them the preparation of heterogeneous metal catalysts are very simple and involve only readily available starting materials like Raney catalysts and tartaric acid as a modifier.

Chapter 4 contains the background of the development of effective modified Ni catalysts, discusses the methods of preparation of different types of stable and active metal catalysts, and discusses the selection of effective modifiers and the most suitable substrate molecules having practical interests. On the basis of these studies a reaction mechanism for the new effective catalytic systems was suggested and experimentally examined. The Chapter discusses the preparation variables for the development of this new type of effective chiral modified Ni catalyst, the supported metal catalysts, the chiral modified bimetal and multimetal catalysts including rare earth metals, and the new chiral modified nickel-ruthenium and palladium catalysts. Attempts are undertaken to elucidate the mechanism of enantioselectivity and to reveal the general regularities of asymmetric actions.

4.1. Background and elaboration of Ni catalysts modified with amino and hydroxy acids

Studies of heterogeneous enantioselective hydrogenation of 2-oxocarboxylic acid esters over chirally modified Ni catalysts were developed first in the early 60's by Izumi's group (Osaka, Institute of Protein Researches) and by Klabunovskii's group (Moscow, Zelinskii Institute of Organic Chemistry) and have been summarized in monographs and reviews [1-31]. Later, experiments were continued by the groups of Smith (SIU Carbondale), Sachtler (Leiden, Evanston), Yasumori (Tokyo), and Wells (Hull). Then during the 70-80's experimental and theoretical works were developed toward improving Ni catalytic systems [32-41,109-113], involving the elaboration of methods of chiral modification of catalysts [42-60,120]; searching for the best suitable substrate molecules besides the 2-oxo-systems [61-66,117-119], and elaborating new methods for increasing enantioselectivity in modified catalysts [67-70,115,116,218-222,235-253], including the search of new bimetal and multimetal modified catalysts based on nickel [120-183,193-198] and supported catalysts [198-206]. Mechanistic studies were reported [75-79,84-112,184-187,208-217,223-234]. Methods of enhanced durability of modified catalysts and their use in practice (see also Chapter 7) were considered in other reports [71-74,80-83,188-192].

A well known fact was that some inorganic and organic compounds, that possess strong complexing ability, can affect the stability and selectivity of metallic catalysts. The effect of non-reacting compounds (so called *modifiers*), on the rates of hydrogenation of unsaturated compounds have been shown by Balandin [234]. He found that the rates depend not only on the nature of the solvent but also on the nature of added compounds that are not able to react under the reaction conditions. Such compounds can modify the catalyst surface and strongly influence the rate and selectivity. If such compounds possess chiral characteristics they might influence the asymmetric action of the catalyst. Probably the earliest example of the combination of a chiral modifier (fructose) and a heterogeneous catalyst (ZnO) was described by Erlenmeyer (see Chapter 5). Certain, but very small enantioselective actions of heterogeneous metal catalysts in the presence of optically active compounds were observed during the hydrogenation of 2-methylcinnamic acid over Raney Ni in a solution of D-glucose resulting in the formation of optically active product with an *ee* of not more than 0.5%. This combination also produced optically active (+)-*N*-phenylalanine from the hydrogenation of 2-acetamidocinnamic acid [9].

The original details of the influence of optically active additives on catalytic reactions were reported by Isoda et al. [32,33]. Hydrogenation of C=O and C=N bonds in diethyl 2-oxoglutarate and diethyl 2-oximinoglutarate were carried out in the presence of modifier compounds (camphor, borneol, amino acids) *in situ* at elevated hydrogen pressures and temperatures.

The most characteristic data are summarized in Table 4.1. and show very striking results: optical yield reached 24% in the presence of several compounds (camphor, borneol) that are not inclined to complex-formation with nickel surface.

The higher optical yields were achieved with the addition of L-amino acids to the reaction mixtures. The addition of L-tyrosine resulted in the formation of phenylalanine with an *ee* of 50%. Unfortunately, these data were not noticed for some time and only in 1963 did the group of Akabori and Izumi [35-41] attempt to reproduce these results; however, the first results were of modest success even using other methods of preparation of the catalysts.

In the early studies of Izumi's group (Osaka) [1-8] the new method of modification of Raney Ni by preliminary treatment of the catalyst with modifier solution was elaborated.

Modifier	Reaction conditions °C	bar	solvent	Product *ee* %
		Diethyl 2-oxoglutarate		
(+)-Camphor	80	63		24.4
(+)-Camphor	100	60		6.7
(+)-Camphor	75	60	Hexane	7.0
(+)-Camphor	50	60	Ethanol	7.0
(+)-Borneol	100	10	Hexane	1.1
(+)-Borneol	80	10	AcOEt	1.1
(-)-Borneol	80	60	AcOEt	23.3
L-Tyrosine	70	74		7
L-Glu	80	56		6
L-Leu	85	30		2.5
L-Phe	85	16		5
		Diethyl 2-oximinoglutarate		
(+)-Camphor	120	20	AcOEt	1.1
(+)-Borneol	120	20	AcOEt	1.1
L-Tyr	60	70		3
		2-acetamidocinnamic acid		
L-Tyr	50	80		50
L-Tyr	60	80		36
L-Tyr	65	74		18

Table 4.1. Enantioselective hydrogenation of the C=O bond in diethyl 2-oxoglutarate to diethyl 2-hydroxyglutarate, of the C=N bond in diethyl 2-oximinoglutarate to diethyl glutamate, and of the C=C bonds in 2-acetamidocinnamic acid to phenylalanine on Raney nickel catalysts, modified with optically active ketones, alcohols and amino acids (according to mainly Isoda et al.[32,33]).

The first results showed that optically active amino acids are not very effective as modifiers and showed that the most effective modifiers are hydroxycarboxylic acids, such as tartaric and malic acids. Also, it was found that the treatment of Raney Ni catalysts with these modifying solutions diminished the catalytic activity for the hydrogenation of the C=O bonds in *beta*-keto esters but not for the hydrogenation of C=C bonds in olefinic substrates (Scheme 4.1.).

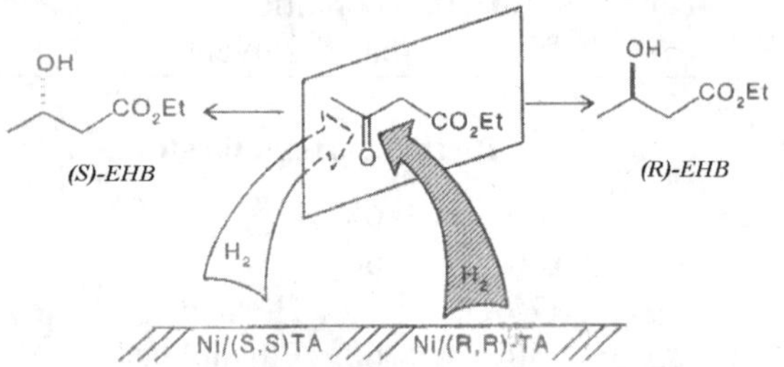

Scheme 4.1. Formation of a chiral center from a prochiral center.

Balandin and Klabunovskii's group (Moscow) [9,10,113] and Smith's group (Carbondale) [220-222,263] examined this type of chiral catalysts using the new method of preliminary modification of the catalysts in solutions of optically active compounds and confirmed that these catalysts produce the hydrogenation products; however, they possess very small optical activities if modifiers are amino acids or amino alcohols [for example (-)-ephedrine or (-)-1-(4-nitrophenyl)-2-aminopropan-1,3-diol], and they found good results using ethyl acetoacetate instead of methyl acetoacetate as the substrate.

The next important finding was that: the best catalyst is Raney Ni modified with tartaric acids (TA) rather than amino acids (AA) using the esters of 2-oxocarboxylic acid (mainly methyl acetoacetate, MAA) as model substrates. Using RNi-TA produced MHB with an *ee* above 50% from the hydrogenation of MAA in the liquid phase at higher temperatures, 50-70°C, and higher hydrogen pressure, 50-100 bar.

A very important discovery was that enantioselective hydrogenation can be realized only with *beta* -keto-esters rather than *alpha*-keto esters. The asymmetric hydrogenation of *alpha* -keto esters demands other types of chiral catalytic systems (see Chapter 5). Hydrogenation of monoketones or 2,3-diketones over RNi-TA gave only modest results, whereas hydrogenation of 2,4-diketones, e.g. acac, gave good results.

Preparation of catalysts

Because catalytic activity and enantioselectivity of modified metal catalysts strongly depend on their preparation procedures, some examples of the preparation procedures of Raney Nickel catalysts modified with TA (MRNi-TA-NaBr) or AA and of supported Ni catalysts (HNi-AA-aerosil) are described below.

Smith and Musoiu [218,264] prepared RNi-TA catalyst by leaching 0.3g Ni-Al alloy at 100-105°C for 2-3h with 15 mL of 20% aqueous NaOH solution. The prepared catalyst was centrifuged, washed with water (7x30ml) with shaking each time for 5-10 min. Modification was carried out by soaking the catalyst in 100 mL of 2% aqueous solution of TA at a pH of 5.0-

5.1. After modification the catalyst was centrifuged, washed with water (2x2ml), and methanol (2x2ml).

Groenewegen and Sachtler [211] prepared supported Ni catalyst using two methods: impregnation and precipitation. The impregnation of the aerosil support was carried out with a solution of nickel nitrate followed by drying at 100°C, heating at 500°C and, finally, reduction in flowing hydrogen at 300-500°C. In the precipitation method, aerosil was mixed with an aqueous solution of Ni-salt and ammonia was added slowly at such a rate that with increasing pH the metal precipitated on the surface of support rather than in the solution. The resulting precipitated catalyst was heated at 500°C and reduced in flowing hydrogen at 300-400°C. By these preparation methods, catalysts containing 17% Ni on aerosil were prepared. Modifications of catalyst were carried out by subliming the appropriate optically active amino acids in vacuum at 100°C.

An efficient RNi-TA-NaBr catalyst was prepared by Harada and Izumi [105] according to the following method: 1.9g of pulverized Ni-Al alloy (42:58) was added in small portions into a solution of 4.5g NaOH in 20 mL water over a 2 min period. The resulting mixture was heated at 110°C for 1h. The liquid was decanted, and the resulting catalyst was washed 15 times with 30 mL of water. Modification was carried out with a solution of 1g TA and 10g NaBr in 100 mL of water. The pH of the solution was adjusted to 3.2 by the addition of 1N NaOH. Modification of 0.8g of Ni was carried out by soaking in this solution, then heating at 100°C for 1 h. The resulting modified catalyst was washed with 10 mL of water. This process was repeated three times. Then the catalysts was washed with methanol (2x50ml) and then with 25 mL of methyl propionate. The hydrogenation of MAA was carried out with this catalyst in a solution of methyl propionate with small amounts of acetic acid. The reaction rate was 8.5 mol/h/g Ni, the optical yield of MHB reached 88.6% (without addition of AcOH; under these conditions the *ee* was only 39.2%). Hydrogenation of acetylacetone (acac) on this catalyst resulted in a diol with an *ee* of almost 100%.

A very effective Ni catalyst was prepared from Ni-acetylacetonate supported on crystallized *alpha*-alumina (Japanese origin "Sumico Rundum") and modified with (2*R*,3*R*)-tartaric acid + NaBr solution. A maximum *ee* of 87% was obtained from the hydrogenation of MAA [110,111,256].

The effectiveness of enantioselective hydrogenations on chiral modified RNi catalysts depends on many factors. It is influenced by variables of modification procedure, pH of solution modifier, structure of modifier, concentration of modifier, temperature of solution, period of action of modified solution, and presence of additional components. In the case of modifier (2*R*,3*R*)-tartaric acid, it proved very important to add NaBr into the modifying solution. The amount of TA adsorbed on the catalyst determined the effectiveness of the resulting MRNi catalyst.

Modification variables

Enantioselective hydrogenation of MAA to MHB was studied as a model reaction and it was found that the hydroxy-dicarboxylic acids, like tartaric acid, were the most effective modifiers. 2-Methyltartaric acid was a very effective modifier, the *threo*-acid showed the best results in this case: RNi catalyst modified with this acid at pH 5.0 resulted in producing MHB with an *ee* of 51%, while the *erithro*-acid had almost no effect [42].

Much attention was spent on the problem of the influence of the nature of alkali cations when adjusting the pH value of the modifying solution. For the hydrogenation of MAA on RNi-TA only NaOH produced good results when used in the TA modifying solution to adjust to a definite pH value (mostly 5.2 for the modification with TA). Using KOH, LiOH or RbOH diminished *ee* values of the resulting MHB. It was suggested that the Na cation in the solution reacts with the free COOH group of tartaric acid during the process of modification, whereas the other COOH group and the OH group are coordinated with the surface of the nickel catalyst. The nature and amounts of other amines or alkalis added to the TA modifying solution exhibited small effects on the *ee's* of the resulting MHB. On the other hand, modifications of RNi with mono-hydroxycarboxylic acids are not sensitive to the nature of the alkali cation in the modifying solution, which indicates that the cation does not take part in the adsorption of these modifiers. A similar behavior was observed when RNi was modified with amino acids [55]. From these data the following conclusion was drawn: in acidic modifying solutions one COOH group of the modifying dicarboxylic acid is linked with the Ni surface and the other is connected to the alkali cation in the solution.

The temperature, concentration, and pH of the modifying solution have strong influences on the effectiveness of the resulting modified catalyst. Increasing the time of modification and the concentration of modifier at optimal temperature and pH enhances enantioselectivity of the catalyst up to an optimal value due to corrosion of the nickel surface with formation of nickel-tartrate chelates coming into the solution. Therefore, for the best enantioselective effect the amounts of TA on the surface of Ni must reach an optimal value.

Instead of the conventionally used modification by pre-immersion of the catalyst in a solution of modifier, Osawa et al. [58,255] used an *in situ* modification during the enantioselective hydrogenation of MAA. Fine nickel powder modified with (2*R*,3*R*)-tartaric acid was used and sodium salt was added to the reaction media. By this method the optical yield was increased up to an *ee* of 79%. Improvement of this method [257] consists in modification *in situ* of finely reduced Ni-powder by addition of (2*R*,3*R*)-tartaric acid and NaBr to the reaction media. In this case, an *ee* of 89% was obtained in the hydrogenation of MAA. The addition of small amounts of NaBr to the reaction media increased both the *ee* and the reaction rate, while, in contrast, the rate decreased with the addition of NaBr to the modification solution in the conventional modification method. Addition of NaBr to the reaction media for the *in situ* modification would have both of the following roles:

i) Na^+ increases *ee* and rate
ii) Br^- increases *ee* and decreases rate.

According to Osawa et al. [258] the *in situ* modified and reduced Ni catalyst showed high durability during repeated use in enantioselective hydrogenation of MAA. In comparison with Raney Nickel used earlier, reduced Ni from nickel-oxide and fine Ni powder were the best precursors for the preparation of effective catalysts. Nickel hydroxide and nickel carbonate were used as the precursors of the nickel oxide [262]. They were calcined to nickel oxides, then reduced to nickel metal, and finally modified with Tart-NaBr modifiers. The temperature of calcination of the precursors was an important variable for attaining high *ee* values in the hydrogenation of methyl acetoacetate. 1100°C was found to be the best calcination temperature, at which the nickel was produced with crystalline sizes of 40-70nm. One of the possible reasons for the lower durabilities of the conventionally pre-modified Ni catalyst with repeated use could be desorption of tartaric acid from the catalyst surface during repeated hydrogenations.

Structure of modifier molecule
Among many classes of organic compounds the best results were obtained with the following combination: a dihydroxy-dicarboxylic acid as modifier and a *beta*-ketocarboxylic acid ester (or *beta*-diketone) as substrate.

Further investigations showed that the means of improving enantioselectivity consists not in searches for new modifier structures but in elaboration of the stereochemical mechanism of reaction based on concepts of 'dual sites', the existence of chiral modified and non-stereospecific centers, on the surface of catalysts (see, Sachtler[209-217], Klabunovskii [9-15], Smith [263-265]).

Effect of solvents and additives
The solvents and additives employed in the reaction system exert a significant influence on optical yield and reaction rate. In the early works (see Lipgart et al.[120]) it was reported that the use of a polar solvent such as methanol, ethanol, or isopropyl alcohol increased *ee's* in the hydrogenation of EAA on RNi-TA in comparison with hydrogenation of neat EAA. Addition of water to the ethanol solvent in the amounts of 25% and 50% diminished the *ee* values of EHB from the original 16% to 4.4% and 0.4%, respectively [120]. But it was found later that aprotic solvents increased the *ee* of MHB and; that either THF or methyl propionate proved to be the most suitable solvents [8,105]. On the other hand a pronounced effect of additives containing carbonyl groups, carboxylic groups or unsaturations was noted. Moreover Ninomiya [63,65] and Higashi et al.[64] found that the addition of water or reducible components to the reaction mixture exert significant influences on *ee* during the hydrogenation of MAA.

Thus Higashi [64] observed that increasing the amounts of added water to the reaction mixture during the hydrogenation of MAA on MRNi, modified with L-amino acids like Ala, Glu and Val, resulted in (*R*)-(-)MHB with *ee* values from 0.4 to 17%. Besides, the addition of a very small amount of water sharply diminished the *ee* and even changed the configuration of the MHB: instead of the (*R*)-(-)-enantiomer the (S)-(+)-MHB was formed. But when RNi was modified with (2*R*,3*R*)-tartaric acid, no significant drop of *ee* was observed after addition of water, and the inversion of the configuration of MHB was not found [65].

The studies resulted in the conclusion that water affects the asymmetric center of the catalyst rather than the keto-enol tautomerism of MAA.

It is of interest to point out that in the enantioselective hydrogenation of ethyl pyruvate on a Pt-alumina-cinchonidine system (see Chapter 5, Wehrli [48]) the presence of water improved the asymmetric effect, which indicates a quite different enantioselective mechanisms in that case.

As mentioned above, the presence of additives containing carbonyl or carboxyl groups in the reaction mixture exert large effects on enantio-selectivity. Ninomiya [63] found that dehydroacetic acid (**1** in Scheme 4.2.), methyl (*E*)-3-acetoxycrotonate (**2**) and diketene (**3**) were always present in small amounts as impurities in commercial MAA. These impurities affected the *ee* values of the MHB during hydrogenations of MAA with MRNi catalysts. For example, the addition a small amount (4%) of dehydroacetic acid to the reaction mixture during the hydrogenation of MAA on MRNi, modified with L-Glu or L-Val, resulted in an increase in the *ee* of MHB from 20.9% up to 32.7%.

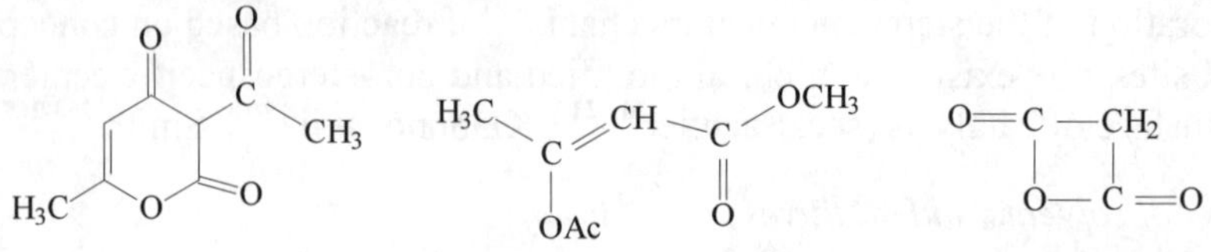

1 **2** **3**

Scheme 4.2.

Temperature effect

It is very difficult to draw general conclusions about the influence of temperature factors on *ee* because the reaction temperature affects the *ee* of MHB in different ways depending on the nature of the modifier. For example, when MRNi is modified with L-amino acids (Glu,Val), the *ee* values of the resulting MHB reached a zero value at 40°C and the configuration changed from (*R*)-(-)-MHB to (*S*)-(+)-MHB as a function of temperature, whereas when MRNi is modified with (2*R*,3*R*)-tartaric acid, the resulting MHB showed a maximum *ee* value at 40°C (Osaki [56,102]).

Effects of the structure of substrates

As mentioned above, the best substrate in the enantioselective hydrogenation on modified Ni catalysts proved to be *beta*-keto esters and *beta*-diketones, compounds that are capable of forming an intermediate six-membered chelate with a Ni atom, or a 'Ni-center' on the surface. Optical yields drop drastically if a monoketone substrate is used.

Besides the acetoacetic acid esters, many studies were devoted to enantioselective and diastereoselective hydrogenation of ketones, especially alkan-2-ones [70,78-80,87] and diketones, such as pentan-2,4-dione (acetylacetone, acac). The latter reaction consists of two steps, the first is the formation of a hydroxyketone, 4-hydroxypentan-2-on (Scheme 4.3.)

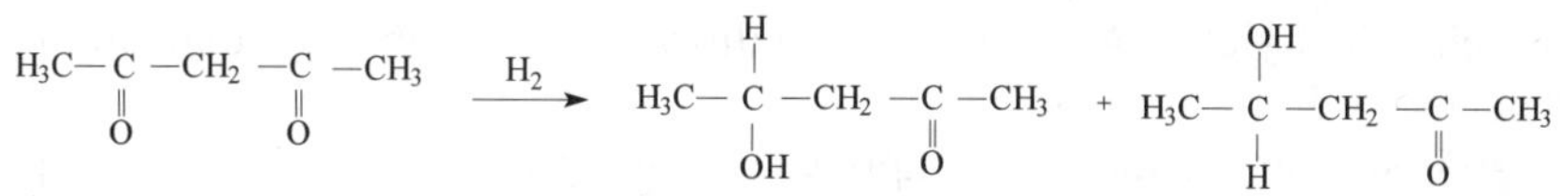

Scheme 4.3.

The second step is the desorption of the hydroxyketone into solution. It is followed by its readsorption and the final hydrogenation into the diol. During the hydrogenation of *acac* the chiral (2*R**,4*R**)-pentan-2,4-diol and the achiral *meso*-diol can also be formed (Scheme 4.4.).

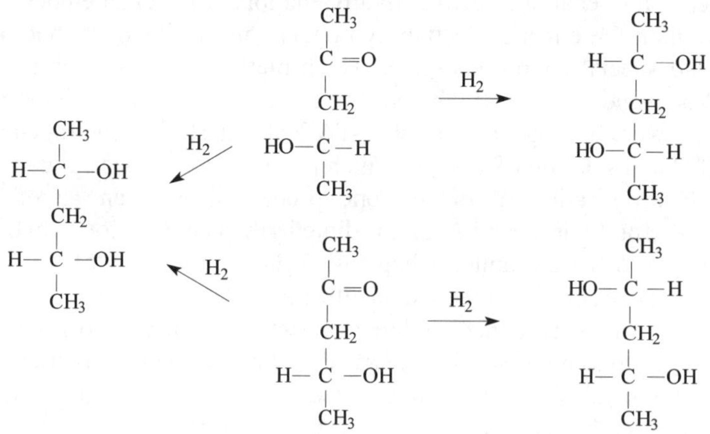

Scheme 4.4.

If the reaction is not selective, the product contains the hydroxyketone, the achiral *meso*-diol, and the chiral (2*R**,4*R**)-pentan-2,4-diol with different optical activities. On chiral catalysts, the first step, the formation of the 4-hydroxypentan-2-one, can proceed enantioselectively, and the second step can proceed diastereoselectively, which can explain the high effectiveness of the enantioselective hydrogenation of *beta*-diketones.

Acac is a strong complex-forming reagent, therefore, a part of the Ni catalyst is corroded and transformed (even up to 75%) into the solution as the [Ni(acac)$_2$] complex. Saturation of Ni with hydrogen and also modification with TA diminishes the solubility of Ni and its ability to form complexes with *acac*. These facts support theoretical suggestions about the formation of the triple complex [Ni - Modifier - Substrate] on the surface of modified Ni catalyst as the reaction proceeds [9,113].

Tanabe [91] hydrogenated *acac* over RNi modified with L-Glu and (2*R*,3*R*)-TA and found that the reaction proceeded according to Scheme 4.3. and Scheme 4.4. A very effective hydrogenation of *acac* on RNi catalyst modified with TA + NaBr gave the diol with high optical yield, 87%, with a 65% chemical yield. After one recrystallization from ether the optically pure pentane-2,4-diol was obtained.

A similar method was applied to the preparation of a number of optically pure diols (Scheme 4.5.) (Ito et al. [88]).

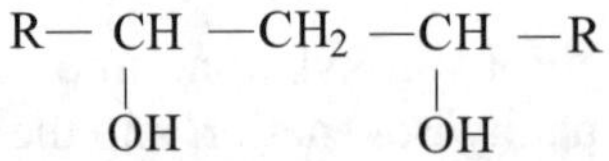

R = Et, Pr, Bu or Ph

Scheme 4.5.

In general the enantioselective hydrogenations of monoketones are less effective than the enantioselective hydrogenation of diketones, but Osawa et al.[79] found special conditions under which methyl-ketones, such as acetone, butan-2-one and 3,3-dimethylbutan-2-one, produced optical yields of 74%, when they were hydrogenated on RNi-TA-NaBr at 100 bar hydrogen pressure and 100°C in a solution of THF, to which pivalic acid has been added.

Hydrogenation of octan-2-one proceeded with an *ee* of 80% at 40°C [71,70,87] and hydrogenation of 2,6-dimethylheptan-3,5-dione gave the corresponding (3*S*,5*S*)-2,6-dimethylheptan-3,5-diol with an *ee* of 99.5% and a 29% chemical yield after three recrystallizations [71].

It is of interest that diethyl oxalacetate (Scheme 4.6.), that can be considered both as an *alpha*- and a *beta*-keto ester, cannot be hydrogenated on the RNi-TA catalyst. This demonstrated that small changes in the groups around the carbonyl group may result in difficulties in asymmetric hydrogenation (Tanabe et al.[92]).

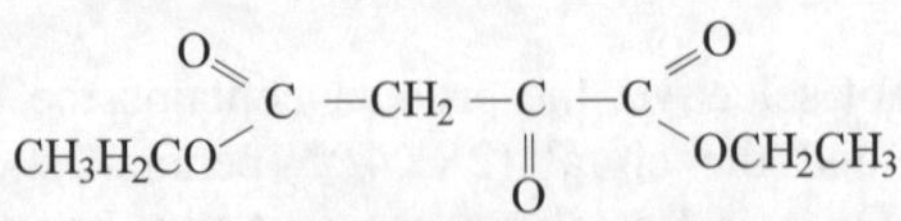

Scheme 4.6.

4.2. Preparation variables for new nickel catalysts modified with tartaric acid

The important question of the effect of the composition of the original Ni-Al alloys on enantioselective ability was not studied in detail. But Zubareva et al. [119] studied samples of Ni catalysts prepared from commercially available Ni-Al alloys with different contents of intermetallic $NiAl_3$ and Ni_2Al_3 with results shown in Table 4.2. An increase of $NiAl_3$ in the alloy facilitated etching of the alloy and diminished the contents of Al in the resulting Ni catalysts.

Sample No.	Chemical analysis (%)				Phase analysis (%)		
			Ni- Al alloys				
	Ni	Al	Cr + Ti		$NiAl_3$		Ni_2Al_3
1	50	50	-		10.0		90.0
2	15.2	51.2	3.6		25.1		74.9
3	38.3	59.7	2.0		51.2		48.8
			RNi catalysts				
	Ni	Al	Cr + Ti	Rest	Ni	Al	Bayerite
2	70.6	15.8	2.8	10.8	91.6	5.4	3.0
3	73.0	6.8	0.12	20.1	97.9	0.0	2.1

Table 4.2. Chemical and phase compositions of Ni-Al alloys and RNi-catalysts (according to Zubareva et al. [119]).

RNi No. 3, prepared from the commercial NiAl alloy No. 3, containing 2% Cr+Ti and small amounts of other additives, contained 6.8% Al, approximately half of the Al in RNi sample No. 2, which was prepared from the NiAl alloy No. 2. Impurities during the preparation of RNi No. 3 were almost completely removed. Table 4.2. and Fig. 4.1. show that decreasing the Ni_2Al_3 phase in the alloy and increasing the contents of $NiAl_3$, promoted an increase in both the general activity and the enantioselectivity of the resulting RNi catalysts.

Harada et al. [103] have found that the presence of traces of aluminium or Al-compounds in Ni catalysts are unfavorable for enantioselective activity. Therefore the use of Ni catalyst prepared by reducing nickel oxide instead of etching Ni-Al alloy resulted in a MHNi that exhibited optical yields up to 85%.

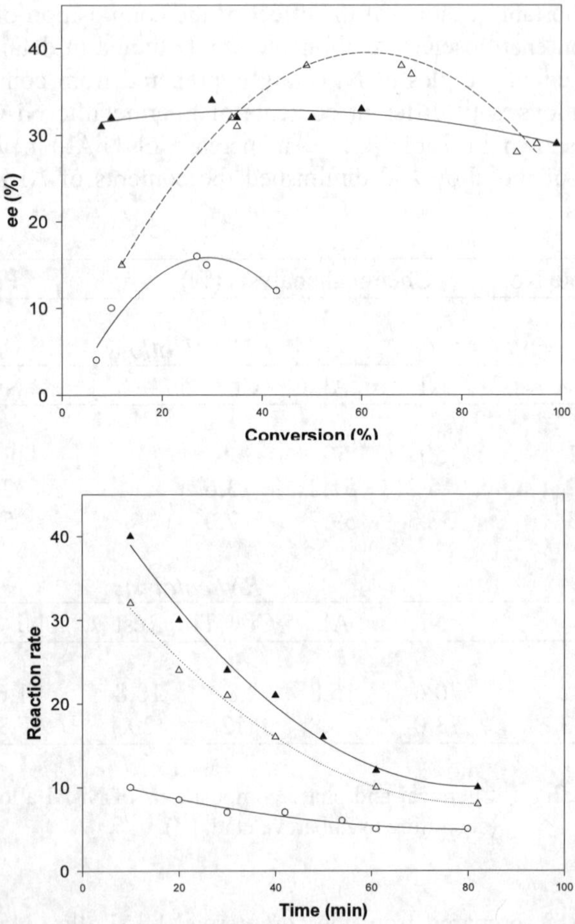

Figure. 4.1. Influence of the composition of Ni-Al alloy on the catalytic properties of RNi catalysts, modified with (2*R*,3*R*)-tartaric acid, in the enantioselective hydrogenation of ethyl acetoacetate. Sample numbers 1(o), 2(Δ) and 3(▲) are in Table 4.2. (according to Zubareva et al. [119]).

It was found that the addition of 1-2% AcOH to the reaction mixture is very favorable. Table 4.3. shows that commercially available NiO reduced for 1 h at 350°C in flowing hydrogen gives the best results (81.6% *ee*) in the hydrogenation of MAA.

Source of Activated Ni	Reaction Temperature	Solvent	Optical Yield of MHB
Raney Ni	60	neat	44
Ni-formate	100	neat	54.8
NiO	120	THF + AcOH	68
NiO	120	THF + AcOH	81.6

Table 4.3. Effect of different kinds of activated nickel catalysts modified with tartaric acid in the enantioselective hydrogenation of MAA (adapted from [103].

The catalytic activity and enantioselectivity of the catalysts depend on the mode of preparation of RNi from the original alloy. According to Gross and Rys [223] etching the Ni-Al (1:1) alloy with 20% NaOH aqueous solution under different conditions (temperature, time, concentration of alkali solution) results in RNi with different activities and stabilities. The catalysts prepared at a low etching temperature showed higher hydrogenation activities and aluminium contents than catalysts prepared at a high etching temperature; moreover, after modification with TA solution the low etching catalysts produced higher *ee* values in the hydrogenation of MAA. Most catalytically active catalysts, such as W-6, exhibit high activities but low enantioselectivities, while one of the least active catalysts, W-1, exhibits high *ee* values. It follows that Ni catalysts that are least contaminated with remaining Al (or with other components) may provide catalysts with high enantioselectivities. Therefore the nickel catalysts prepared by decomposition of nickel salts exhibited higher enantioselectivities than Raney Ni catalysts.

Vedenyapin et al. [116] used mild reduction conditions in the preparation of HNi catalysts by heating Ni-nitrate in flowing He at 450°C followed by reduction of the resulting NiO in a H_2-He mixture (1:10) at 280°C. Eight grams of the resulting Ni was modified in air with 100 mL of 0.5% solution of (2*R*,3*R*)-(+)-tartaric acid at pH 5.1 and 50°C for 1h. Hydrogenation of EAA (neat) with 0.5g of this Ni at 100°C and 100 bar gave (-)-EHB with an *ee* of 71%, a value that exceeded an earlier value [218].

In general, nickel catalysts prepared by reduction of NiO are very sensitive to preparation variations, even to the mode of manufacture of commercial NiO (*ee* of obtained MHB can change from 68 to 86%). Modifications of HNi with tartaric acid + NaBr solution were presented in Figures 4.2.-4.4. This catalyst was characterized by high *ee*'s but exhibited the disadvantage of slow reaction rates; therefore, the reaction had to be performed at temperatures above 100°C, in which case by-products were formed.

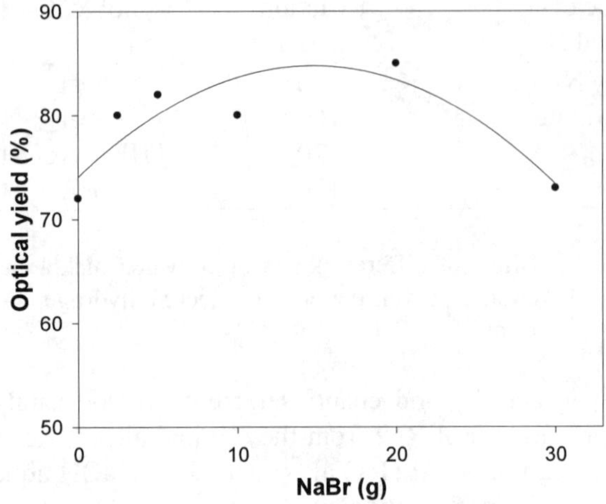

Figure 4.2. Relationship between the amounts of NaBr in the modifying solution and the optical yield of MHB on MNi-Tart-NaBr (conditions are the same as in Figure 4.4.) [59].

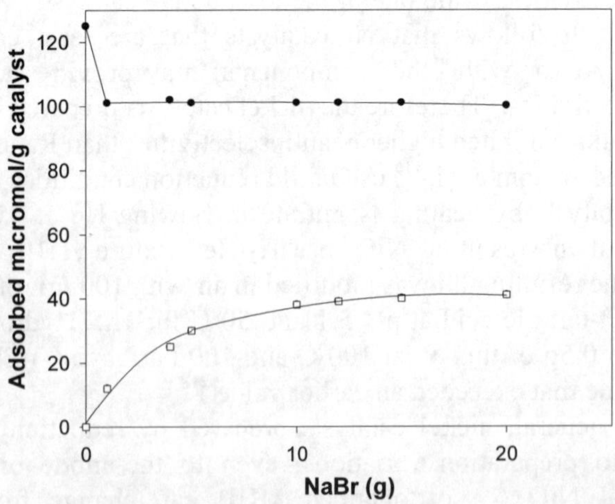

Figure 4.3. Relationship between the amounts of NaBr (o) and adsorbed tartaric acid (•) in solutions modifying activated Ni (conditions are the same as in Figure 4.4.)(adapted from [59,103]).

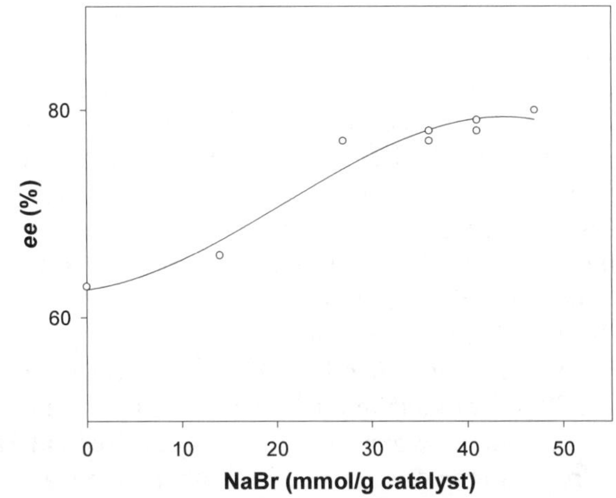

Figure 4.4. Relationship between the amount of adsorbed NaBr on the surface of Tart-NaBr-MRNi catalyst and *ee* in the hydrogenation of methyl acetoacetate.
Conditions: 110 bar, 100°C and the ratio of volumes of MAA to solvent to AcOH are 11.5 : 23 : 0.2 (Harada [59]).

As has been shown in Figures 4.2.-4.4., adsorption of NaBr on non-specific Ni-centers can enhance *ee* in this hydrogenation as a result of blocking of "racemic" centers, which produce racemic MHB and result in increases in optical yields [59].

Inorganic salts	No. of modification procedures	Optical yield of (*R*)-MHB
None	1	39.2
NaBr	1	83.1
NaBr	2	86.1
NaBr	3	88.6
NaCl	1	72.1
Na_2SO_4	1	56.4

Table 4.4. Effects of inorganic salts, added to the (2*R*,3*R*)-tartaric acid modifying solution, on enantioselectivity in the hydrogenation of MAA on MRNi (mainly according to Harada [105]).

It was frequently noticed that addition of water and some other reducible additives to the reaction system during the reaction have a significant effect on *ee* [63-65]. Also it was found that during the preparation of MRNi, addition of some inorganic salts to the modifying solution sharply increased enantioselectivity during the hydrogenation of MAA. Table 4.4. summarizes the effect of addition of different salts. The greatest effect occurred with the addition of NaBr. This mode of modification with (2*R*,3*R*)-tartaric acid + NaBr was used in further studies by Izumi's group [1-8].

The basic modification procedure consisted of the following: 10g of Raney Ni were soaked in the modifying solution at 100°C. The mixture was kept for 1h with shaking, the solution was removed by decantation and the residue was washed with 10 mL of deionized water, then the catalyst was washed with 50 mL of methanol and with 25 mL of methyl propionate. This procedure was repeated two or three times (results seen in Table 4.4.).

Hydrogenation was carried out after removal of the solvent by decantation, suspending the MRNi in 23 mL of MAA, followed by hydrogenation in an autoclave at 90 bar and 100°C with shaking until hydrogen uptake ceased (about 10h).

Another method of improving enantioselectivity of modified Ni was reported by Tai et al. [67-69].

The idea of the existence of two kinds of centers on the surface of modified Ni catalysts, MRNi, was realized by using ultrasound irradiation in the treatment of the catalyst *before* its modification [67,68]. A sample of W-1 type RNi (according to Rys [223]) was prepared from 38 g of Raney Ni-Al alloy (Ni:Al=42:58). This alloy was etched in 20% NaOH solution followed by further washing with 400 mL deionized water. Then a suspension of the catalyst was subjected to ultrasound irrradiation in an ultrasonic generator (118W, 48kHz) for 3 min and the turbid supernatant suspension was separated by decantation from paramagnetic Ni-powder with a magnet. This operation was repeated several times to afford RNi-U catalyst. The particles in the supernatant were collected by centrifugation. Sixty-nine mg of particles (0.3% of the original 38 g RNi alloy) were obtained. The same procedure was repeated with HNi catalyst for preparation of HNi-U. Modifications of the catalysts were performed in a solution of 24 g (2*R*,3*R*)-tartaric acid + 240 g NaBr in 2400 mL water.

Table 4.5. compares results using traditional MRNi catalysts as references and results using catalysts treated with ultrasound. The new catalysts were characterized by an increase in reaction rates, by factors of 2 to 4, and an increase in *ee* of 5-8%. The highest *ee*, 94%, was obtained in the preparation of (*R*)-methyl 3-hydroxytetradecanoate (No.13). In the hydrogenation of MAA, a new catalysts proved to be very efficient for the large scale production (10 kg) of (*R*)-MHB with an *ee* of 79% (No.4). It is important to note that ultrasound treatment should be used before the modification procedure. When RNi was ultrasonicated in a modification solution, the resulting catalysts showed poor enantioselectivity (No.5, *ee* 58.6%) and the solution thereby contained was contaminated with aluminium tartrate, which had originated from aluminium in the Ni-Al alloy. The use of MHNi instead of MRNi resulted in no significant changes from the ultrasonic treatment (No.6 and No.7). This fact indicated that the ultrasonic treatment is only effective on the fragile aluminium contaminations in Raney Nickel catalysts.

Sample No.	$R-C(=O)-CH_2CO_2CH_3$ R	Catalyst	Reaction time (h)	Method	ee (%)	Chemical yield,% [69]
1.	Me	MRNi-U	2.5	A	83.9	70(>98)
2.	Me	MRNi	5.0	A	78	
3.	Me	MRNi-U	3.0	B	83.9	
4.	Me	MRNi-U	12	C	79	
5.	Me	MRNi-U	3	A	58.6	
6.	Me	MHNi-U	20	A	81	
7.	Me	NiHNi	20	A	81	
8.	Et	MRNi-U	4	A	92	
9.	$CH_3(CH_2)_6$	MRNi-U	36	A	89	79 (>98)
10.	$CH_3(CH_2)_6$	MRNi	144	A	83	
11.	$CH_3(CH_2)_8$	MRNi-U	48	A	91	80 (100)
12.	$CH_3(CH_2)_8$	MRNi	192	A	86	
13.	$CH_3(CH_2)_{10}$	MRNi-U	48	A	94	81(>98)
14.	$CH_3(CH_2)_{10}$	MRNi	192	A	86	
15.	$CH_3(CH_2)_{12}$	MRNi			87	82
16.	Me	MRNi			83	

Table 4.5. Enantioselective hydrogenations of methyl 3-oxoalkanoates over ultrasonicated (U) and untreated Tart-NaBr-MRNi and Tart-NaBr-MHNi catalysts at 100 bar and 100°C (adapted from Tai [67-69]).
Reactant A (10 g), catalyst (1.9 g), $EtCO_2Me$ (20 mL), AcOH (0.2 mL),
Reactant B (200 g), catalyst (10 g), $EtCO_2Me$ (200 mL), AcOH (2 mL),
Reactant C (10 kg), catalyst (100 g), $EtCO_2Me$ (10 L), AcOH (15mL).

Ultrasonicated catalyst, RNi-U, proved to be very effective besides the hydrogenation of 3-oxoalkanoates, in the enantioselective hydrogenation of 1,3-diketones (pentan-2,4-dione, 2,3-dimethylheptan-3,5-dione) into diols, too, with *ee*'s of 86-91%. For example, this new type of catalyst was used for the preparation of optically pure (*R*)- and (*S*)-methyl 3-hydroxyalkanoates after hydrogenation in high enantioselectivity followed by crystallization of their dicyclohexylammonium salts [69] (see Table 4.5, Chemical yields).

Recently Tai's group [66] demonstrated a new example of high efficacy of enantioselective hydrogenation in heterogeneous catalysis, which is comparable to the best cases of the enantioselective reactions of homogeneous chiral metal-complex catalysts. The MRNi-TA-NaBr catalyst was treated with ultrasonic irradiation in water, then it was modified with a solution of tartaric acid + NaBr at 100°C and at pH 3.2 (the pH of the solution was adjusted with NaOH) for 1h. A series of *beta*-keto esters was hydrogenated on this catalyst at 100°C and 60°C. The results showed that the higher *ee*'s were obtained at the lower temperatures. The hydrogenation reactions were carried out both on untreated and on ultrasonicated catalysts (Table 4.6.).

	untreated catalyst *ee* (%)	ultrasonicated catalyst *ee* (%)
methyl (3-oxobutanoate)	86	86
methyl (3-oxopentanoate)	91	94
methyl (4-methyl-3-oxopentanoate)	88	96
methyl (5,5-dimethyl-3-oxohexanoát)	84	96

Table 4.6. Hydrogenation of methyl (3-oxoalkanoates) on untreated and ultrasonicated MRNi-TA-NaBr catalyst.

The most important result was that the highest *ee* value was obtained in the hydrogenation of methyl (3-cyclopropyl-3-oxopropanoate) into the optically very pure prospective synthon, methyl (3-cyclopropyl-3-hydroxypropanoate). The starting material was shaken in 4g of THF and 0.2 mL of AcOH at 100 bar hydrogen in an autoclave with 0.9g ultrasonicated MRNi-TA-NaBr and resulted in producing (*S*)-(+)-methyl (3-cyclopropyl-3-hydroxypropanoate) with 100% chemical yield. In one experiment the optical yield was found to be 96% at 100°C and 98% at 60°C. In other experiments reproducibilities of the *ee*'s of the (*S*)-(+)-hydroxyester at 60°C were 98.1%, 98.3% and 98.6%. It is clear that hydrogenation under the same conditions with a catalyst modified with (2*S*,3*S*)-tartaric acid gives (*R*)-(-)-methyl (3-cyclopropyl-3-hydroxypropanoate) with an *ee* of 98%. This was the maximum optical yield received in enantioselective hydrogenations on heterogeneous metal catalysts based on Raney Ni.

The idea of two kinds of centers on the surface of nickel in the modified MRNi catalysts was strengthened by results from the treatment of catalyst *before* its modification with ultrasound irradiation [67,68]. Theoretically though, a more effective method might be selective poisoning of the non-enantioselective centers in chirally modified catalysts.

According to the model of active surfaces of modified catalysts two kinds of centers, selective and non-selective, exist. It may be possible to improve enantioselectivity of hydrogenation if non-modified centers producing racemic MHB can be blocked with a poison, while modified centers are being protected with tartaric acid modifier and therefore are not poisoned. The effect of poisoning depends on the strength of adsorption of the poison in comparison with that of the modifier and ester to be hydrogenated. Yasumori [225] attempted to increase the *ee* in enantioselective hydrogenation of MAA on MRNi-Tart by adding catalytic poisons to the reaction. Figure 4.5. shows the effects of added poisons on the resulting *ee* of MHB.

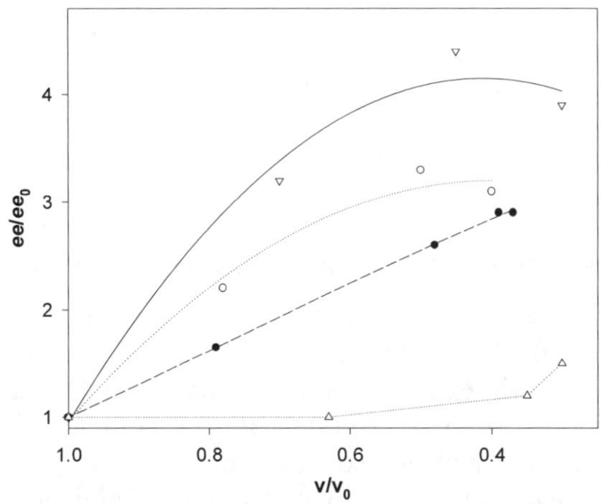

Figure 4.5. The effect of poisoning on enantioselectivity in the hydrogenation of MAA at 80°C on Tart-MRNi catalysts.
The plot represents the relative optical yields, ee/ee_o, versus the relative reaction rates, v/v_o, where v_o and ee_o (11%) are the rate and *ee* for the unpoisoned catalyst. Poisons are butylamine (∇), triethylamine (o), ammonia (•) or pyridine (Δ) (mainly according to Yasumori [225]).

Poisons applied in [225] were butylamine, triethylamine, ammonia and pyridine, all of which diminished the reaction rate but increased the optical yield .

Another method was elaborated by Chernysheva et al.[186]. They studied the enantioselective hydrogenation of EAA (neat) on MRNi-Tart in the liquid phase at 75°C and 1 bar, using two methods of poisoning.

The first method started with modification and it was followed by poisoning (M-P),

The second method started with poisoning, and was followed by modification (P-M).

Poisons were thiophene and pyridine. In method P-M, the addition of thiophene *prior to* modification, inhibited the most active racemic EHB producing centers of the catalyst, resulting in the increase of *ee* from 11% to 15% at the 20% conversion level. Addition of pyridine by the same method, increased the *ee* from 2.35% to 13.5% at 10% conversion. However, upon further hydrogenation the *ee* decreased.

In method M-P the addition of thiophene after modification increased the *ee* value from 6.8 to 11% at 10% conversion. Thus the inhibition effect of the poison in the P-M process appears to be more effective than in the M-P process. This suggests that in the latter case all of the active centers are already occupied by the tartaric acid modifier molecules, which is more easily displaced from the less active chiral sites than from the racemic sites.

It was found that the relative adsorption coefficinents (poison/EAA), calculated according to Balandin's theory of hydrogenation [234], are 25 for pyridine and 14.6 for thiophene (in the M-P process), while in the P-M process they are much higher (1400).

4.3. Bimetallic and multimetallic modified hydrogenation catalysts

4.3.1. Supported Ni catalysts promoted with noble metals

In 1976 Orito et al. [193-197] discovered that nickel catalysts deposited on Kieselguhr (Kg) (1:1), alloyed with 1% noble metals, and modified with (2*R*,3*R*)-tartaric acid exhibited an exceptionally high enantioselectivity in the hydrogenation of MAA to (*R*)-(-)-MHB. According to Orito [197] the optical yield of MHB reached an extremily high value, 99.4%, using Ni-Pd-Kg catalyst. However, Orito's group at that time determined the value of *ee* using polarimetric measurements and adopted the specific rotation value of 20.9° for the optically pure MHB (neat). But as it was later established, the specific rotation of neat MHB is 22.95° (Harada et al.[105]), therefore the highest value of *ee* in the above studies should have been reported as 79.9-90.5 %, which are still extremely high values. The importance of these results led us to examine methods of preparation of the catalysts, the methods of carrying out experiments on the hydrogenation of MAA [193], and discuss the reason of such high enantioselectivity.

Kieselguhr (manufactured by the National Chemical Laboratory, Japan) was used without any additional purification or treatment (according to the personal communication by Prof. Orito in 1980 to one of the authors of this book, E.K.) in the solution of nickel nitrate and a platinum metal salt such that the ratios of Ni : Pt : Kg = 100 : 1 : 100 were attained. The mixture was allowed to stand for 1h and then was treated with a solution of sodium carbonate. The precipitate was throughly washed and dried for 24 h at 110°C. The resulting 3g of catalyst was heated by quartz lamp in a stream of hydrogen (7 L/h) up to 300°C for 1h (formation of alloy?), after which it was cooled and added to 100 mL of a 1.5% aqueous solution of (2*R*,3*R*)-tartaric acid at pH 4.1 under bubbling nitrogen. The temperature of the solution gradually increased to 85°C. The catalyst was then centrifuged and washed initially twice with water (with 25 mL in each case), then with methanol, and finally with the solvent in which the hydrogenation was to be carried out. The reaction was performed in an autoclave with stirring under a hydrogen pressure of 60-80 bar with a substrate to catalyst ratio of 14:1. The reaction temperature was raised from 20°C to the required level at a rate of 0.7° per min. When the hydrogen uptake was complete, the catalyst was filtered off and the product distilled. The compilation of data in Table 4.7 is based on references 193-197 and. shows the maximal optical yields of MHB under different conditions (the *ee* values were corrected in accordance with the 22.95° specific rotation of MHB [105]).

By carrying out the reaction in THF and by adding small amounts of carboxylic acids into the reaction mixture the *ee* values of the products can be increased by 6-10%.

The method of synthesis of MHB with an *ee* of up to 90.6% was developed for the Ni-Pd-Kg catalyst, on which modification with tartaric acid is carried out at pH 4.1-4.5 and 84°C. Before reaction, the catalyst must be treated for 4h in THF without stirring, having added 1.5% of acetic acid at 109°C and maintaining a hydrogen pressure of 55 bar (Table 4.7., experiment No. 10). The reaction mixture (MAA in 44 mL of THF to which 1 mL of AcOH has been added) is kept initially without stirring for 1h at the same temperature and pressure and is then stirred for 10-20 h. The optical yield of MHB thus obtained reached a maximum value of 90.6% (Table 4.7., experiment No. 11). Instead of acetic acid, formic or benzoic acid may be added. The optical yield of the product could be increased if the added monocarboxylic acid has a pK_a, lower than that of the modifying agent, (2*R*,3*R*)-tartaric acid ($pK_a = 2.89$).

It follows from Table 4.7. that alloying of the nickel catalyst with platinum metals significantly increases enantioselectivity and the overall rate of the reaction. The deposited Ni-Ru-Kg catalyst is more effective (producing an *ee* of 50%) than the Raney catalysts, RNi-Ru, without a support, on which *ee*'s of only 4% and 17% were achieved, for the MAA and EAA hydrogenations, respectively [9]. An increase in the temperature of the modifying TA solution from 20°C to 80-90°C helps to increase the *ee* values of MHB. When the reaction is carried out in alcohol solutions, the *ee's* of MHB diminish but the overall rate of reaction increases. The best solvents proved to be THF and ethyl acetate with addition of a carboxylic acid (acetic, formic or benzoic). The antiparallel changes of *ee* and the overall reaction rate observed earlier by Orito [194-197] were confirmed by Klabunovskii and Vedenyapin [9]. Amongst the supports investigated (activated charcoal, alumina, silica gel, Kieselguhr) the best results were obtained with Kieselguhr (but only according to the data of Orito's group because, unfortunately, there are no literature data on the reproducibility of these results). Nevertheless, a question arises concerning the reason for the high enantioselectivity of nickel catalysts alloyed with other metals and especially concerning the specific role of supports. Bimetallic catalysts based on nickel and copper with added platinum metals (ruthenium, palladium), according to the reference 9, did not exhibit an appreciable enantioselectivity in the EAA hydrogenation reaction (see next Parts of this Chapter 4 and references 9 and 238).

Number of experiment	Catalyst	Solvent	Rate	$[\alpha]_D^{20}$	*ee* (%)
1[a]	Ni-Kg	none	5.9	12.20	53.2
2	Ni-Rh-Kg	none	7.8	13.75	59.9
3	Ni-Pt-Kg	none	9.7	14.20	61.9
4[b]	Ni-Pt-Kg	THF	4.7	15.00	65.4
5[c]	Ni-Pt-Kg	THF	7.2	18.05	78.6
6[d]	Ni-Pd	none	3.8	9.15	39.8
7	Ni-Pd-Kg	none	8.7	14.65	63.8
8[c]	Ni-Pd-Kg	THF	6.9	18.22	79.4
9[c]	Ni-Pd-Kg	THF	2.8	20.10	87.6
10[c]	Ni-Pd-Kg	THF	2.5	20.44	89.1
11[e]	Ni-Pd-Kg	THF	2.5	20.79	90.6
12[f]	Ni-Pd-Kg	THF	2.6	20.49	89.3
13	Ni-Pd-C	none	5.6	13.02	56.7
14	Ni-Pd-Al_2O_3	none	5.2	11.13	48.5
15	Ni-Ir-Kg	THF	4.9	16.50	71.9
16[b]	Ni-Ru-Kg	none	8.8	11.35	49.5
17	Ni-Ru-Kg	none	7.8	12.35	53.8
18[g]	Ni-Ru-Kg	none	9.3	12.05	52.5
19[c]	Ni-Ru-Kg	THF	6.7	16.75	73.0
20[h]	Ni-Ru	none	1.6	4.10	17.0

Table 4.7. Rate of formation of (*R*)-(-)-MHB (mmol h^{-1} g^{-1}) and *ee* values in the enantioselective hydrogenation of methyl acetoacetate on deposited nickel-kieselguhr catalysts, promoted with 1% noble metals and modified with (2*R*,3*R*)-tartaric acid (according to summarized data of Orito et al.[193-197]).
Composition: Ni:Pt:Kieselguhr = 100:1:100
Samples: 2.3g catalyst, 2.3g MAA, 44 mL solvent
Modification: 1.5% aqueous solution of TA, at pH 4.3 and 84°C
Hydrogenation: 55 bar H_2, temperature 21-130°C
(a) Ni:Kg = 1:1; (b) catalyst was modified at 20°C; (c) 1 mL AcOH was added; (d) Ni:Pd =100:1; (e) 0.05g HCOOH was added;
(f) 0.5g PhCOOH was added; (g) Ni:Ru:Kg = 100:1:100;
(h) Ni:Ru =100:1.
Modification conditions: pH 4.5, 20°C. Samples: 0.2g of the catalyst, 41 mL ethyl acetoacetate. The hydrogenation was carried out at 60°C (these data are given for comparison, according to Klabunovskii[9])

Nitta et al.[205] investigated in details modified nickel catalysts deposited on different supports including Kieselguhr. They exhibited high enantioselectivities in many cases, but Kieselguhr as a support revealed no special properties. Apparently the addition of small amounts of noble metals to the nickel catalyst preparation stage promotes a more complete reduction of nickel salt and the formation of nickel crystallites of optimum size [205]. The NiO reduction process is influenced significantly by Cl- and Br-anions. The addition of $PdCl_2$ to the initial NiO-SiO_2 system promotes the reduction of NiO but has little influence on enantioselectivity [205].

Contrary to the data of Orito [193-197], the deposition of nickel on Kieselguhr of the Japanese origin, "Shimalite brand" or on alumina with subsequent modification by (2*R*,3*R*)-tartaric acid does not make it possible to attain a value of *ee* in excess of 54-68% from the hydrogenation of MAA, even in the presence of high metal contents on the support (see Harada et al.[59]). Similar *ee* values were also obtained when nickel was deposited on TiO_2 (*ee* of 70%) and ZrO_2 (*ee* of 68%) at 90 bar hydrogen pressure and 20°C [59].

The higher *ee* values attained by Orito [193-197] appear to be partly accounted for by the high nickel content on the support (50%), which actually converts the deposited catalyst into nickel black catalyst and also by the especially favorable conditions for formation of the catalyst created by the addition of platinum metals. It is also possible that the Kieselguhr employed in the above studies contained promoting but unknown admixtures. Nitta [205] concluded that the high effectiveness of the nickel catalysts [193-197] may be accounted for by the special condition in catalyst formation. According to Nitta [205], the introduction of 1% Pd into the Ni-SiO_2 catalyst during the reduction stage of the $NiCO_3$-SiO_2 precursor promotes the reduction of the salt, allowing for a decrease in the process temperature, and forms larger nickel crystallites. To better understand the reason for the high enantioselectivity of noble metal-promoted nickel-support catalysts detailed studies of the preparation process of modified nickel catalysts will be needed.

Conditions	Nitta et al.			Orito et al.
	A	B	C	
$Ni(NO_3)_2 : Na_2CO_3$	1	1.2	0.83	1
Deposition temperature (K)	348	358	273	298
Ageing time (min)	15	30	5	60
Drying temperature (K)	383	393	323	383
Drying time (h)	24	45	20	24
$NiCO_3$ content in deposit (%)	35-40	7-10[a]	60-65	-

[a] $Ni(OH)_2$ content

Table 4.8. Conditions of preparation methods of Ni-SiO_2-TA catalysts for the hydrogenation of MAA (ratio of Ni:SiO_2 = 1:1) adopted from Nitta et al.[205] and Orito et al. [193].

Nitta [205] attempted to reproduce the data of Orito [193-197]. The Ni-SiO_2 catalysts were prepared by three methods differing in the condition of deposition of catalyst and its reduction. The role of Kieselguhr (60-200 mesh) as a support and the influence of the introduction of Pd on the formation of the catalysts and on their enantioselectivity were also investigated.

Table 4.8. presents the conditions for the preparation of nickel catalysts described by Nitta [205] and, for comparison, the conditions quoted by Orito [193]. In that work [205] Nitta used silica gel with a specific area of 600 m^2/g and Kieselguhr with a specific area of 5.7 m^2/g supplied by the Japanese company Yoneyama Yakuhin Co. The suspension of the support was impregnated with a solution of nickel nitrate and precipitated with a solution of carbonate at 75°C. After drying at 100°C the catalytic mass was reduced in a stream of hydrogen for 1h at 500°C. In order to obtain the Ni-Pd-SiO_2 system, $PdCl_2$ was added to the catalytic mass before its reducion. The Pd was introduced in amounts to prepare a ratio of Ni:Pd:support =100:1:100 [205]. As can be seen from Table 4.8., methods A and C differ in the severity of the conditions during the deposition and treatment of the catalyst mass. The catalyst obtained by method C contained an excess of $NiCO_3$. Method A is distinguished by milder conditions, similar to those adopted in Orito's experiments [193-197].

Preparation method [a]	Catalyst [b]	T_r (K)	D (nm)	*ee* (%)
A	Ni-Kg	523	31	32.5
A	Ni-Pd-Kg	523	16	26.3
A	Ni-Kg [c]	573	31	12.1
A	Ni-SiO_2	573	-	23.8
		673	9	50.6
A	Ni-Pd-SiO_2	573	-	27.8
		673	12	41.6
B	Ni-SiO_2	673	4	29.8
B	Ni-Pd-SiO_2	673	5	20.1
C	Ni-SiO_2	573	13	43.4
C	Ni-Pd-SiO_2	573	13	30.1
C	Ni-SiO_2 [c]	573	13	36.8
C	Ni-Pd-SiO_2 [c]	573	13	34.6
C	Ni-Pd-Kg	573	12	27.7
	Ni-Pd-Kg [c]	573	-	63.8 [d]
	Ni-Pd-Kg	573	-	89.1 [e]

Table 4.9. The influence of the catalysts preparation method, the nature of the support, the reduction temperature (T_r) and the added palladium on enantioselectivity in the hydrogenation of methyl acetoacetate (MAA) to (*R*)-(-)methyl hydroxybutyrate (according to Nitta et al . [205]).
(a) see Table 4.8.; (b) Ni-Pd/support =100:1/100, modification with (2*R*,3*R*)-tartaric acid at pH 5.1; (c) modification at pH 4.1 (adapted mainly from Orito [193]).

Method B affords a deposit containing less effectively reduced $Ni(OH)_2$, the reduction of which gives rise to large crystallites, even in the presence of palladium [205]. Depending on the composition of the initial mixture, the catalysts differ significantly in their enantioselectivity in hydrogenation of MAA in AcOEt at 60°C and a hydrogen pressure of 10 bar. It was shown [205] that nickel catalyst with added Pd, deposited on silica gel or Kieselguhr are usually less effective than the catalyst without added Pd, which conflicts with the data of Orito's group [193]. Admittedly, the introduction of palladium into the catalyst obtained by method A promoted some increase in enantioselectivity as a consequence of the increase in the average size of the nickel crystallites (see Table 4.9.).

The use of Kieselguhr did not increase but actually somewhat decreased *ee's* even when modification was carried out by the method of Orito [193] at pH 4.1. The reaction rate on all samples of Ni-SiO_2-TA catalysts were similar and amounted to 1.4 - 0.8 mmol/min/g Ni. The introduction of Pd decreased somewhat the average size of the crystallites, which increased the rate of reaction, but did not increase *ee's*. However, according to Orito [193] the catalysts on Kieselguhr with added Pd are more active (the rates of the reactions were 2.5-8.7 mmol/min/g Ni) and more enantioselective, with *ee*'s 63.8 - 89.1%. Thus, contrary to Orito's data [193-197] the introduction of Pd and the use of Kieselguhr as a support does not lead to the formation of especially effective catalysts [205].

Increasing the Ni content above 25% in the fraction of $NiCO_3$ during the preparation method increased the average size of the nickel crystallites (Figure 4.6.).

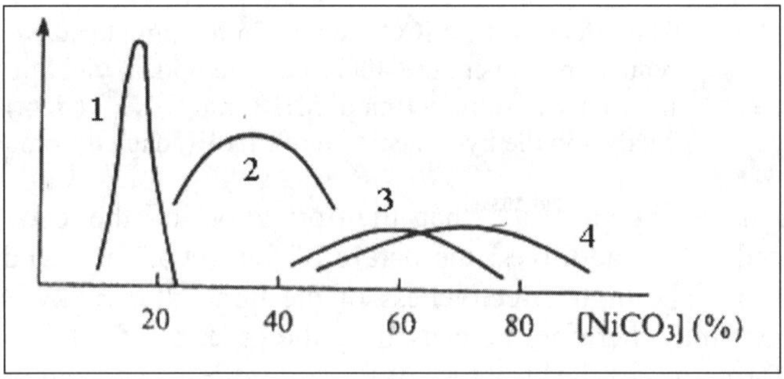

Figure 4.6. Influence of $NiCO_3$ content in the initial catalytic mass (catalyst reduction temperature 400°C) on the average size of the nickel crystallites.
Relative amounts of crystallites with 3.5 (1), 8.0 (2), 16.0 (3), and 18.0 (4) nm average size, respectively (according to Nitta [203]).

The optical yield passed through a maximum corresponding to the catalyst obtained from the catalyst mass containing 37% of $NiCO_3$ and reduced at

500°C. Increasing the fraction of $NiCO_3$ to 58% makes it possible to lower the reduction tempearture to 300°C. This leads to an increase in mean diameter of crystallytes, D, to 15 nm, which promotes an increase in *ee*. Hence, it follows that the studied enantioselective hydrogenation of MAA is a structure-sensitive reaction and may be used even as a method for the monitoring of the distribution of nickel crystallites. Figure 4.7. presents the variations of the catalytic activity and the optical yields from MAA hydrogenations on $Ni\text{-}SiO_2$ with the fraction of $NiCO_3$ in the catalytic mass.

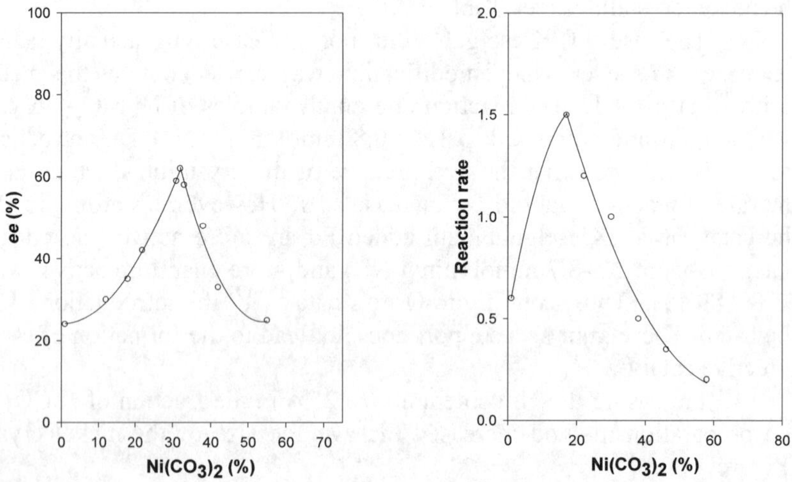

Figure 4.7. The effect of the $NiCO_3$ content in the initial catalytic mass (catalyst reduction temperature 400°C) on the optical yield of MHB (left) and the catalytic activity (mmol MHB min^{-1} gNi^{-1}) (right) on $Ni\text{-}SiO_2$-TA catalyst in the hydrogenation of MAA (adapted from Nitta [203]).

Studies have shown [201,203] that the formation of the catalyst is greatly influenced by the additives, the nature of the initial salts, and the reduction conditions. The high effectiveness of the bimetallic catalysts deposited on Kieselguhr may therefore be caused by the precise influence of these factors on the formation of the structure of the catalytic surface and primarily on the size of the nickel crystallites [198,199.200-203].

The introduction of Pd into the catalyst apparently affects the degree of reduction of nickel which occurs at lower temperatures. This influence depends greatly on the nature of the initial salts; the presence of Pd promotes the reduction of $NiCO_3$, which is formed on precipitation with Na carbonate, but has little effect on the reduction of $Ni(OH)_2$. The presence of the support accelerates the partial transformation of $NiCO_3$ into $Ni(OH)_2$, but reduction of the hydroxide requires more severe conditions.

According to Orito [193-197], the addition of AcOH significantly increases *ee* in the hydrogenation of MAA on Ni-Pd-Kg catalyst. The addition of certain salts into the modifying solution of tartaric acid also helps to increase *ee*.

The nature of the anion plays an important role, and NaBr proved to be the most effective modifying additive to RNi-TA catalyst preparations (Harada et al.[105]). High concentrations of halide salts in the modifying TA solution have a negative effect on the rate of reaction and on *ee* (see Sachtler [209]), but the addition of NaI or KI at very low concentrations tends to increase the optical yield. It has been found that NaI reduces *ee* sharply at a concentration of just 0.002%, whereas NaBr is effective even at higher concentrations by a factor of 100. This is associated with the fact that the iodide is adsorbed on nickel much more strongly than the bromide. Thus the anion affects *ee* much more than the cation [105].

The nature of cation is important in the etching of the Ni-Al alloys and in the establishment of a suitable pH for the modifying tartaric acid solution. Best results were achieved when Na^+ ions were employed at a NaBr concentration of 8% and a tartaric acid concentration of 1%. The optical yield in the hydrogenation of MAA was 70.2%. On the other hand, Li^+ and Rb^+ ions hardly affect *ee*. The cations are coordinated to the oxygens in the TA and MAA molecules adsorbed on the nickel. Sodium and potassium participate in the octahedral coordination which intensifies the template effect in the stereochemical interaction of TA with MAA (or MHB). The anions are by themselves incapable of showing the template effect, particularly at high salt concentrations; their function consists of fixing the appropriate amounts of cation on the catalyst surface needed for coordination to the TA-MAA (or MHB) complex. The influence of NaBr is also manifested by the fact that the salt is a weak poison for the nickel catalysts. It blocks the most active centers on the nickel surface increasing the fraction of selective centers and thereby intensifying the enantioselective effect of the modified centers (see, Sachtler et al.[209-217]).

According to Nitta [205] the introduction of chlorides or bromides into the catalyst salt mass during the formation of deposited catalyst, promotes an increase in *ee* by virtue of the increase in D, but the overall rate of the reaction decreases under these conditions. Additions of $NiCl_2$, NaCl, $FeCl_2$, and NaBr or of HCl or HBr to the catalyst precursor mass prepared by method A (see Table 4.8.) increases *ee* from 36.1% to 49.7-57.5%, and also increases D from 13 to 20-30 nm while narrowing the size distribution of the nickel crystallites [205].

The increase in *ee* after the addition of $NiCl_2$ during the preparation process is of interest. The introduction of 29.8% of this salt into the catalyst mass prepared by methods A, B, and C (see Table 4.8.) sharply diminishes the specific surface of the reduced metal, which indicates an increase in D. Depending on the method employed, the *ee* increased from 36.1% to 54.7% in method A, from 19.3% to 51% in method B, and from 39.4% to 53.2% in method C. The sharp decrease in the specific surface area indicates not only the growth of D but also the presence of surface chloride residues, which hinder the adsorption of hydrogen during reaction. At the same time, the Cl- ions remaining on the catalyst surface after its reduction may act as promoting additives; that is, they may behave similarly to NaBr in the modifying solution (Harada, Izumi [105]).

According to these authors, the RNi-TA-NaBr catalyst was prepared by leaching the Ni-Al alloy (42:58) at 100°C. The catalyst was washed with portions of deionized water 15 times and was modified with an 1% aqueous solution of (2*R*,3*R*)-tartaric acid, containing 10g NaBr, at 100°C for 1h. After modification, the catalyst was washed again with water and then with methanol. MAA was hydrogenated in a solution, containing EtCOOMe + 0.2 mL AcOH at a hydrogen pressure of 90 bar and a temperature of 10°C. The reaction rate was 0.2 mmol/min/g.Ni. A threefold repetition of the modification procedure increased *ee* from 83.1% to 88.6%. Under the same conditions, but without the addition of NaBr and AcOH, the *ee* was only 31.2%. The introduction of NaBr at a concentration of 0.01-0.04 mmol/g Ni, did not affect the degree of adsorption ofTA on nickel and the optical yield increased in proportion with the NaBr concentration. This has been attributed to the presence of two types of centers on the surface of the modified catalyst [9]: selective centers, modified with TA on which the (*R*)-(-)-MHB is formed, and nonselective centers on which the racemic MHB is produced. This method of co-modification with NaBr in addition to tartaric acid also has its negative aspects. For example, NaBr diminishes the activity of the nonselective centers and lowers the overall reaction rate. In order to compensate this effect, it is necessary to increase the pressure and temperature of the reaction, and this entails a decrease in regioselectivity and the appearence of side products, in particular the methyl (5-hydroxy-3-methyl-4-oxoheptanoate).

The behaviour of the deposited nickel catalysts are somewhat different from that of the Raney catalysts. The presence of residual amounts of Cl^- ions in the $Ni\text{-}SiO_2$ catalyst imparts stability during repeated hydrogenations. On the other hand, the $Ni\text{-}SiO_2$ (1:1) catalyst modified with TA with added NaBr is not quite so effective [205] as RNi [105]. The modification by NaBr increases *ee* in the MAA hydrogenation reactions from 36.1% to 46.0%, but after repeated hydrogenations on the same portion of the catalyst, the *ee* fell to 11.0%, while the catalyst promoted by the addition of $NiCl_2$ to the catalyst mass before reduction retained its stability in three consecutive experiments (chemical yields are 90%; 79%, and 81% and *ee*'s are 52.7%, 55.0%, and 54.9%, respectively). This can be explained not only by the presence of residual amounts of the halogen on the nickel surface, but also by a more uniform size distribution of the nickel crystallites. Promotion by halides (nickel chloride) of the initial catalyst mass increased the degree of reduction of NiO to metal [205]. This effect is manifested to a particularly notable extent in reduction of the catalyst mass obtained by method B (Table 4.8.) and containing $Ni(OH)_2$, which is reduced with difficulty as a consequence of the strong interaction with the support. As a result of this, the catalyst obtained has higher activity but lower enantioselectivity.

The low enantioselectivity of the Ni-alumina catalyst, in the preparation of which NaOH is used for the hydroxide preparations, can be accounted for by the strong interaction of the metal with the support and the formation of Ni-Al compounds on the surface. However, similar catalysts formed under circumstances which rule out the formation of small nickel crystallites, have the same enantioselectivity as Ni-SiO_2 (Zubareva et al.[163]).

A series of studies of Izumi's group [1-4,103,105] were devoted to the development of methods for increasing the enantioselectivity of the RNi-TA catalysts in the hydrogenation of MAA. Increasing the leaching temperature of the Ni-Al alloy increases optical activity of the product, presumably due to the decrease in the residual amount of Al present in the catalyst as aluminium hydroxide. The OH groups present in TA and $Al(OH)_3$ compete with one another for the formation of hydrogen bonds with MAA and thereby reducing *ee*. Elimination or blocking of the aluminium-enriched nickel centers increases the fraction of chiral centers and increases *ee*. This view is confirmed by the fact that the nickel catalyst, HNi-TA, obtained from reduction of NiO in a stream of hydrogen (8 L/h) at 350°C and modified with a 1% solution of TA at 85°C and pH 4.1, exhibits higher enantioselectivity than the RNi-TA Raney catalyst. The overnight hydrogenation of MAA (11.5 mL) in a solution of EtCOOMe (23ml) + AcOH (0.2 mL) at 110 bar initial hydrogen pressure and 120°C, leads to the formation of (*R*)-(-)-MHB with an *ee* of 81.6%, whereas the optical yield on the RNi-TA catalyst under approximately the same conditions leeds to only an *ee* of 44-46% [55,103].

4.3.2. Enantioselective lanthane-nickel intermetallic compounds

Multicomponent metallic hydrogenation catalysts, based on intermetallic compounds (IMC) of rare-earth elements with nickel, copper, cobalt, and other *d*-metals and their hydrides are of interest for asymmetric catalysis as models of clusters quite similar to bimetallic systems. Most studies were devoted to two structural systems LnM_3 and LnM_5, where Ln = La, Sm, Gd, Ce, Pr, and Nd and M = Ni (see Klabunovskii, Konenko's group [13,28,171-183,251,252]). Comparison of $LnNi_5$ catalysts with Ni catalysts supported on oxides of Ln, show higher activities of the IMC's and their hydrides in hydrogenation of propene (100°C, 1 bar), where $LaNi_5$ proved to be the most active catalyst [173].

Specific activities of hydrides of IMC, such as $LaNi_5$ or $LaCo_5$, are ten times higher, than activities of supported catalysts, such as 5% Ni-CeO_2 [13].

Originally intermetallic compounds and their hydrides like $LaNi_5H_6$ and $La_{1/2}Pr_{1/2}Ni_5H_5$ treated with aqueous solution of (2*R*,3*R*)-tartaric acid at pH

4.8, produced *ee's* in the hydrogenation of EAA into (*R*)-(-) EHB at 80°C and 100 bar hydrogen pressure of not more than 9.8 and 5.7%, respectively (Konenko et al.[171]). Using the catalyst 5% Ni-CeO_2 produced the same result: (*R*)-(-)EHB with an *ee* of 5.7%. This fact probably indicates that the responsible component for enantioselectivity in the IMC's is Ni, which as the result of segregation, enriched the surface of the IMC's and formed during the process of modification with tartaric acid chiral clusters. However, it was found that after modification the IMC $LaNi_3$ produced an *ee* of 11.9% in spite of the fact that segregation in this sample became difficult or even impossible (Konenko [175]). Substitution of the fraction of Ni atoms for other metals, such as Cu in $LaNi_{(5-x)}Cu_{(x)}$ increased the stability of IMC and its hydrides (Starodubtseva et al.[174 - 176,252]).

In accordance with Balandin's multiplet theory of heterogeneous catalysis [234] the best catalytic activity will occur in catalytic systems with optimal bond energies between catalyst substrate molecules. Indeed, most catalytic and enantioselective activities in a series of IMC's of the general formula $LaNi_{(5-x)}Cu_{(x)}$ reveal $LaNi_3Cu_2H_{2.7}$ producing an *ee* of 15.6% and occupying a position of stability between $LaNi_5$ and $LaNiCu_4$ [174]. Thus isostructural substitution of one or two atoms of Ni with Cu in $LaNi_5$ leads to more effective chiral catalysts; so with a content of Cu 28.7 atomic % in $LaNi_3Cu_2$, an *ee* of 17% was accomplished [174]. This fact indicated that the catalytic and enantioselective properties of IMC resulted from the whole system rather than from the segregation of Ni into a separate phase on the surface of the IMC's [13].

The influence of tartaric acid modifying solution pH on the *ee's* of products from hydrogenation also influence the catalytic activity of the entire IMC cluster. Thus for IMC's of the type $LaNi_4Cu$, the optimal pH is 5, whereas for $SmNi_4Cu$ and $GdNi_4Cu$ the optimal pH is 9. In accordance with values of stability constants of complexes of Ln-TA, modification at pH 9 of the IMC of composition $LnNi_3Cu_2$, where Ln = La, $La_{(0.5)}Sm_{(0.5)}$, and $Sm_{(0.5)}Gd_{(0.5)}$, lead to increasing optical yield: 4.4% to 6.1% and 9.9%, respectively. Hydrogen pressure and reaction temperature also influene the value of *ee*. The most effective catalyst, $LaNi_4Cu\ H_{6.2}$, produced *ee* values of 11.5% at 80°C and 40 bar hydrogen pressure, 27% at 80°C and 80 bar, and 9.7% at 40°C and 80 bar [13].

Thus chiral properties of intermetallic catalysts are determined both by the ratio of Ni to Cu in the crystalline lattice of the hydride and by the nature of the atoms of different rare earth elements in the clusters.

Most effective catalysts were prepared by introducing cobalt into the matrix of the hydride of $LaNi_5$ [13,178]. These catalysts revealed increased enantioselectivity and catalytic activity in comparison with $LaNi_{5-x}Cu_x$. For $LaNi_4CoH_{3.5}$ an *ee* of 26.6% was obtained. This value is twice the value obtained with IMC catalysts containing Cu.

Figure 4.8. shows the dependence of *ee* and conversion on the amounts of cobalt atoms in $LaNi_{5-x}Co_x$. For comparison, the data of IMC's containing Cu are also shown. As can be seen from Figure 4.8. for both types of IMC's the *ee* and conversion of EAA depends on the number of Co or Cu introduced into the intermetallic $LaNi_5$ [13].

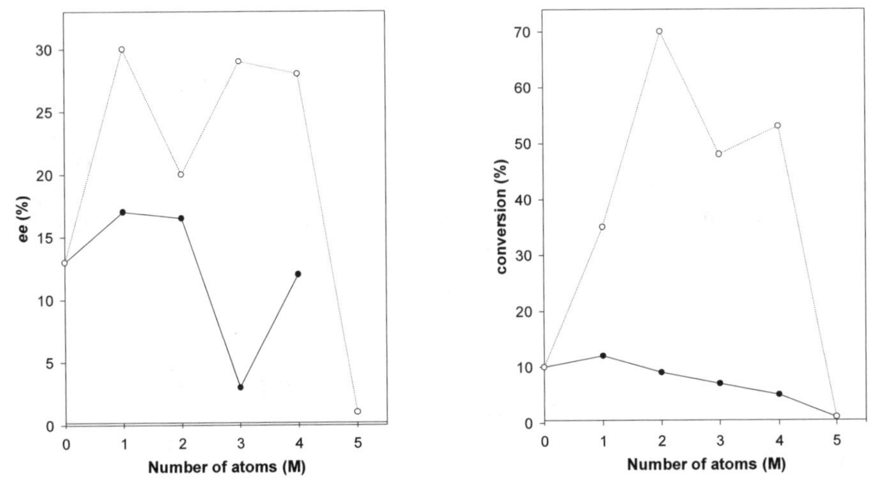

Figure 4.8. Dependence of optical yield (left) and conversion (right) in the enantioselective hydrogenation of EAA into EHB on the number of atoms (M) in the crystalline lattice of $LaNi_{(5-x)}M_{(x)}$.
M = Co (white circles) and Cu (black circles), pH 4.5.
EAA hydrogenated at 80°C, 80 bar, 7h (according to Klabunovskii [13]).

Upon introduction of Cu into $LaNi_5$, *ee* value did not exceed 17%, but upon introduction of Co, *ee* increased to 30%, and after modification of intermetallic $LaNi_3Co_2$ with a solution of tartaric acid at pH 4.5 with addition of trimethylacetic acid, *ee* increased to 45% (Figure 4.9.) [13]. It is interesting to note that the introduction of NaBr in the modifying solution of tartaric acid was very effective for the preparation of modified RNi catalysts, but had no influence in the modification of IMC's [13].

Introducing into the IMC other metals like Mn, Cr, Ti, V, or Al instead of Co, $LaNi_5$, increased its stability for forming hydrides, but decreased enantioselectivity. Only in the case of $LaNi_4MnH_2$ and $LaNi_4CrH_{2.5}$ were moderate enantioselectivities obtained with *ee's* of 20% and 16%, respectively [13].

Interesting results were obtained from partial substitution of Co for Cu in $LaNi_{5-x}Co_xH_n$ [13]. In general both the catalytic activity and the enantioselectivity decreased in the hydrogenation of EAA, and the $LaNi_3Co_2H$ system produced a rather high *ee* of 52% as shown in Table 4.10. (Starodubtseva et al.[180]).

Intermetallic catalyst	Conversion (%)	*ee* (%)
$LaNi_3Co_2\,H_{3.5}$	100	52
$LaNi_2CoCu\,H_{2.7}$	96	9
$LaNi_2Co_2Cu\,H$	20	27
$LaNi_2CoCu_2\,H_{3.8}$	30	17
$LaNiCo_2Cu_2\,H_{3.8}$	12	13

Table 4.10. Conversion and *ee* values on different intermetallic catalysts.

Intermetallic catalysts without Ni exhibited low activity, successive and complete substitution of Ni for Co leads to sharp decrease in catalytic activities and the enantioselectivity. Partial introduction of Cu atoms into $LaCo_3$ did not give increasing activities [13].

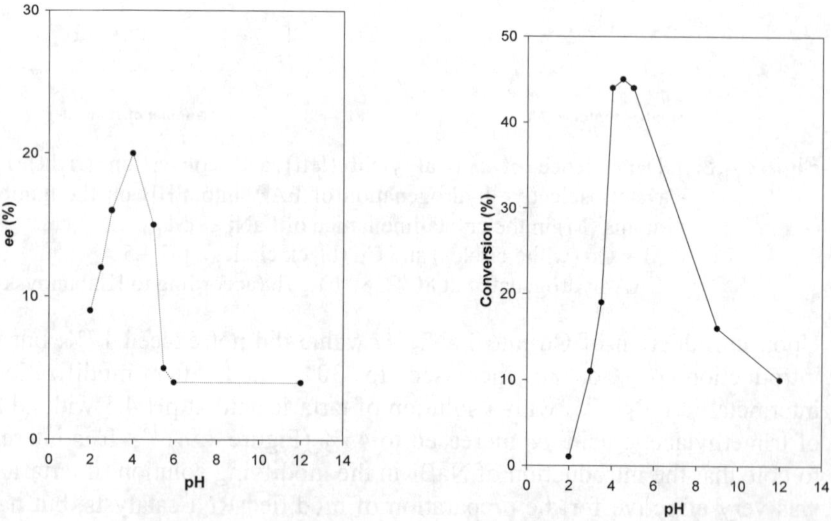

Figure 4.9. Dependence of optical yield (left) and degree of conversion (right) on the pH of the modifying solution of (2*R*,3*R*)-(+)-tartaric acid in the $LaNi_3Co_2H_{3.5}$ catalyzed hydrogenation of EAA [13, 177].

The greatest enantioselectivity, *ee* = 60%, was observed on the system $LaNi_{(5-x-y)}Co_{(x)}Cu_{(y)}$ which can be explained by the formation of mixed complexes $[LaCo_{(x)}Cu_{(y)}$-TA] [13]. Such behavior of intermetallic catalysts based on systems $LaNi_{(5-x-y)}M_{(x)}M_{(y)}$ is significantly different from Ni-Co catalysts modified in the same way as IMC; on binary Ni-Co catalysts, the *ee* values continuously increased with increasing Ni content of the catalysts.

Klabunovskii et al.[146] have been shown that in the hydrogenation of EAA and acetylacetone into ketol and diol, enantioselectivities are diminished upon using a series of metal catalysts, like Ni, Cu, Co, modified with TA or amino acids and in the case of Co-Ni the *ee* was not above 5-8%. But IMC's contained La, Ni, Co components were much more active and stereospecific than binary Ni-Co catalysts, which can be explained by the nature of clusters that involved atoms of three different elements: La, Ni and Co (or Cu) (see Starodubtseva et al. [177]).

Modifying at the optimal range of pH, intermetallic catalysts were more stable during the course of catalytic reactions, than was suggested by their ferromagnetic properties (Faraday method). Thus the higher *ee*'s corresponded to the less degradable hydrides during the modification and catalytic processes. The catalytic properties of intermetallic catalysts depend also on the strength of IMC-hydrogen interaction. Therefore the introduction of Co in IMC increases the strength of the IMC - H bond and increases the stability of the hydride. This behavior is especially characteristic for the catalyst $LaNi_3Cu_2$, for which the maximal values of magnetic moments corresponds to the maximal density of hydrogen atoms in lattice of hydride and optimal values of bond energies IMC -H (Konenko et al.[178]).

For intermetallic catalysts the dependence of *ee* on pH of the modifying TA solution results in the same extrenal character and effect as exhibited by the modification of Cu and Co catalysts[13].

For the hydrogenation of EAA, Figure 4.8. shows the extrenal dependence of *ee* and degree of conversion on the initial value of pH of the modifying TA solution with a maximum at pH 4.5 for the catalyst $LaNi_3Co_2$ $H_{3.5}$. The high catalytic activity of the IMC's containing Co is connected to the stability of the original hydrides. The isotherms of adsorption of hydrogen by the IMC's showed that replacing Ni for Co diminishes their stability [177]. Comparing the stability of the two IMC's, $LaNi_{(5-x)}Co_x$ H_n and $LaNi_{(5-x)}Cu_x$ H_n, showed that hydrides in the lattic containing Co are more stable than those containing Cu. This indicates the formation of chiral complexes involving tartrate on the surface of the catalysts [9].

Composition of intermetallides (%)	La	Ni	Co	Cr
IMC, original	14.7	60.0	15.8	9.5
IMC-hydride	17.5	64.6	7.9	10.0
IMC $H_{4.2}$, after modification at pH 4.5	8.3	65.2	7.8	18.7
IMC $H_{4.2}$, after modification at pH 7.0	22.3	44.5	12.2	21.2
Bulk composition of IMC original hydride: $LaNi_3CoCrH_{4.2}$	32.3	40.9	13.7	12.1

Table 4.11. Composition of surface layers of intermetallic compounds.

In addition to the enantioselective effect from complex formation, also observed was the enhancement effect of pH and the evolution of hydrogen during modification of the hydrides of the IMC's. During these processess the presence of components (especially La) in the modifying solution resulted in enrichment of IMC surface Ni. Data on Table 4.11. from X-ray photoelectronic spectra show the composition of surface layers of a number of IMC's [13].

The substitution of La for Sm or Gd in the $LaNi_3CoCr\text{-}H_{4.2}$ IMC resulted in a sharp decrease in catalytic and enantioselective activity, which indicates an involvement of the rare-earth metals in chiral clusters besides the involvement of the *d*-metals [172]. This decrease of *ee* values can be attributed to the increased stability of the tartrate complexes in the case of IMC's containing Sm or Gd in comparison with IMC's containing La [13]. Substitution of La in the $LaNi_5H_4$ IMC [13] by other rare-earth metals like Pr, Sm, and Gd, also leads to decreases in stability of the hydrides [183] and this forces an increase in hydrogen pressure during the preparation of IMC-hydrides. In contrast, substitution of Ni for other metals in $LaNi_5$-hydride increases the IMC-H bond strength. Therefore, catalysts based on the most stable hydrides like $LaNi_4M$ (M=Al,Ti,V) exhibit high catalytic activity, but low enantioselectivity [13]. Table 4.12. summarizes some intermetallic hydride's catalytic activities in the hydrogenation of EAA [13,177,179,181,182].

As can be seen from Table 4.12., using multimetallic IMC catalysts with 5 and 6 components do not exhibit any preferences for enantioselective hydrogenation compared to the simpler structures.

Chiral intermetallic, heterogeneous and heterogenized catalysts were put to practical use; for example in the diastereoselective hydrogenation of D-fructose into mannitol and sorbitol [13, 192, 251].

Inulin, a polysaccharide containing one D-glucose and 10-50 D-fructose units, when hydrolized and hydrogenated in a one pot approach, would be an attractive D-mannitol feedstock. With this goal in mind, the hydrogenation was studied [192a] with fructose as a model compounds using a water soluble catalyst, $[Ru(TPPTS)_2]$ (where TPPTS is the tris(3-sulphonatophenyl)phosphine ligand, Scheme 4.7.), at 90°C. The addition of HCl increased selectivity and activity in the preparation of D-glucitol and D-mannitol.

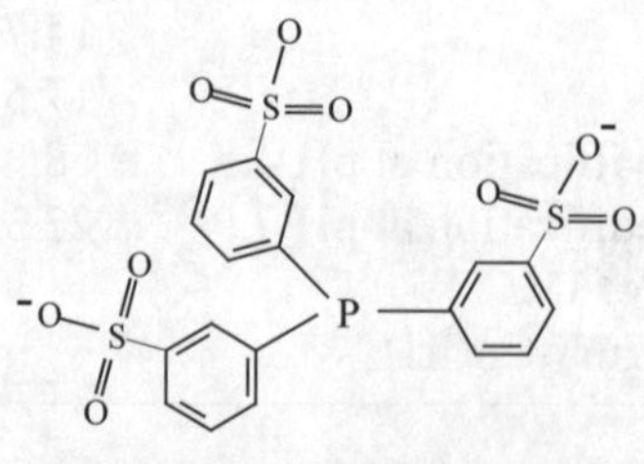

Scheme 4.7. The TPPTS ligand.

Intermetallic hydride	conversion (%)	*ee* (%)
$LaNi_5$-H $_{6.0}$	14.90	12.7
$LaNi_4Co$-H $_{4.8}$	48.80	26.6
$LaNi_3Co_2$-H $_{3.7}$	100.00	20.0
$LaNi_2Co_3$-H $_{3.9}$	52.90	28.0
$LaNi_2Co_2Cu$-H $_{3.9}$	10.00	21.0
$LaNiCo_4$	0.00	0.0
$LaCo_4Cu$	19.30	3.5
$LaCo_3Cu_2$	14.40	0.0
$LaCo_2Cu_3$-H $_{3.0}$	0.00	0.0
La $_{0.5}$ Sm $_{0.5}$ Ni_4Co	27.70	19.7
Sm $_{0.5}$ Gd $_{0.5}$ Ni_4Co	0.00	0.0
$CeCo_3$	5.60	0.0
20% Ni-80% Co	22.00	11.0
$LaNi_4Al$-H $_{4.0}$	60.90	6.3
$LaNi_4Ti$-H $_{3.5}$	93.40	1.1
$LaNi_4V$-H $_{3.5}$	95.4	5.2
$LaNi_4Cr$-H $_{4.0}$	96.0	16.3
$LaNi_4Mn$-H $_{5.0}$	40.70	23.4
$LaNi_4Cu$-H $_{5.8}$	17.50	17.0
$LaNi_3Cu_2$-H $_{3.7}$	13.30	15.6
$LaNi_2Cu_3$-H $_{3.5}$	9.50	4.3
$LaNi_3CoCr$-H $_{4.2}$	58.20	13.5
$LaNi_3CuCr$-H $_{3.9}$	64.70	16.0
$LaNi_4Co$ $_{0.5}$ Cu $_{0.5}$-H $_{4.2}$	10.30	0.0
$LaNi_3CrMn$-H $_{3.9}$	48.20	3.8
$LaNi_3Co$ $_{0.5}$ Cu $_{0.5}$ Cr $_{0.5}$ Mn $_{0.5}$-H $_{4.2}$	36.60	6.8
$LaNi_2Co$ $_{2.5}$ Mn $_{0.5}$-H $_{5.3}$	65.40	20.9
$LaNi_2Co_2Mn$-H $_{3.5}$	77.80	7.2

Table 4.12. Dependence of conversions and optical yields in the hydrogenation of ethyl acetoacetate on intermetallic catalysts (at 80°C and 80 bar for 7h) on the composition of hydrides of IMC, modified with (2*R*,3*R*)-(+)-tartaric acid at pH 4.5 [13,177, 179,181,182].

Veksler et al.[192 b] found that to accomplish selective hydrogenation of D-fructose to the preferential formation of mannitol, the pH of the modifying solution of TA is important. This process was studied using RNi modified with (2*S*,3*S*)-TA > L-Glu > (2*R*,3*R*)-TA > RNi [192b]. RNi-(*SS*)-Tart proved to be the best catalyst among the other modified catalysts. Hydrogenation of an aqueous solution of fructose at 80°C and 90 bar hydrogen resulted in D-glucitol with a yield of 65.7% and a ratio of sorbitol to mannitol of 1:9.

For this process a number of isostructural IMC's, $LaNi_3Co_2$, $LaNi_3Cu_2$, and $LaNi_3M'M''$ were used, where M'M" are CoCu, CoCr, MnCr, or $Co_{0.5}Cu_{0.5}Cr$, modified with (2*R*,3*R*)-(+)-TA. It was found that the crucial significant factors for catalyst selectivity were the pH values of the modification process.

At pH 7-8 for the intermetallic catalysts $LaNi_5$, $LaNi_3Co_2$, and $LaNi_3Cu_2$ and at pH's 4.5-5.5 for the catalyst $LaNi_3CoCr$ at 80-90°C and 50 bar hydrogen pressure the excess of D-mannitol was found to be 46% and of D-glucitol 39% with the optimal catalyst being $LaNi_3Co_2$-TA. The study of the mechanism of this reaction showed that mainly the *alpha*-D-fructopyranose and *alpha*-D-fructofuranose stereomers were responsible for the formation of D-mannitol, [13].

It is of interest that the above mentioned IMC's, based on La, Ni, and Co, are very stable and active catalysts for the vapour phase decomposition of methanol into hydrogen and carbon monoxide (Klabunovskii et al. [188,189]).

4.4. Modified nickel and bimetallic catalysts

In general, chiraly modified bimetallic catalysts exhibited low to moderate enantioselectivity in the hydrogenation of EAA. A number of studies were carried out by Klabunovskii's group to investigate new metal catalysts, modified mainly with tartaric acid and amino acids, which were active in enantioselective hydrogenation of ethyl acetoacetate and acetylacetone.

They were catalysts containing Cu [140,141,143-148,150,184,242,246,247], Co [114,140-142,146,161,162, 184,239,242, 247], Ru [121-124,,235,243,244], or Pd [157,160,190,220,221,250,253], and bimetallic catalysts containing Ni-Cu [149-160,245,249], Ni-Co [140,142,163-165,183], Ni-Ru [133,134,235], Ni-Pd [170,193-197,253], Cu-Ru [133-139,169,235,245,250], or Cu-Pd [166-170,190,250], and multimetallic catalysts based on intermetallide Ln-Ni [13,171-183,188,189,251,252]. The mechanism of the reaction on bimetallic catalysts was studied in [9,14-19,21-30,145,146,184-187,238].

4.4.1. Nickel - Copper Catalysts

Chiral Ni-Cu powder catalysts without support [149-160,245,249] were rather active in the liquid phase hydrogenation of EAA. These catalysts were prepared by three procedures used by Chernysheva et al. [159,248]:

a) by reduction of mixed oxides of Ni and Cu ("Ni-Cu-1")
b) by reduction of their hydroxides ("Ni-Cu-2")
c) by reduction of their carbonates ("Ni-Cu-3").

Changes in the nature of the original salt led to different phases and surface compositions of the catalysts and to different relationships of the reduced and remaining non-reduced forms of the metal compounds, which influenced the activities and enantioselectivities of catalysts after their modification with TA.

Table 4.13. demonstrates the effect of various methods of preparation on the phase composition of Ni-Cu catalysts modified with tartaric acid and on their catalytic and enantioselective properties during hydrogenation of EAA (according to Zubareva et al.[154]).

As can be seen from Table 4.13., the Ni-Cu-1 catalyst consists of two phases of solid solutions, Ni-Cu enriched with Ni or Cu and small quantities of NiO. This catalyst is characterized by high *ee*'s but low catalytic activities. In the 'precipitated' catalysts, Ni-Cu-2 and Ni-Cu-3, copper and nickel separated as phases resulting in sharp increases in hydrogenation activities but lower enantioselectivities. Reduction of precursors of these three catalysts in flowing gas (hydrogen + nitrogen) produced catalysts that gave enantiomeric excesses of 50.0, 14.4, and 21.3%, respectively, in the hydrogenation of EAA.

Catalyst	S (m^2/g)	Rate (mmol/ $m^2 \cdot min$)	Optical yield (%)	Phase I		Phase II		Cu	Ni	NiO	CuO
				Cu	Ni	Cu	Ni				
Ni-Cu-1	9.3	0.6	50.0	23.3	76.7	79.4	20.6	-	-	5.1	-
Ni-Cu-2	1.5	8.5	14.4	-	-	-	-	49.8	26.0	22.4	1.8
Ni-Cu-3	1.3	11.8	21.3	23.3	76.7	85.4	14.6	23.0	35.2.	-	1.6

Table 4.13. The effect of preparation methods on the catalytic and enantioselective properties and phase compositions of Ni-Cu catalysts modified with tartaric acid in the hydrogenation of EAA (according to Zubareva et al.[154]).

Apparently, the presence of small amounts of NiO on the surface of Ni-Cu-1 was favorable for increasing complex formation with TA and hence for increasing optical yields; however, a further increase in the amounts of NiO or CuO in the Ni-Cu-2 and Ni-Cu-3 catalysts led to diminishing *ee*'s.

In order to further elucidate the reasons for higher enantioselectivities, consideration will be given below to the x-ray photoelectronic spectra of the TA modified Ni which proved to be enantioselective in the hydrogenation of MAA in the gas phase, see Figure 4.10. (according to Yasumori [225]).

The electronic state of the Ni catalyst surfaces of different natures were studied by x-ray photoelectron spectroscopy in which binding energies were calibrated with those of $4f_{5/2}$ and $4f_{7/2}$ from peaks of Au taken as the standard.

Spectrum (a) for the modified surface exhibits a characteristic peak 1 at 852.7 ev, which is close to that for the Ni film at 852.5 ev. Peaks 2 and 4 correspond to Ni_2O_3 and peak 3 at 858.5ev corresponds to the satelite peak in the spectrum of a Ni film. None of the Ni surfaces covered with adsorbed oxygen and oxides gave rise to any peaks near peaks 1 and 3. In the spectra of Ni-tartrate (c) and Ni-oxide coated with TA (d) peak 1 did not appear. In the modified surface there are two peak maxima at 1.1 and 1.2ev which indicate that the modified surface is protected from oxidation by TA.

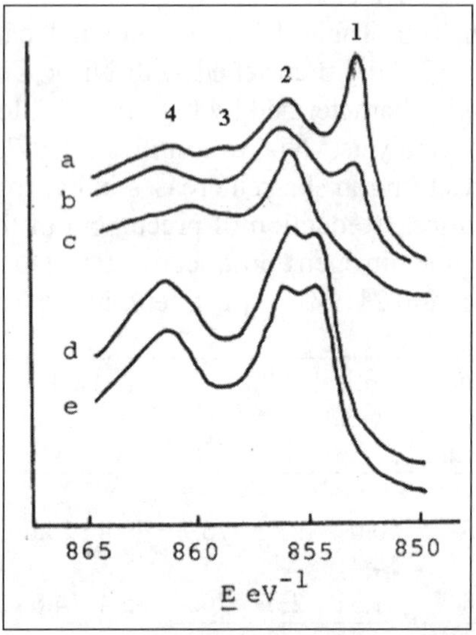

Figure. 4.10. X-ray photoelectron spectra of Ni $2p_{3/2}$. The plots of intensity (I) of characteristic peaks (1, 2, 3, and 4) versus binding energy (E). (a) modified Ni, (b) unmodified Ni, (c) nickel tartrate, (d) Ni-oxide modified with TA, (e) Ni-oxide (according to Yasumori [225]).

Zubareva et al.[154] studied the X-ray photoelectron spectra of a Cu-Ni-TA catalysts (Table 4.13.) in the region of Ni3p and Cu3p of the original Ni-Cu-1 catalyst (Figure 4.11., a, curve 1). The ratio of suface concentrations of Ni:Cu was near 1.5. In the Ni-Cu-2 catalyst the surface layer was enriched with Cu^{o} (curve 2). After modification of both catalysts with TA, their surfaces became enriched with Ni in such a way that for Ni-Cu-1 the Ni:Cu ratio was 1.3 and for Ni-Cu-2 this ratio was 2. Spectra of Ni2p and Cu2p in Ni-Cu-1 (Figure 4.11., b and c) showed that Cu and Ni are in the reduced state. And from Figure 4.11.,b, curve 3, it follows that a portion of Ni in the modified Ni-Cu-1 catalyst was transformed into the Ni^{2+} oxidized state, which corresponds to the NiO state.

Thus maximal enantioselectivity of catalyst Ni-Cu-1 (Table 4.13.) can be explained by the presence of NiO, CuO, and Ni^{2+} centers on the surface. These conclusions were confirmed with data of Nitta et al. [199] for Ni-silica catalysts modified with TA that were active in the enantioselective hydrogenation of MAA and showed [199] that centers of Ni and Ni^{2+} are on the surface of the catalysts.

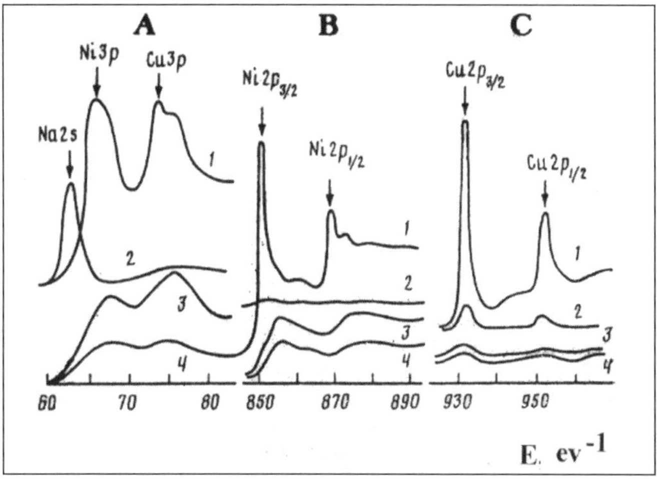

Figure 4.11. X-ray photoelectronic spectra of Ni3p and Cu3p (a), Ni2p (b) and Cu2p (c) of different Ni-Cu catalysts.
(1) original Ni-Cu-1 catalyst; (2) original Ni-Cu -2 catalys;
(3) Ni-Cu-1 modified with TA; (4) Ni-Cu-2 modified with TA (according to Zubareva et al.[154]).

The data on differential thermal desorption spectra of the reduction of NiO and CuO proved to be very important. Despite the differences in reduction temperatures of individual oxides (320°C and 270°C), a maximum *ee* was observed at that reduction when mixture of oxides were reduced at a lower temperature, 220°C, because Ni-Cu catalysts contain two phases: a Cu phase and a Cu-Ni-alloy phase. With increase of the copper content, the Ni-Cu phase is enriched in the bulk with Ni and on the surface with Cu. According to Klabunovskii et al.[151] the Ni-Cu (70:30) catalyst is five times more active than the pure Cu catalyst, but after modification of this catalyst with TA, its enantioselectivity in the hydrogenation of EAA was lower than either the modified pure Cu catalyst (*ee* 50%) or pure Ni-TA catalyst (*ee* 33%) under the same conditions. Thus the bimetallic Ni-Cu catalysts revealed synergism in catalytic activity but not in enantioselectivity.

Increasing the copper content of the mixed Ni-Cu, inactivates the catalysts, because copper, as an inactive component and dilutes the Ni centers responsible for enantioselectivity. The specific activity related only to the Ni-Cu alloy phase in the concentration range of 20-80 mol% Cu (Figure 4.12.), so the *ee* and the specific rate remained constant corresponding to the constancy of the surface concentration of the active component in the Ni-Cu catalysts.

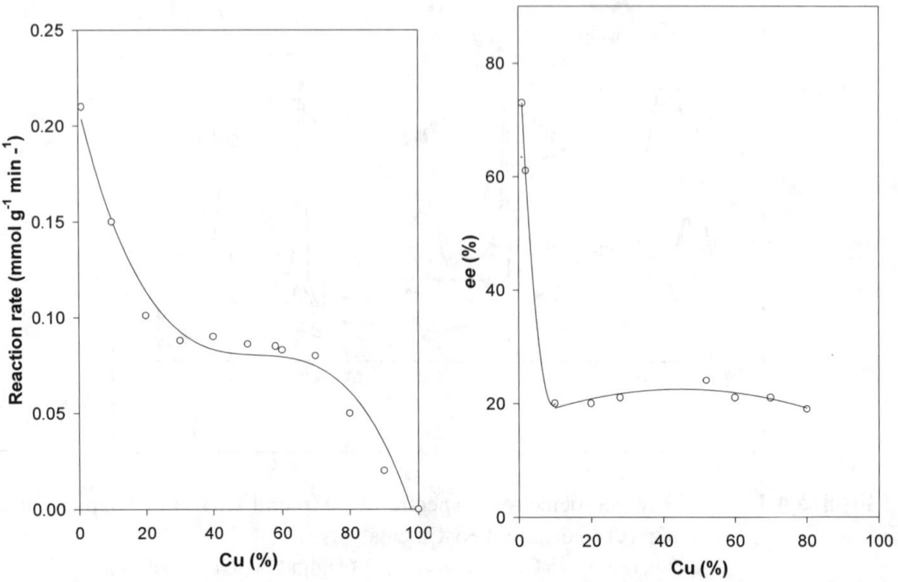

Figure 4.12. Effect of the composition of Ni-Cu catalysts, modified with TA, on the reaction rate (left) and on the *ee* (right) in the hydrogenation of EAA [13].

Increasing the fraction of Ni in the Ni-Cu catalysts leads to segregation of Ni on the surface, which was confirmed by Chernysheva et al.[149] in the hydrogenation of EAA on Ni-Cu catalysts modified with (*S*)-phenylalanine, *S*-Phe. Indeed, on the Cu-*S*-Phe catalyst the product (*S*)-(+)-EHB was obtained, whereas on the Cu-Ni catalyst, (90:10) modified with *S*-Phe, another enantiomer, the (*R*)-(-)-EHB, was obtained. The latter product was also observed on Ni-*S*-Phe. This fact confirmed that Ni was segregated on the surface of the Ni-Cu catalysts with the maximum Ni contents of 20-40% (see Figure 4.13.) [13].

The effect of modifying solution pH on the *ee* during hydrogenation on Ni and Ni-Cu catalysts was displayed by the appearance of two *ee* maxima at pH 6 and pH 10.5 (Figure 4.14.). For the pure Ni-TA catalyst the resulting *ee*'s were 57% and 71% at pH's of the modifying solutions of 5 and 10, respectively. However, for the Ni-Cu-TA catalyst *ee* values were 20% at pH 5 and 22% at pH 10 (Figure 4.14.).

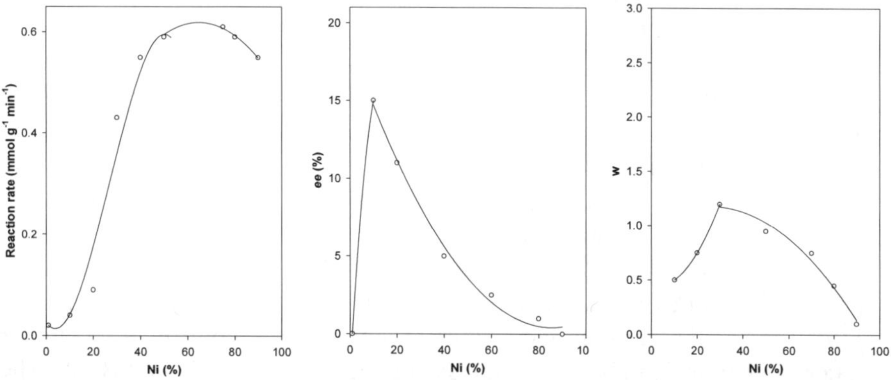

Figure 4.13. Dependence of reaction rate (left) and optical yield (middle) and the rate of formation of the excess of (-)-enantiomer (w) (w = v x *ee*) (right) on the composition of Ni-Cu catalysts, modified with *S*-phenylalanine (acccording to Chernysheva et al.[149]).

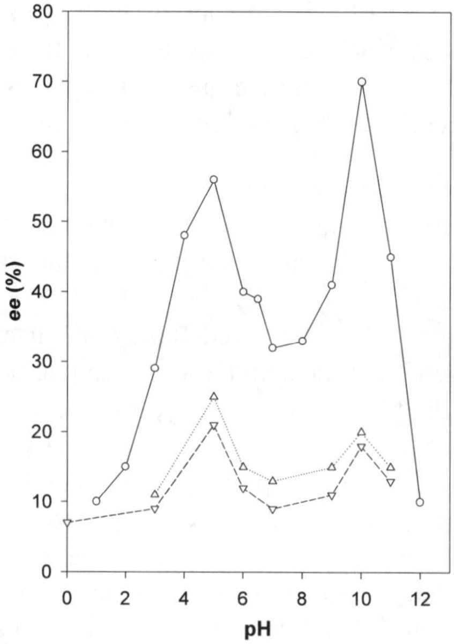

Figure 4.14. Dependence of *ee* of (-)-EHB on the pH of the modifying solution of (2*R*,3*R*)-(+)-tartaric acid.
Catalysts are Ni (o), Cu-Ni (50:50) (Δ), Cu-Ni (20:80) (Δ) (according to Klabunovskii [13]).

Such dependence of *ee* on the pH of the modifying solution was significantly different from data obtained on Ni [149], over which only one maximum was found, indicating the formation of two kinds of complexes on the catalyst surface.

Introduction of 10-15% Cu into a Ni-TA catalyst, sharply diminished the degree of conversion and *ee*. This indicates that Cu is not an indifferent support but can influence the catalytic properties. Probably the high Cu content in the catalyst destroys the chiral clusters, consisting mainly Ni [13], which are responsible for asymmetric effects.

It was found that tartaric acid as a modifier forms complexes on the surface of Ni-silica and Cu-silica catalysts in which IR spectra reveal quite close coordination between the COOH groups. This is confirmed by similar values of *ee* in which Ni-silica catalysts are modified at pH 5 and pH 10. The similarity of the effect of pH of the modifying solution on Ni and Ni-Cu catalysts indicate that the structure of chiral clusters on the surface of both catalysts remained unchanged. Probably additional introduction of copper into Ni-Cu catalyst did not produce any ligand effect and did not change the behaviour of chiral clusters, but only changed the number of such clusters ('cluster effect').

The dependences of *ee* and rate of hydrogenation of EAA on the composition of Ni-Cu catalysts indicate that the centers of enantioselective hydrogenation contain four times more chiral species (atoms and clusters) than centers of "racemic" hydrogenation leading to racemic EHB.

4.4.2. Cobalt and cobalt-nickel catalysts

Informations provided by chiral cobalt catalysts modified with tartaric acid and amino acids are very important for the theory of enantioselective hydrogenation.

Neupokoev et al. [161,242] prepared Raney cobalt catalyst modified with tartaric acid and found that the active catalyst can be prepared by etching the Co-Al alloy with 20.3% Co content. Hydrogenation of EAA gave product with only an *ee* of 4%. However, Raney cobalt catalyst (RCo) prepared from Co-Al alloy with 35.5% Co content and modified with amino acids was more active in the hydrogenation of EAA and acetylacetone. Hydrogenation with an RCo-S-Leucine catalyst resulted in EHB with an *ee* of 8.8% [162].

Zubareva et al. [165] found that some chiral bimetalic Raney Ni-Co catalysts modified with TA in hydrogenation of EAA also produced relatively low enantioselectivities, *ee* of only 5-8%. It is of interest that intermetallic catalysts (IMC) based on $LaNi_5$ M'M" (see Parts 4.3. and 4.5.), containing additional metals, M' and M" and modified with TA, revealed significant enantioselectivities after building the cobalt component into this catalyst. In this case *ee* increased up to 52% [180].

Cobalt assisted the segregation of Ni and increased its concentration on the surface of IMC's. In order to elucidate the reasons for such enhanced enantioselectivity of the Ni-Co catalysts, samples of powdered Ni-Co catalysts modified with TA were prepared by decomposition of acetates or nitrates of Ni and Co at 450°C followed by reduction of the resulting oxides in a stream of hydrogen or a mixture of hydrogen + helium at 350°C. Modifications with tartaric acid were carried out at pH 4.5-5.0 (see Zubareva et al. [164,179]). Hydrogenations of EAA were carried out at 20-80°C and 25-100 bar hydrogen in AcOEt solution containing AcOH. Under these conditions the Co-catalyst did not produce any enantioselectivity; however, after introduction of 20% Ni, the *ee* reached 3.6% and after addition of 80% Ni the *ee* reached 15% [179]. Changes of the catalyst composition and the reaction conditions resulted in an increase in *ee* of up to 46-50% with Ni-Co (50:50)-TA catalyst. Under the same conditions RNi-TA catalyst resulted in a 65% *ee*. This indicated that the main component responsible for enantioselectivity is the Ni component, and its increasing concentration in the Ni-Co catalyst sharply increased the *ee* values. Thus, these catalysts, based only on Ni-Co components, significantly differ from IMC catalysts, which include La, Ni, and Co components, in that the increasing content of Co rather than Ni (e.g. in $LaNi_3Co_2$-$LaNi_2Co_3$ systems) leads to increasing *ee*'s in the hydrogenation of EAA.

Chiral Ni-Co catalysts were examined in an industrially important reaction, in the hydrogenation of the equilibrated mixture of menthone and isomenthone (70:30) into menthol stereoisomers. The use of the Ni-Co-(90:10)-TA catalyst resulted in an optical yield of 54% (Klabunovskii et al.[183]). On an RCo-TA catalyst the unavailable neoisomenthol was obtained with a yield of 80% (Zubareva et al.[142]).

4.4.3. Supported nickel-copper and nickel catalysts

Supported chiral Ni and Ni-Cu catalysts [149-160,245,248,249] are of special interest because they allow the elucidation of the nature of the metal-support interaction and the asymmetric adsorption of modifier and substrate molecules by use of the IR spectra of adsorbed molecules.

In general, supported metal catalysts are less effective than Raney metal modified catalysts, and the enantioselectivities of supported catalysts are near to those of the Raney catalysts only in the cases when large amounts of metal are found on the surface of the support [198,200]. This can be explained, at least in the cases of bulk metal catalysts, as a consequence of an increase in crystallite sizes [207,208] and diffusion of the tartaric acid modifier into the pores during modification of the catalyst (Sachtler [209]). Detailed consideration of this problem is in discussed in Chapter 5.

The supported chiral catalyst, Ni-Cu-TA, was supported on aerosil [28], and proved to be active in the hydrogenation of acetylacetone (acac) (1) into hydroxyketone (2) and diol (3) (Scheme 4.8.).

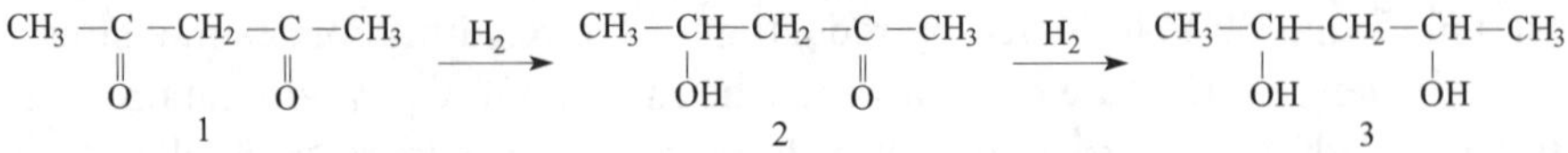

Scheme 4.8.

Under the conditions used (*R*)-(-)-4-hydroxypentan-2-one (2) was produced chemoselectively with an *ee* of 9.1% without production of diol (3).

The Ni-Cu-TA-aerosil and Ni-TA-aerosil catalysts were also compared in the enantioselective hydrogenation of EAA [28]. It was found that increasing amounts of metals on the surface of the support, increases *ee* and conversion. Increased also were the rate of formation of the excess (-)-enantiomer, w, (where w = v times *ee*) and the rate of formation of racemic EHB (v_r =v - w), which goes through a maximum (Figure 4.15.).

This can be explained by an increase of Ni on the surface of the support accompanied by an increase in the general surface of the unmodified part of the metallic phase.

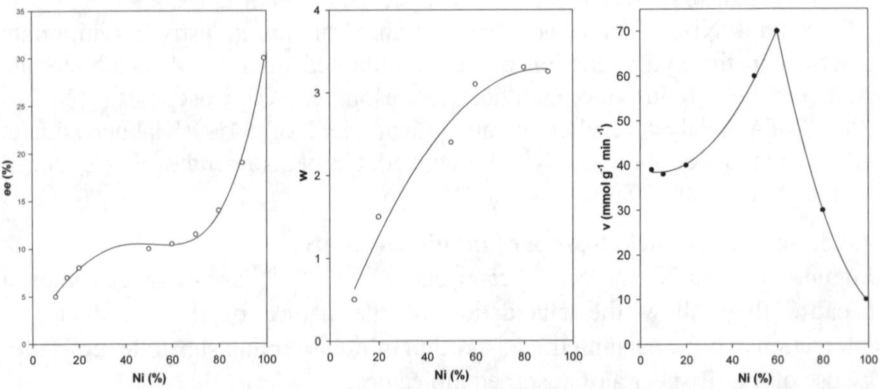

Figure 4.15. The effect of Ni content on the surface of the support on the optical yield (left), on the rate of formation of excess (-)-EHB (w) (middle) and on the rate of formation of racemic EHB (v) (left) in the hydrogenation of ethyl acetoacetate [28].

With the increase of Ni content the surface area begins to decrease as a result of consolidation of small Ni crystallites into larger Ni particles that are less active in enantioselective hydrogenation.

This process was shown both for Ni [198] and for Ru-TA catalysts [9]. Thus the optical yield, which was not very high in the case of diluted supported catalysts, increased with increasing metal content on the surface of the supports [199]. This explained the high effectiveness of catalysts with large amounts of metal on the supports.

The nature of the support also exerts a very strong influence on enantioselectivity of the supported catalysts [199]. Thus in the enantioselective hydrogenation of MAA over a series of catalysts in which the Ni to support ratio was 1:1, diminishing optical yields (*ee* %) were exhibited in the following series: Ni without support (57) > SiO_2 (56.2) > zeolite (45.6) > Al_2O_3 (43.7) > kieselguhr (29.1) > MgO (6.1) > TiO_2 (3.1). A probable explanation is that the degree of interaction of Ni with these supports increases in this series and inhibits the growth of large Ni crystallites during the catalyst preparation process by means of an interaction that does not occur in the case of the Ni-aerosil [199].

Keane et al. [254] have found that the gas phase hydrogenation of butan-2-one to butan-2-ol can be accomplished with an *ee* of 31% at 70°C over a 2.2% Ni-Zeolite Y catalyst modified with 0.008 mol/L solution of L-tartaric acid. The optimum particle size of the catalyst was found to be 3 nm (this size seems to be important in catalysis [263]). Modification of catalysts with solutions of L-Val or L-Glu had a negative results on enantioselectivity.

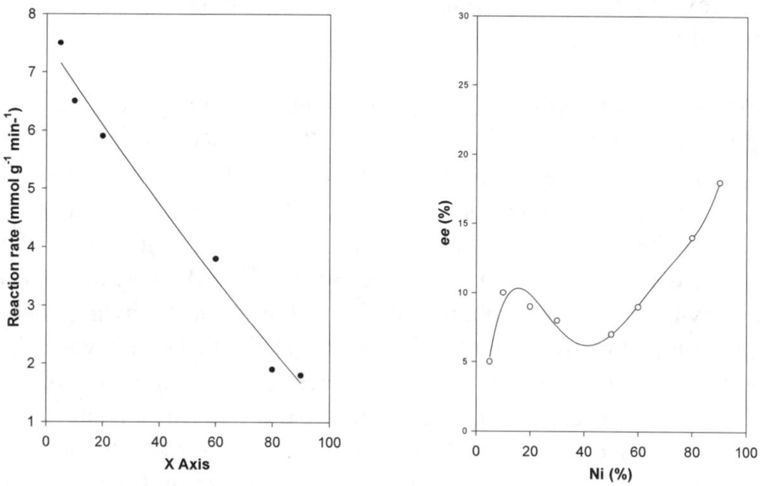

Figure 4.16. The effect of Ni content on the reaction rate (left) and *ee* values (right) in the hydrogenation of EAA on Ni-TA-aerosil catalyst (according to Vedenyapin et al.[155]).

Vedenyapin et al.[155] studied catalysts prepared by soaking supports like silica, aerosol, and kieselguhr with solutions of Ni salts followed by reduction in flowing hydrogen at 380°C. The enantioselectivity of such catalysts after modification with TA depended on the nature of the support and diminished in the series of supports: aerosil > kieselguhr > silica gel. In this series the mean size of crystalline particles decreased as 22, 17, and 12 nm, respectively. Five percent copper in the 5% Ni-aerosil catalyst resulted in a decrease of crystalline size to 2.3 nm and a concomitant decrease of *ee* from 4.1% to 3.3%. Increasing Ni content on the support from 10 to 60% increased the mean crystalline size from 8-11 to 12 nm and upon further increase of Ni the *ee* value increased but the reaction rate fell (Figure 4.16.).

Under the same reaction conditions, powdered Ni-TA catalyst produced *ee* values above 50-60%, which showed the complicated character of the dependence of enantioselectivity; therefore, the increase of *ee* can not be connected only with the increase of Ni crystallite size. Ni supported on aerosils produced higher dispersions of the metallic phase and decreasing *ee* values. Such inhibiting influence of the support was observed up to 80% Ni content.

In the case of Ni-Cu-aerosil catalysts *ee*'s and reaction rates underwent sharp decreases upon adding small amounts of copper into a Ni catalyst. Infrared spectra of supported 20% (Ni-Cu)-TA catalysts of different Ni-Cu compositions showed the interaction of the Ni-Cu phase with both COOH groups of tartaric acid but in different states depending on the ratios of Ni:Cu. A lower covalent character of binding of the TA with metals was observed for Ni-Cu catalysts of composition 1:1 [150].

4.4.4. Modified ruthenium and ruthenium-copper catalysts

Chirally modified Ru [121,130,132,134,192,235,243,244] and Ru-Cu [133-139,169,235,245,250] catalysts, active in the enantioselective hydrogenation of EAA, were studied for the first time in a number of papers by the Klabunovskii's group [121-139]. A Ru-silica catalyst modified with TA at pH 5.5 produced only an *ee* of 3% in the hydrogenation of EAA at 75°C and 80 bar [122]. A skeletal Ru catalyst modified with an aqueous solution of TA was used in the hydrogenation of EAA in the liquid phase at 80°C and 100 bar. The optical yield was only 2% [121]. On these Ru-TA catalysts it was observed for a first time that the optical yield depends on the crystallite size of Ru (see the details in Chapter 5).

The Ru-Cu catalysts modified with TA also proved to be not very effective. Skeletal Cu-Ru catalysts (1:1) were prepared by leaching a Cu-Ru-Al alloy (15:15:70) at 60°C with a 20% aqueous NaOH solution. Modification of the Cu-Ru (1:1) catalyst was carried out in a 4% aqueous solution of (2*R*,3*R*)-tartaric acid at pH 4.4. Hydrogenation of EAA was carried out without solvent at 30°C and 140 bar for 5h. For comparison the hydrogenation of EAA using skeletal Ru and Cu catalysts were also examined under the same conditions. Results are summarized in Table 4.14.

Catalyst	Conversion (%)	Rate (mmol/min)	ee (%)
Cu-Ru	78	102	3.5
Cu-Ru	91	120	2.3
Cu-Ru	100	130	2.4
Cu	6	7.9	19.6
Ru	100	130	1.4

Table 4.14. Catalytic and enantioselective activities of Raney Cu-Ru, Cu and Ru catalysts in the hydrogenation of ethyl acetoacetate (according to Kuznetsova et al. [137])

Promotion of Ru -TA catalyst with Cu, preserved the high overall reaction rate characteristic for pure Ru catalyst and somewhat increased the *ee* from 1.4% to 3.3% for Cu-Ru catalysts. This revealed a synergism in the action of mixed bimetal catalysts.

In other conditions the *ee*'s of 17-20% for pure Cu-TA catalyst dropped to an *ee* of 0.6% after introduction 0.5% Ru into the catalyst and then increased after further increasing Ru content up to an *ee* of 2.5%, which is characteristic for the pure Ru-TA catalyst.

Studies of the hydrogenation of EAA at 90°C and 100 bar using supported 10% (Cu-Ru)-aerosil-TA catalysts with different Cu-Ru compositions showed that there is an antiparallel dependence between the optical yield and the reaction rate. This was especially true at the 4 to 6 Ru:Cu ratio, in which the *ee* dropped from the 4% of the Ru-Cu catalyst to almost zero, while the rate incresed linearly [135,136].

It is very interesting that under these conditions the Cu-TA catalyst were not active in the enantioselective hydrogenation of EAA, while the skeletal Cu-TA produced an *ee* of 17% [148], and enantioselectivity occurred at the Cu-Ru-aerosil catalyst with a composition of Cu to Ru of 4:6, indicating synergism in the Cu-Ru-TA catalysts as a results of activation of the Cu in mixed bimetal catalysts. This suggestion for the origin of the enantioselective properties exerted by Cu atoms in the Cu-Ru catalysts corresponds to the characteristic properties of skeletal Cu-Ru catalysts in their dependence of *ee* on the pH of the modifying TA solution because optically active (*R*)-(-)-EHB with an *ee* of 12% is obtained only in the pH interval of 3-7, and the best results were at pH 5 [153]. It is of interest that for Cu-Ru-TA and for Ni-TA catalysts the resulting *ee*'s depend on the general content of the metal on the support. Increasing Cu-Ru content from 5% to 15% diminished *es'e* from 25% to 4% and then to zero; however, a constant and rather high reaction rate was maintained. These data indicate that on the surface of bimetal catalysts there are two kinds of Cu-metal centers; one is less active catalytically but enantioselective, while the other is highly reactive but non-stereoselective. Despite the absence of mutual solubilities of Cu and Ru in supported catalysts, both metals are in close interaction, probably in the form of bimetallic clusters.

In skeletal Cu-Ru-TA catalysts, Cu affects the Ru but the reverse action of Ru affecting Cu was not observed, a fact that can be explained by the difference of structures in supported and skeletal catalysts. In the latter case there are the structures of Cu phase covered with Ru phase. After etching of the Ru-Al alloy in the preparation of Raney Ru catalyst, some admixtures of bayerite were found in the resulting catalyst. Ru atoms interacted with the hydrated form of alumina, which provided high dispersion of Ru and presented the possibilities of chiral interaction with the modifier [128] when the Ru-alumina catalyst could not be modified chirally [13]. These conclusions were confirmed by studies of hydrogen spillover on skeletal Ru catalyst [131,132]. Indeed, in skeletal catalysts a more homogeneous distribution of the metal particles and OH groups were found and their interaction were higher than in supported catalysts.

4.4.5. Palladium, palladium-copper and palladium-nickel catalysts

Palladium [157,160,190,220,221,253], palladium-copper [166-170,190,250] and palladium-nickel [170] catalysts were studied in a number of papers.

Orito et al. [193-197] found that small amounts (1%) of noble metal additives, like Ru, Pt, or Pd, in Ni catalysts supported on Kieselguhr significantly increased enantioselectivity in the liquid phase hydrogenation of MAA. Therefore, it was of interest to study bimetallic catalysts based on Pd and mixed with Cu and Ni with greater concentrations of Pd than 1% (see works of Klabunovkii's group [166-170,250]).

Skeletal Cu-Pd of different compositions were prepared [190]. Catalysts were obtained by etching Cu-Pd-Al alloys (70% Al). The prepared catalysts contained Cu and Pd phases and their solid solutions and up to 40-58% bayerite. Linearly increasing the contents of Pd in the catalysts from 5 to 100%, increased the rate of hydrogenation of EAA at 100°C and 100 bar. More complicated dependence on concentrations of Pd was found below 5%. In the interval of 1-5% Pd the *ee* values went through a maximum and at 2% Pd in Cu-Pd the optical yield reached the maximal value of 4-5% [168]. Vedenyapin et al.[167,238,250] used an electrochemical method and found decreasing adsorption of hydrogen on Cu-Pd with Pd contents less than 10%. This can be explained [167] by the involvement of a fraction of the Pd phase (or Cu-Pd alloy) in the Cu phase and was confirmed with the results of enantioselective hydrogenation of EAA. Indeed, only the catalysts containing less than 5% Pd produced enantioselectivity and at increasing concentrations of Pd from 1 to 5% the *ee* falls from 9 to 1%.

In hydrogenations over a Cu-Pd-TA containing 2.7% Pd two optimal pH values of 4 and 8 were found for the TA modifying solution [168]; these produced (*R*)-(-)-EHB's with 4.1 and 4.6% *ee*'s, respectively (Figure 4.17.).

Likewise, as was the case with the Cu-Ru catalysts, the violation from additivity was observed upon the introduction of small amounts of Pd into the Cu catalyst. At 5% Pd the specific rate related to m^2 of surface area of the Pd phase increased up to 150 times the rate of hydrogenation on the pure Pd catalyst. This indicated sharp changes of the properties of copper in bimetallic catalysts after mixing it with palladium.

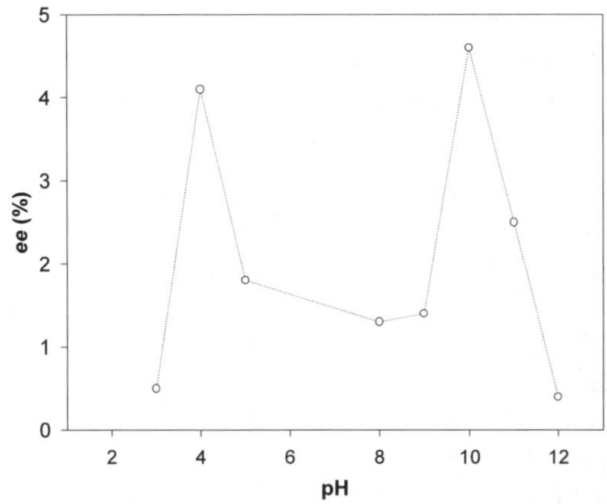

Figure 4.17. Effect of the pH of the (2*R*,3*R*)-tartaric acid modifying solution on the optical yield of (*R*)-(-)-EHB on Cu-Pd (97.3:2.7) catalyst (according to Vedenyapin et al.[168]).

Comparison of the chiral bimetallic catalysts, Cu-Pd-TA and Cu-Ru-TA, showed significant differences. In the case of Cu-Ru-TA catalyst, introducing 0.1-0.5% Ru into Cu-TA leads to almost complete loss of enantioselectivity, while in the cases of Cu-Ru and Cu-Pd catalysts such chiral deactivation proceeds only after introduction of more than 5% Pd. The general catalytic activity of the Cu-Ru-TA catalysts increased with increasing Ru content, while the Cu-Pd catalysts exhibited a synergism of catalytic activity, which was explained by a peculiar structure of the active center and by invoking a ligand effect [167,168]. A similar effect for skeletal Cu-Ru-TA catalysts was observed.

Because Cu did not adsorb hydrogen, the exposed surface of Ru in Cu-Ru catalysts can be determined by adsorption of hydrogen using galvanostatic and potentiodynamic measurements in KOH and H_2SO_4 media [13].

Comparison of the properties of Cu-Ru-TA and Cu-Pd-TA catalysts using electrochemical methods showed that the strength of adsorption of hydrogen on these catalysts diminished with a weaking of the Ru-H and Pd-H bonds as a result of a ligand effect found by Vedenyapin et al.[169]. Unlike the sandwich structure of the Cu-Ru-TA catalysts (the surface of Cu-Ru catalyst is covered with a layer of Ru atoms), electrochemical data show the Cu-Pd catalysts' surfaces are enriched with copper below 5% Pd [168], and this causes a synergistic catalytic action resulting from a ligand effect. In this way the special behavior of the Cu-Pd catalysts and their peculiar dependence of optical yield on the of pH's of the TA modifying solutions can be explained.

Supported Cu-Pd and Ni-Pd catalysts were prepared by impregnation of aerosil and alumina supports with solutions of Cu and Ni salts followed by reduction of these precursors in flowing a hydrogen-helium mixture (1:10) at 360°C and modified with a solution of TA + $PdCl_2$ at pH 5. The Ni-Pd-silica catalyst produced an optical yield of 7% from the hydrogenation of EAA. Table 4.15. shows the results [170].

Catalyst	Optical yield (%)	Rate mmol/min g.cat
10% Pd-SiO_2	0.0	18
10% Ni-SiO_2-TA	7.1	9
(2:8) Pd-Cu-SiO_2	0.0	5
(1:10) Pd-Cu-SiO_2-TA	1.0	5
(1:9) Pd-Ni-SiO_2-TA	7.0	8
(1:10) Pd-Ni-SiO_2-TA	4.6	11

Table 4.15. Reaction rate and enantioselectivities of Cu-Pd and Ni-Pd supported catalysts in the hydrogenation of ethyl acetoacetate (according to Kuznetsova et al.[170]).

Under the conditions studied it was found that the supported Ni catalysts also produced low enantioselectivities, and changing the silica support for alumina completely inhibited enantioselectivities.

4.5. Mechanism of enantioselective hydrogenation

4.5.1. Enantioselectivity as a function of the bond strength in intermediate surface complexes

Balandin, Klabunovskii et al. [9,28,113] published the first ideas about the structure of the intermediate complexes in enantioselective hydrogenation. According to their idea, the reaction proceeds through an intermediate complex, consisting of atoms of the catalyst on the surface (C) or an active center of the catalyst (C-C-C), the amino acid or hydroxyacid chiral modifier molecule (M), and the prochiral substrate molecule (S), (predominantly carbonyl-compounds, such as 3-oxocarboxylic acid esters or *beta*-diketones) which can also coordinate to the metallic catalyst, and after addition of two hydrogen atoms produces the hydrogenated product (P) (Scheme 4.9.).

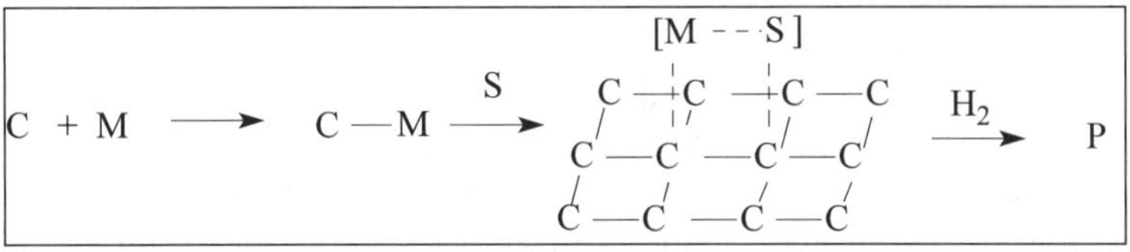

Scheme 4.9.

Scheme 4.9. does not give the detailed picture of the intermediate complex and does not precisely delineate the structure of catalytic center. But after modification, this scheme was applied to the hydrogenation of ethyl acetoacetate on Ni catalyst modified with an amino acid (Scheme 4.10.a.) and (2*R*,3*R*)-(+)-tartaric acid (Scheme 4.10.b.) (Klabunovskii et al. [25-28,153,184])

Scheme 4.10. corresponds to the "doublet" scheme of the multiplet theory of catalysis by Balandin [204] but the Ni' centers should not be applied as centers of the multiplet theory but used in accordance with the principle of microscopic reversibility [28,204].

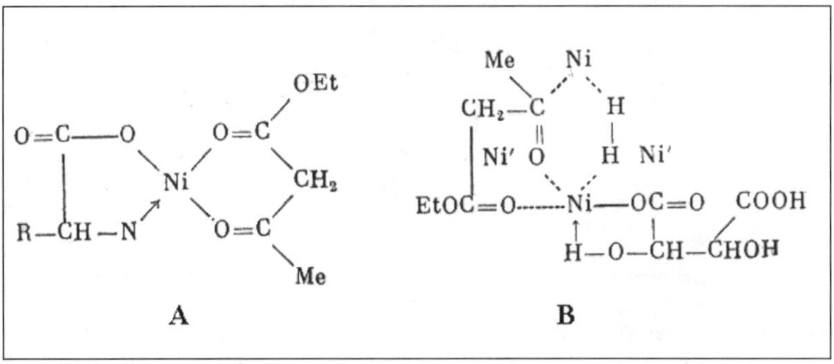

Scheme 4.10. Possible structures of the intermediate complexes in the enantioselective hydrogeantion of ethyl acetoacetate on Ni, modified with an amino acid (A) or with tartaric acid (B) [9,28].

The general idea that the enantioselective reaction proceeds through a triple complex, including the catalytic center (C), the modifier (M) and the substrate molecules (S) was elaborated in several models about the intermediate complex formation on the catalyst surface.

Klabunovskii et al.[28] suggested that hydrogenation of EAA or acetylacetone would be perfomed through reaction of a heteroligand complex which was composed of the intermediate produced from the substrate and the modifier adsorbed on the catalyst as shown in Scheme 4.10.

The conformations of EAA or acetylacetone (acac) in the complex were studied by means of circular dichroism and magnetic rotatory dispersion spectra (Faraday effect) of the complexes prepared from modeling the species formed on the surface of the catalysts [239,247,260]. There was good agreement between the carbonyl ligand conformation and the results of enantioselective hydrogenation. The model is presented in Figure 4.18., in which the complex is disposed on the edge of crystalline catalyst particle modified with (2*R*,3*R*)-tartaric acid (a) or an amino acid (b) (Klabunovskii,Vedenyapin [9], Ollis [31]). According to Balandin [234,259] the Ni catalysts revealed a peak height of 1 nm.

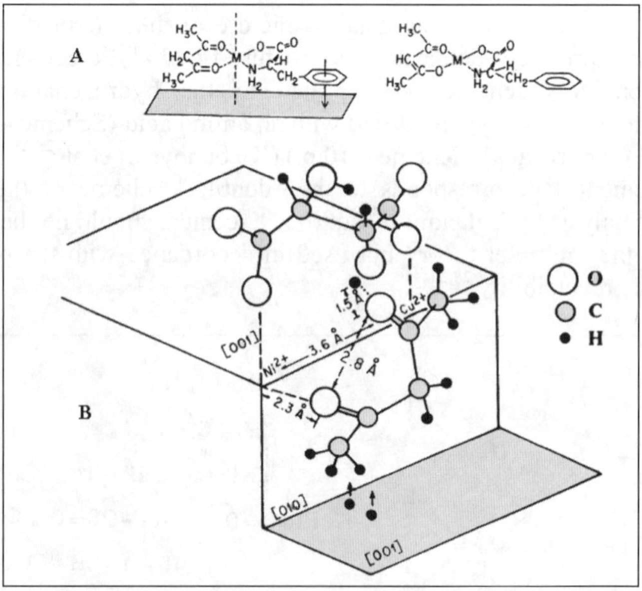

Figure 4.18. Stereochemical models of intermediate complexes in the hydrogenation of ethyl acetoacetate on crystalline Ni modified with (2*R*,3*R*)-tartaric acid (a) and with amino acid (b) (according to Klabunovskii,Vedenyapin [9]).

Using IR spectroscopy Sachtler et al. [208-217] developed models of the intermediates for the adsorption of complexes on Ni-silica catalysts that are modified with amino acids and hydroxyacids and active in the gas phase hydrogenation of MAA. These are presented in Schemes 4.11. and 4.12.

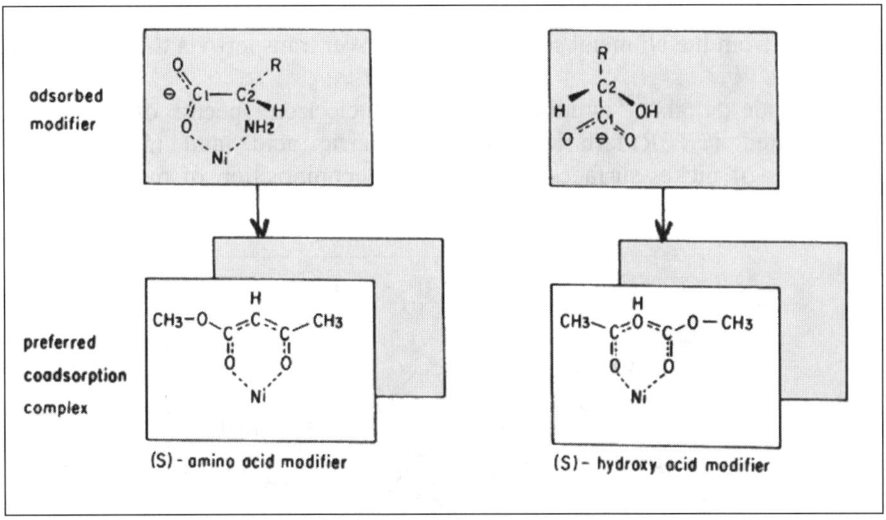

Scheme 4.11. Preferred arrangements of the coadsorbed MAA molecules on Ni-silica catalysts, modified with amino acids and hydroxyacids (according to Sachtler, Yasumori [209-217,224-226]).

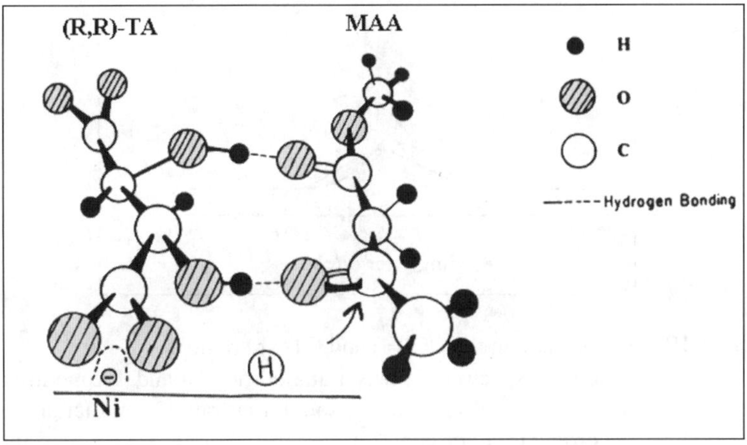

Scheme 4.12. Stereochemical model of the interaction of the modifier, TA, with the substrate, MAA (according to Sachtler, Yasumori [209-217,224-226]).

They proposed the models shown in Schemes 4.11. and 4.12., based on the interaction of MAA with amino acids and hydroxyacid with MRNi, in which each atom of Ni tetracoordinated with TA. One of the remaining two coordination sites of Ni participates in the hydrogenation, and the structure of the Ni complex acts as a template. The hydrogen used for the reaction is supplied from the Ni metal surface by a spillover transport via the tartaric acid ligand.

Inoue et al. [226] studied X-ray photoelectron spectra of MAA and coordinated (2*R*,3*R*)-tartaric acid, (*S*)-aspartic acid, and (*S*)-alanine as modifiers of nickel surfaces, prepared by decomposition of nickel formate, DNi (Figure 4.19.).

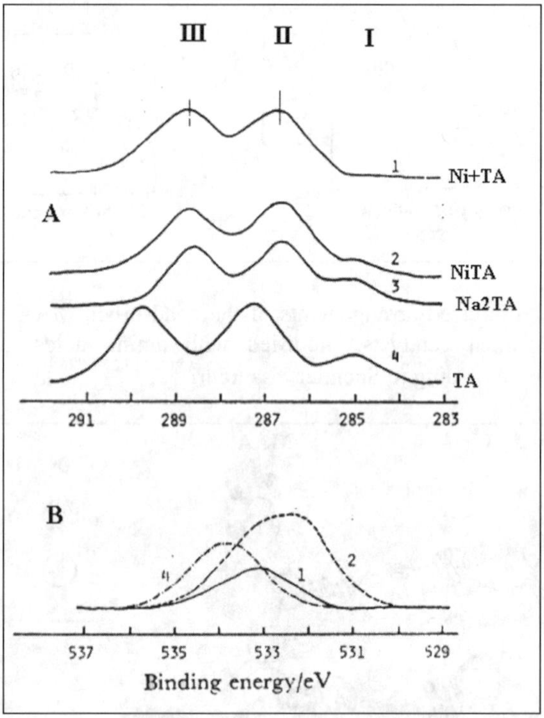

Figure 4.19. XPS spectra in the C1s (a) and O1s (b) regions. Difference spectrum between unmodified Ni and TA modified Ni, at pH 5 (1), nickel tartrate (2), sodium tartrate (3) and tartaric acid (4) (according to Inoue et al.[225]).

Figure 4.19.a. shows two peaks, III and II, that have been assigned to the C atom of the carboxy group and to the *alpha*-C-atom of the tartrate molecule. A small peak at 285 ev (I) is the carbon peak of a contaminant. The binding energies of peaks III and II for adsorbed TA (288.7 and 286.7 ev) are much closer to those for NiTA (288.7 and 286.7 ev) and Na_2TA (288.6 and 286.6 ev) than those for free TA (287.3 ev and 287.3 ev).

Figure 4.19.b. shows the O1s peak for TA, adsorbed on Ni (1) which appears at ca. 533 ev. The peak for NiTA (2) appears at a binding energy lower than that of adsorbed TA, whereas the peak for free TA (4) appears at a higher binding energy. The broad O1s peak for TA was assigned to OH and COOH groups in the formation of the Ni complex, since the C1s spectrum for TA, adsorbed on Ni, is similar to that for NiTA and Na_2TA, while the O1s spectrum of TA, adsorbed on Ni, differs from that of NiTA. Therefore, the OH probably is not in contact with the surface of Ni.

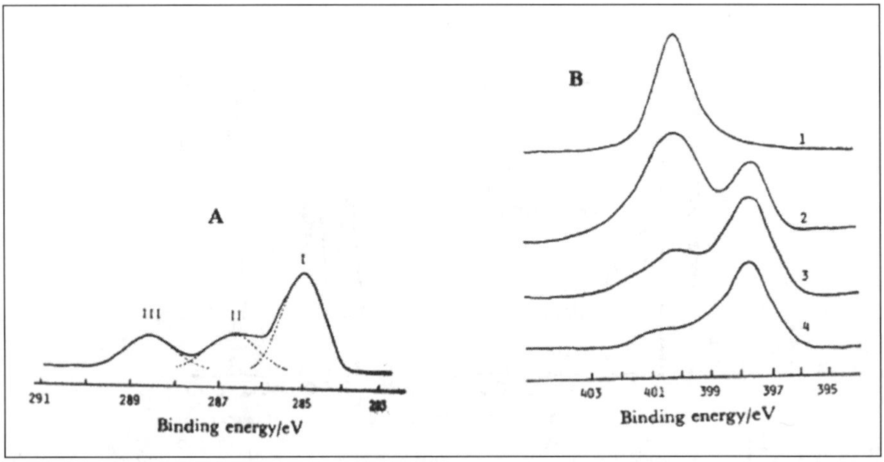

Figure 4.20. XPS spectra in the C1s region of MAA, adsorbed at room temperature on nickel surface (a), and in the N1s region of adsorbed amino acids (b).
(1) aspartic acid, evacuated at room temperature
(2) aspartic acid, exposed to 0.008 bar of H_2 at 90°C
(3) alanine, exposed to 0.013 bar of H_2 at 100°C
(4) aspartic acid, exposed to 0.008 bar of H_2 at 280°C
(according to Inoue et al.[225]).

The XPS method was also applied to the study of adsorption of aspartic acid and alanine on Ni, and the conclusion was made that amino acids were coordinated on Ni in a manner similar to the corresponding Ni chelate complex [226]. Figure 4.20.a. shows the N1s spectra of adsorbed amino acids. The adsorption of MAA at RT on a clean Ni surface provids three peaks in the C1s region (285.0, 286.6 and 288.6 ev). Peak III can apparently be assigned to C atoms of a carbonyl group, whereas peak I is associated with C atoms of a hydrocarbon chain, and peak II is due to the carbon atom in the methoxy group. Peaks II and III originates from the keto-form of MAA and should provide the intensity ratio of 1:2, so the observed equal intensity is derived on the basis of the assumption that about half of the adsorbed MAA is in the enol-form, which is confirmed by IR spectra [224]. Figure 4.20.b. confirm a model (Sachtler [211] and Hatta [106]) in which the amino acid is linked to a surface Ni atom through both amino and carbonyl groups.

Yasumori et al.[224-226] proposed a stereomodel of the interaction of MAA with HNi on the surface of a nickel catalyst prepared by the reduction of NiO (marked as "HNi") based on IR spectra data which showed that the hydrogen on C_2 in MAA had been replaced by deuterium during the deuteriumation of MAA on HNi in the gas-phase reaction.

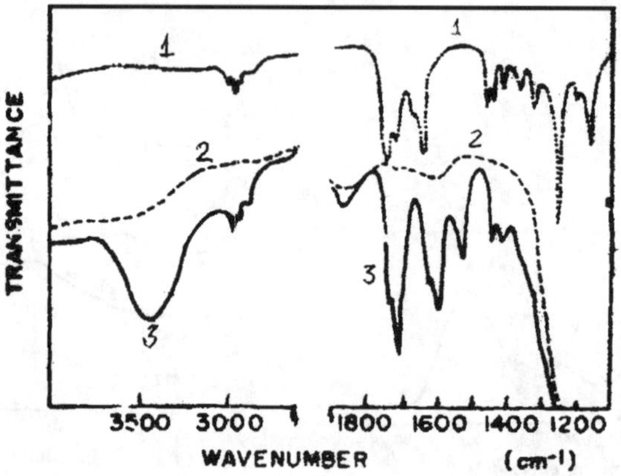

Figure 4.21. IR spectra of gaseous and adsorbed methyl acetoacetate on Ni catalyst.
Gaseous MAA (1), Ni supported on silica gel (2), MAA adsorbed on Ni (3), (according to Yasumori et al. [224]).

IR spectra of Ni supported on silica gel before and after adsorption of MAA are shown in Figure 4.21. The band at 3100-3200 cm^{-1} was assigned to the stretching vibration of the enolic hydroxyl group and the two adsorption bands at 1740 and 1710 cm^{-1} were attributed to *beta*-carbonyl and carboxyl groups, respectively. The band at 1620 cm^{-1} is indicative of the interaction of the C atom of carbonyl group with Ni. Therefore, it was concluded[224] that the adsorbed form of MAA is in the enol form, a fact that was confirmed by deuterium exchange experiments.

Kishi et al.[104,106] also found adsorption bands of *acac* on Ni which are similar to forms in Ni(*acac*)$_2$ chelate in which it is considered that *acac* adsorbs in the enol form probably with formation of a pseudo-aromatic ring (according to UV spectra [107,261,263]).

Thus, according to Yasumori [224], reaction in gas phase proceeds through the enol form of MAA. But this scheme probably does not apply to liquid phase reactions because enantioselective hydrogenation can be accomplished with 2,2-dimethyl-3-oxobutyrate, which cannot have an enol form.

Enantioselective hydrogenation of 2,2-dimethyl-3-oxobutyrate resulted in an *ee* of 30% (modified at a pH of 5.0 at 0°C with the reaction at

60°C and 80 bar) compared to the hydrogenation of MAA in solution, which gave an *ee* of 38% (according to Smith et al. [218]). Thus the enol formation does not play an exclusive role in liquid phase enantioselective hydrogenation in systems like MAA-MNi. Analysis of the experimental data from hydrogenations of MAA, EAA, and *acac* on Ni, Cu, and Co catalysts that have been modified with chiral amino acids and hydroxyacids, revealed the regularities connecting enantioselectivity with the stabilities of the intermediate complex [C-M-S] formed on the surface of the catalyst during reaction.

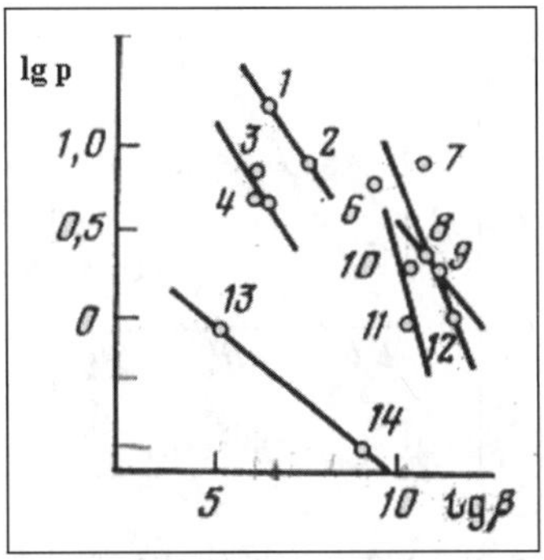

Figure 4.22. Dependence of optical yield (lg p) on stability constants (lg β) in the hydrogenation of methyl acetoacetate on Ni, modified with amino acids. (1) Glu, (2) Asp, (3) Phe, (4) Leu, (6) Lys, (7) Trp, (8) Meth, (9) Cyst, (10) Thr, (11) Ser, (12) Pro, (13) Arg, (14) Hist.

As can be seen from Figure 4.22., data on the hydrogenation of MAA on RNi modified with chiral amino acids published mainly by Izumi's group (see Izumi [1-8]) show the linear dependences between enantioselectivities (ln p) and stability constants, lg β, of the complex [Ni-amino acid] formed on the surface of the catalyst according to equation (1):

$$C + nM \overset{\beta}{\leftrightarrow} [CnM] \qquad (1)$$

Figure 4.22. shows good correlation of the experimental data as a number of curves unite the amino acid modifiers of similar structures.

The next step in the evaluation of correlations of enantioselectivity with stability constants of the intermediate complex was made when the bond strength of substrate molecule [S] with chiral surface comlexes [C-M], was measured by the constant $K_{CM\text{-}S}$ in accordance with equation (2), where AA are amino acids.

$$[\text{Cu-AA-EAA}] \overset{K_{CM\text{-}S}}{\leftrightarrow} [\text{CuAA}]^{+} + \text{EAA}^{-} \quad (2)$$

In the large interval of changes of constants, the dependence of lg p on lg $K_{CM\text{-}S}$ resulted in a curve similar to the "volcano-shape" of multiplet theory (Figure 4.23.).

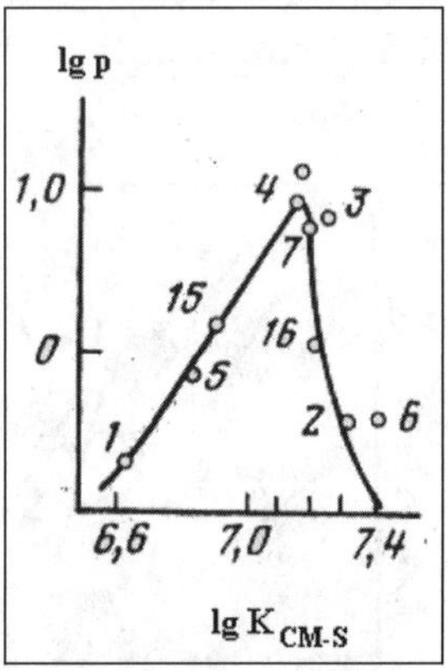

Figure 4.23. Dependence of optical yield (lg p) on stability constant, $K_{CM\text{-}S}$, in the hydrogenation of ethyl acetoacetate on Cu modified with amino acids, (1) Glu, (2) Asp, (3) Phe, (4) Leu, (5) Ala, (6) Lys, (7) Trp, (15) Val, (16) Tyr .

For the correlations of EAA and acetylacetone (acac) data were used from the hydrogenations of EAA and *acac* on Cu and Co catalysts modified with amino acids (AA).

In the case of the cobalt catalyst, only a linear dependence was obtained, which can be considerd as one branch of the same extreme dependence "volcano shape" curve [204].

A more general approach was developed by Vedenyapin, Fridman, and Klabunovskii [144-146] in which the measure of complex stability the co-proportional constant, K_d, was taken for the reaction:

$$[\text{Cu(AA)}_2] + [\text{CuS}_2] \overset{K_d}{\leftrightarrow} 2\,[\text{Cu (AA) (S)}] \quad (3)$$

In this approach, equilibration on the surface of catalyst was taken as equilibration in the reaction mixture between the complexes that can form from homoligand complexes in solution, and then the connection of *ee* (p) with K_d was obtained [144-146]

$$\text{Lg p} = nK_d + C \quad (4)$$

These data were summarized in Figure 4.24.

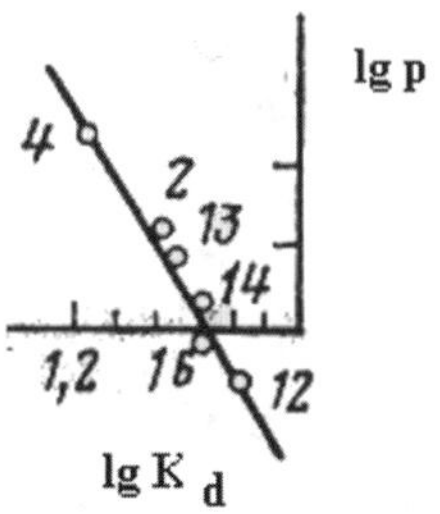

Figure 4.24. Dependence of optical yield (lg p) on coproportional constant, K_d, in the hydrogenation of ethyl acetoacetate on Co, modified with amino acids. (2) Asp, (4) Leu, (12) Pro, (13) Arg, (14) Hist, (16) Tyr.

Instead of optical yield in such calculation, it is more correct to use the value of "enantioselectivity", E , equal to the relation of enantiomers [R] / [S]. (The connection of p with E has linear character only at small values of p and E and in general reveal a complicated dependence).

The general equations were given [144-146] for the connection of E with the stability constant $K_{CM\text{-}S}$ and with constant co-proportionation, K_d according to equations (5) and (6):

$$\lg E = a + \beta \lg K_{KM\text{-}S} \quad (5)$$

$$\lg E = b + 2\beta (\delta\text{-}1) \lg K_d \quad (6)$$

where coefficient β characterizes the degree of asymmetrizing action of the modifier molecule and δ-accounts the degree of influence of modifier on the strength of the C-S bond, the metal atom in the catalyst with the substrate molecule, involved in the heteroligand complex [M-C-S].

Values of coefficients for eqs. (5), (6) and (7) were calculated from the experimental data (Table 4.16.).

Value [ε] is sensitive to changes of the bond strengths of both ligand S (MAA, EAA, acac) with complex [C-AA] and AA with [C-S]. With the diminishing of bond strengths CM-S and CS-M, there is a trend of increasing of [ε] as the differential value of the induced Cotton effect in the absorbance band of the chromophore of the substrate molecule.

Complexes	a	β	b	δ	A	B
[Cu-AA-EAA] (AA = Glu, Ala, Val, Trp,Leu, Tyr, Phe)	0.065	0.14	0.09	3.1	0.093	0.036
[Cu-AA-acac] (AA = Glu,Val, Phe)	0.01	0.003	-0.60	1.32	-	-
[Ni-AA-MAA] (AA= Glu,Val, Ala, Ser, Leu, Pro, Asp, Tyr)	-0.035	-0.15	0.005	4.41	0.086	0.085
[Co-AA-EAA] (AA= Glu, Pro, Leu,Tyr, Arg, Hist)	-0.012	-0.03	-0.036	1.33	0.032	0.018

Table 4.16. Coefficients for equations (5), (6), and (7) for different amino acids (AA) and ligand (EAA, MAA, and acac) complexes (according to Klabunovskii [28]).

Electronic interaction in chelate knots of complexes and elucidation of the nature of bonds Cu-N and Co-N in chelates [Cu(AA)(acac)] and [Co(AA)(acac)], were studied using circular dichroism and Faraday effect spectra, in which AA = tyrosine (Tyr), triptophane (Trp) and phenylalanine (Phe).

Table 4.17. presents experimental data showing a good correlation between the conformation of the AA ligand and the configuration of the hydrogenation products MHB (or EHB) and ketol (in the hydrogenation of *acac*). All *ee* values were determined only for the initial stage of the reaction and were taken into consideration only at a definite degree of conversion. For Cu and Ni catalysts the correlation *lambda*-S was observed and for the Co catalyst correlation *delta*-R was observed.

It is of interest that the values of differential dichroic absorbance coefficients correlated with optical yield and enantioselectivity, E, according to equation (7) (A and B values are in Table 4.16.).

$$\lg E = A + B \lg([\varepsilon]) \qquad (7)$$

Thus a quantitative evaluation of chirodiastaltic interactions inside the intermediate complexes [CMS] was made during the asymmetric hydrogenation of carbonyl compounds and correlations were found that characterizes this processes.

Catalysts	Modifier	°C	Δ[ε]	Substrate	Conform.	Product	Config.	*ee* (%)
Ni	R-Tyr	100	0.150	acac	δ	ketol	R	2.3
Ni	R-Trp	100	0.900	acac	δ	ketol	R	3.2
Co	S-Phe	100	4.150	acac	δ	ketol	R	1.0
Co	R-Tyr	100	0.163	acac	λ	ketol	S	0.4
Co	R-Trp	100	2.275	acac	λ	ketol	S	0.7
Co	S-Phe	100	0.172	EAA	δ	EHB	S	1.3
Co	R-Tyr	100	0.074	EAA	λ	EHB	R	0.5
Co	R-Trp	100	0.112	EAA	λ	EHB	R	3.4
Cu	S-Phe	20	0.24	acac	λ	ketol	S	1.2
Cu	R-Tyr	20	0.20	acac	δ	ketol	R	0.2
Cu	S-Trp	20	0.35	acac	λ	ketol	S	1.4
Cu	S-Phe	20	0.21	EAA	λ	EHB	S	0.93
Cu	R-Tyr	20	0.42	EAA	δ	EHB	R	3.4
Cu	S-Trp	20	0.65	EAA	λ	EHB	S	7.1

Table 4.17. Relationship between the conformation of the dicarbonyl-compounds in their heteroligand complexes with amino acids and the configuration of the hydrogenation products.
°C: modification temperature; Δ[ε]: differential dichroitic absorption coefficient; conform.: conformation of the substrate;
config.: configuration of product.

References

1. Izumi, Y., and Tai, A. (1977) *Stereo-differentiating reactions. The nature of asymmetric reactions,* Kodansha Ltd.Tokyo, Acad. Press, N.Y., 334 pp.
2. Izumi, Y., and Tai, A. (1975) *Rittai kubetsuhanno* (Japanese), Kodansha Ltd., Tokyo.
3. Izumi, Y., and Tai, A. (1977) *Stereo-differentiating reactions. The nature of asymmetric reactions* (Russian translation by Klabunovskii, E.I.) MIR Publ., Moscow, 376 pp.
4. Izumi, Y. (1983) Modified Raney nickel (MRNi) catalyst: heterogeneous enantio-differentiating (asymmetric) catalysts, *Advances in Catalysis* **32**, 215 - 271.
5. Izumi, Y. (1971) Methoden der asymmetrischen Synthese - enantioselektive katalytische Hydrierung, *Angew. Chem.* **83**, 956 - 966.
6. Izumi, Y. (1971) Methods of asymmetric synthesis - enantioselective catalytic hydrogenation, *Angew. Chem. Int. Ed. Engl.* **10**, 871-881.
7. Tai, A., and Harada, T. (1986) Asymmetrically modified nickel catalysts, in *Tailored metal catalysts,* Iwasawa Y. (ed.), Reidel Publ., Amsterdam, pp. 265-324.

8. Tai, A. (2002) Asymmetrically modified nickel catalyst (MNi): a heterogeneous catalyst for the enantio-differentiating hydrogenation, Murray Raney plenary lecture in *Chemical Industries Series* (*Catalysis of organic reactions*) (Dekker), Morrell D.G. (ed.).
9. Klabunovskii, E.I., and Vedenyapin A.A. (1980) *Asymmetric catalysis, Hydrogenation on metals* **(**Russian**)**, Nauka Publ., Moscow, 201 pp., *Chem. Abstr*., 1981, **94**, 181525.
10. Klabunovskii, E.I. (1968) *Stereospecific catalysis* (Russian), Nauka Publ. Moscow, 368 pp, *Chem Abstr*., 1969,**70**, B 100520r.
11. Klabunovskii, E.I. (1968) *Stereospecific catalysis in the field of optically active polymers, Dissymmetric modified catalysts* (Russian), Tsniitneftekhim Publ., Moscow, 76 pp.
12. Klabunowski, E.I. (1963) *Asymmetrische Synthese*, VEB Deutscher Verlag der Wissenschaften, Berlin, 213 pp.
13. Klabunovskii, E.I. (1991) Asymmetric hydrogenation on metals, *Uspekhi khimii* **60**, 1920-1945, *Russ. Chem. Rev*. **60**, 980, *Chem. Abstr.* 1991, **115**, 255272a.
14. Klabunovskii, E.I. (1991) On some general regularities in the action of enantioselective hydrogenation catalysts, *J. Organomet. Chem*, **417**, 181-191, *Chem. Abstr*. 1991, **115**, 231385p.
15. Klabunovskii, E.I. (1992) On some general regularities of action of enantiomeric catalysts of hydrogenation, *Kinet. Katal., (Kinetics and Catalysis, russ.*) **33**, 292-303, *Chem. Abstr*. 1992, **117**, 170494r.
16. Klabunovskii, E.I. (1996) Catalytic asymmetric synthesis of *beta*-hydroxyacids and their esters, *Uspekhi khimii (russ.)*, **65**, 350-366, *Russ. Chem. Rev.* 329-344.
17. Klabunovskii, E.I., and Sheldon, R.A. (1997) Catalytic approaches to the asymmetric synthesis of *beta*-hydroxy esters, *Cattech* **1**, 153-160.
18. Klabunovskii, E.I. (1978) Recent advances in enantioselective hydrogenation on metallic catalysts modified with chiral ligands, Gordon Research Conference, Colby-Sawyer, Chairman Smith, G.V.), *Science,* **199**, 1100.
19. Klabunovskii, E.I. (1985) Enantioselective hydrogenation on heterogeneous and metallocomplex catalysts, in *Catalytic Reactions in Liquid Phase,* Proc. 6th All-Union Conf., 1983, Alma-Ata, Publ. Nauka, Acad. Kazakh. SSR., pp. 147-159.
20. Klabunovskii, E.I (1998) Asymmetric catalysis. Concepts and Applications, 4th *Workshop GK and SFB, Structure-properties relationships in Heterocycles,* 1998, Schirgiswalde, Germany, *Abstracts,* p. 3.
21. Klabunovskii, E.I., Petrov, Yu.I., and Vedenyapin, A.A. (1976) On regularities of stereospecific catalysis over heterogeneous hydrogenation catalysts modified with comlexforming optically active ligands, in *"The Role of Coordination in Catalysis"*, Kiev, Pub. "Naukova Dumka", p. 40-45, *Chem. Abstr*. 1977, **87**, 12131c.

22. [a] Klabunovskii, E.I., and Vedenyapin, A.A. (1977) Mechanism of asymmetrizing action of chiral heterogeneous hydrogenation catalysts modified with optically active ligands, *Zh. Fiz. Khim. (J. Phys. Chem., russ.)*, **51**, 3005-3012, *Chem. Abstr.* 1978, **88**, 61803c.
[b] Klabunovskii, E.I. (1973) On some regularities in stereospecific catalysis, *Zh. Fiz. Khim. (J. Phys. Chem., russ.)*, **47**, 1353-1364, *Chem. Abstr.* 1973, **79**, 77560w.
23. Klabunovskii, E.I., Petrov, Yu.I., and Vedenyapin, A.A. (1976) The role of coordination in revealing of asymmetric activity of heterogeneous hydrogenation catalysts modified with chiral ligands, *Koordinatsion. khim. (Coordination chemistry, russ.),* **2**, 737-741. *Chem. Abstr.* 1976, **85**, 93197n.
24. Klabunovskii, E.I. (1966) On the problem of selection of catalysts of asymmetric hydrogenation and hydrogenolysis, in *Problems of kinetics and catalysis, XI., Scientific bases of selections of catalysts of heterogeneous catalytical reactions,* Publ. Nauka, Moscow, 170-174, *Chem. Abstr.* 1966, **65**, 16105e.
25. Klabunovskii, E.I., and Petrov Yu.I. (1967) On mechanism of asymmetric hydrogenation, *Dokl. AN SSSR. (Phys. chem.)* **173**, 1125-1128, *Chem. Abstr.* 1967, **67**, 81620k.
26. Klabunovskii, E.I. (1967) Configurational correlations in asymmetric catalysis, *Dokl. AN SSSR (Phys. chem.)* **173**, 1378-1381, *Chem. Abstr.* 1967, **67**, 82284r.
27. Klabunovskii, E.I. (1982) Advances in the field of enantioselective hydrogenation on chiral complexes of Rh,Pd,Co, *Uspekhi khimii (Russ. Chem. Rev.)* **51**, 1103-1128, *Chem. Abstr.* 1982, **97**, 161818r.
28. Klabunovskii, E.I. (1984) Asymmetric hydrogenation on metallic and metal complex dissymmetric catalysts, *Izv. AN SSSR, Ser. khim.* 505-525, *Chem. Abstr.* 1984, **101** ,54069b.
29. Klabunovskii, E.I., and Polkovnikov, B.D. (1986) Perspectives of industrial realization of asymmetric catalysis, *Khimicheskaya promyshlennost', (Chemical Industry, russ.)*, N9, 518-522.
30. Minachev, Kh.M., Kazanskii, V.B., and Klabunovskii, E.I. (1987) State of development of works on catalysis, Vestnik Acad. Sci. SSSR, N12, 49-57, *Chem. Abstr.* 1988, **108**, 82636x
31. Fish, M.J., and Ollis, D.J. (1978) Heterogeneous catalysis by optically selective surfaces, *Catal. Rev.-Sci. Eng.* **18**, 259 -295.
32. Isoda, T., Ichikawa, A., and Shimamoto, T. (1958) Asymmetric synthesis. I. General conditions of asymmetric reduction of derivatives of oxoglutaric acid, *Repts. Sci. (Res. Inst. Phys. Chem. Tokyo),* **34**, 134 -142.
33. Isoda, T., Ichikawa, A., and Shimamoto, T. (1958) Asymmetric synthesis II, Asymmetric reduction under action of catalytic complexes

Pd-tyrosine and amino acids adsorbed on Raney nickel, *Repts. Sci. (Res. Inst.Phys. Chem. Tokyo),* **34**, 143 - 151.

34. Activation of Raney nickel. Akkumulatoren Fabrik AG, *Belg. Pat. 613,541., 28.02.1962, Germ. appl. 07.02.1962, Chem. Abstr.* 1962, 57, 10572.
35. Akabori, S., and Izumi, Y. (1966) Verfahren zur Herstellung von Katalysatoren fuer asymmetrische Hydrierung von Carbonyl-verbindungen zu Alkoholen, BRD Pat 1,212,944.1966.
36. Fukawa, H., Izumi, Y., Komatsu, S., and Akabori, S. (1962) Studies on modified hydrogenation catalyst. 1. Selective hydrogenation activity of modified Raney nickel catalyst for carbonyl group and C=C double bond. *Bull. Chem. Soc. Jpn.*, **35**, 1703-1706.
37. Izumi, Y., Imaida, M., Fukawa, H., and Akabori, S. (1963) Asymmetric hydrogenation with modified Raney nickel. 1. Studies on modified hydrogenation catalyst, 11., *Bull .Chem. Soc. Jpn.*, **36,** 21-25.
38. Izumi, Y., Imaida, M., Fukawa, H., and Akabori, S. (1963) Asymmetric hydrogenation with modified Raney nickel, 11., Studies on modified hydrogenation catalyst, 111., *Bull. Chem. Soc. Jpn.*, **36,** 155-160.
39. Tatsumi, S., Imaida, M., Fukuda, Y., Izumi, Y., and Akabori, S. (1964) Asymmetric hydrogenation with modified Raney nickel, 111., Studies of the modified hydrogenation catalyst, IV., *Bull. Chem. Soc. Jpn.*, **37**, 847-849.
40. Izumi, Y., Tatsumi, S., and Imaida, M. (1965) Asymmetric hydrogenation with modified Raney nickel, V., Studies on the modified hydrogenation catalyst VI., *Bull. Chem .Soc. Jpn.* **38,** 1206 - 1208.
41. Izumi, Y., Akabori, S., Fukawa, H., Tatsumi, S., Imaida, M., Fukuda, H., and Komatu, S. (1964) Asymmetric hydrogenation with modified Raney nickel, in *Proc. 3th Intern. Congr. on Catalysis,* Sachtler, W.M.H., Scheit, G.C.A., and Zwietering, P. (eds.) North-Holland Publishers. Co., Amsterdam, 1964, pp. 1365 - 1372.
42. Izumi, Y., Tatsumi, S., and Imaida, M. (1966) Asymmetric hydrogenation with modified Raney nickel, VI., *Bull. Chem. Soc. Jpn.* **39**, 361 -364.
43. Izumi, Y., Tatsumi, S., and Imaida, M. (1966), Asymmetric hydrogenation with modified Raney nickel, VII., *Bull. Chem. Soc. Jpn.* **39**, 1087.
44. Izumi, Y., Tatsumi, S., and Imaida, M. (1966) Asymmetric hydrogenation with modified Raney nickel, VIII., *Bull. Chem. Soc. Jpn.* **39**, 2223 -2226.
45. Izumi, Y., Tanabe, T., Yajima, S., and Imaida, M. (1968) Asymmetric hydrogenation with modified Raney nickel IX., *Bull. Chem. Soc. Jpn.* **41**, 941 - 943.
46. Izumi, Y., Matsunaga, K., Tatsumi, S., and Imaida, M. (1968) Asymmetric hydrogenation with modified Raney nickel X., Effect of the

substituent of *alpha*-amino acid as modifying reagent. *Bull. Chem. Soc. Jpn.* **41**, 2515 - 2516.

47. Izumi, Y., Imaida, M., Harada, T., Tanabe, T., Yajima, S., and Ninomiya, T.K. (1969) Asymmetric hydrogenation of C=O double bond with modified Raney nickel, XI., *Bull.Chem. Soc. Jpn.* **42**, 241 - 243.
48. Izumi, Y., Tatsumi, S., and Imaida, M. (1969) Asymmetric hydrogenation of C=O double bond with modified Raney nickel, XII., *Bull. Chem. Soc. Jpn.* **42**, 2373 - 2375.
49. Tatsumi, S. (1968) Asymmetric hydrogenation activity of modified Raney nickel, *Bull. Chem. Soc. Jpn.* **41**, 408 - 418.
50. Izumi, Y., Tatsumi, S., Imaida, M., and Okudo, K. (1970) Asymmetric hydrogenation of C=O double bond with modified Raney nickel, XIII., Modification with peptides, *Bull. Chem. Soc. Jpn.* **43**, 566 - 567.
51. Izumi, Y., and Ninomiya, T. (1970) Asymmetric hydrogenation of C=O double bond with modified Raney nickel, XIV., *Bull. Chem. Soc. Jpn.* **43**, 579.
52. Izumi, Y., Harada, T., Tanabe, T., and Okuda, K. (1971) Asymmetric hydrogenation of C=O double bond with modified Raney nickel, XVII., *Bull. Chem. Soc. Jpn.* **44**, 1418 -1419.
53. Izumi, Y., and Ohkubo, K. (1971) Asymmetric hydrogenation of C=O double bond with modified Raney nickel, XVIII., *Bull. Chem. Soc. Jpn.* **44**, 1330 -1333, 1419 -1420.
54. [a] Harada, T., Imaida, M., and Izumi, Y. (1971) Asymmetric hydrogenation of C=O double bond with modified Raney nickel, XIX., *Bull. Chem. Soc. Jpn.* **44**, 1419 -1420.
    [b] Ninomiya, T. (1972) Asymmetric hydrogenation of C=O double bond with modified Raney nickel, XXII., *Bull. Chem. Soc. Jpn.* **45**, 2548 - 2550.
55. Tanabe, T., Okuda, K., and Izumi, Y. (1973) Asymmetric hydrogenation of C=O double bond with modified Raney nickel, XXV., Contributions of pH-adjusting reagents in the asymmetric hydrogenation, *Bull. Chem. Soc. Jpn.* **46**, 514 - 517.
56. Ozaki, H., Tai, A., and Izumi, Y. (1974) Enantioface-differentiating (asymmetric) hydrogenation of methyl acetoacetate with modified Raney nickel, XXVIII, *Chem. Lett.* 935 - 938.
57. Harada, T. (1975) The asymmetric (enantioface-differentiating) hydrogenation of the C=O double bond with modified Raney nickel, XXIX., *Bull. Chem. Soc. Jpn.* **48**, 3236 - 3238.
58. Osawa, T., Mita, S., Iwai, A., Takayasu, O., Harada, T., and Matsuura, I. (1998) New preparative method of asymmetrically modified nickel catalysts for the enantio-differentiating hydrogenation of methyl acetoacetate, *Stud. Surf. Sci. Catal.*, **118**, 313 - 320.

59. Harada, T., Yamamoto, M., Onaka, S., Imaida, M., Ozaki, H., Tai, A., and Izumi, Y. (1981) The enantioface-differentiating (asymmetric) hydrogenation of the C=O double bond with modified Raney nickel, XXXVI., The development of modified nickel catalysts with high enantioface-differentiating abilities, *Bull. Chem. Soc. Jpn.* **54**, 2323 - 2329.
60. Okamoto, S., Harada, T., and Tai, A. (1979) Enantioface-differentiating asymmetric hydrogenation of keto ester with modified nickel, XXXII., Structural requirements of modifying reagent and substrate, *Bull. Chem. Soc. Jpn.* **52**, 2670 - 2673.
61. Nakagawa, S., Sugimura, T., and Tai, A. (1998) A substrate specific chiral modifier activation on enantio-differentiating hydrogenation over tartaric acid-modified Raney nickel, *Chem. Lett.* 1257 - 1258.
62. Ninomiya, T. (1972) Asymmetric hydrogenation of C=O double bond with modified Raney nickel XXIII., *Bull. Chem. Soc. Jpn.* **45**, 2551 - 2554.
63. Ninomiya, T. (1972) Asymmetric hydrogenation of the C=O double bond with modified Raney nickel, XXIV., The effect of unsaturated compounds on the asymmetric activity of the catalyst, *Bull. Chem. Soc. Jpn.* **45**, 2555 - 2557.
64. Higashi, F., Ninomiya, T., and Izumi, Y. (1971) Asymmetric hydrogenation of C=O double bond with modified Raney nickel, XX., The effect of additives on the the asymmetric activity, *Bull. Chem. Soc. Jpn.* **44,** 1333 -1336.
65. Ninomiya, T. (1972) Asymmetric hydrogenation of C=O double bond with modified Raney nickel, XXI., The effect of water on the asymmetric activity of the catalyst, *Bull. Chem. Soc. Jpn.* **45**, 2545 - 2548.
66. Nakagawa, S., Sugimura, T., and Tai, A. (1997) Almost perfect enantio-differentiating hydrogenation of methyl 3-cyclopropyl-3-oxopropionate over tartaric acid modified Raney Nickel catalyst, *Chem. Lett.*, 859 - 860.
67. Tai, A., Kikukawa, T., Sugimura, T., Inoue, Y., Abe, S., Osawa, T., and Harada, T. (1994) An improved asymmetrically modified nickel catalyst prepared from ultrasonicated Raney nickel, *Bull. Chem. Soc. Jpn.* **67**, 2473 -2477.
68. Tai, A., Kikukawa, T., Sugimura, T., Inoue, Y., Osawa, T., and Fujii, S. (1991) Highly efficient enantio-differentiating hydrogenation over an ultrasonicated Raney nickel catalyst modified with tartaric acid, *J. Chem. Soc., Chem. Comm.* 795 -796.
69. Kikukawa, T., Iizuka, T., Sugimura, T., Harada, T., and Tai, A. (1987) The preparation of optically pure 3-hydroxybutanoic acid and its homologues as the dibenzylammonium salt, *Chem. Lett.* 1267-1270.
70. Osawa, T., Harada, T., and Tai, A. (1990) Enantio-differentiating hydrogenation of 2-alkanones over an asymmetrically modified nickel

catalyst and its application to the preparation of optically pure alkanols, *J. Catal.* **121**, 7- 17.

71. Sugimura, T., Yoshikawa, M., Yoneda, T., and Tai, A. (1990) Preparation of optically pure (3*S*,5*S*)- and (3*R*,5*R*)-2,6-dimethyl-3,5-heptanediol, *Bull. Chem. Soc. Jpn.* **63**, 1080 - 1082.
72. Sugimura, T., Futagawa, T., Yoshikawa, M., and Tai, A. (1989) Diastereo-differentiating Simmons-Smith reaction using a new chiral auxiliary, *Tetrahedron Lett.* **30**, 3807 - 3810.
73. Yoshikawa, M., Sugimura, T., and Tai, A. (1989) Pahutoxin: synthesis and determination of its absolute configuration, *Agricul. Biol. Chem.* **53**, 37 -40.
74. Sugimura, T., Yashikawa, M., Futagawa, T., and Tai, A. (1990) Diastereo-differentiating cyclopropanation of a chiral enol ether: a simple method for the preparation of optically pure cyclopropyl ethers, *Tetrahedron* **46,** 5955 - 5966.
75. Tai, A., Harada, T., Hiraki, Y., and Murakami, S. (1983) Stereochemical investigation on asymmetrically modified Raney nickel catalyst. Mode of interaction between modifying reagent and substrate in the enantio-face-differentiating process, *Bull. Chem. Soc. Jpn.* **56,** 1414 - 1419.
76. Harada, T., Imachi, Y., Tai, A, and Izumi, Y. (1982) Modification of nickel surface with optically active substrate: the method and mechanism, in *Metal-Support and Additive Effects in Catalysis,* Imelik, B. et al. (eds.), Elsevier, Amsterdam, pp. 377 - 384.
77. Nakahata, M., Imaida, M., Ozaki, H., Harada, T., and Tai, A. (1982) The preparation of optically pure 3-hydroxyalkanoic acid. The enantioface-differentiating hydrogenation of the C=O double bond with modified Raney nickel, XXXVII., *Bull. Chem. Soc. Jpn.* **55**, 2186-2189.
78. Osawa, T., and Harada, T. (1982) Enantioface-differentiating hydrogenation of 2-octanone over modified Raney nickel in the reaction media containing organic acid, *Chem. Lett.* 315 - 316.
79. Osawa, T., and Harada, T. (1984) The enantioface-differentiating hydrogenation of the C=O double bond with asymmetrically modified Raney nickel, XXXVIII. The hydrogenation of methyl ketones to optically active secondary alcohols, *Bull. Chem. Soc. Jpn.* **57**, 1518 -1521.
80. Osawa, T., and Harada, T. (1987) The enantioface-differentiating hydrogenation of the C=O double bond with asymmetrically modified Raney nickel, XXXIX. Repeated uses of the catalyst in the hydrogenation of 2-alkanones, *Bull. Chem. Soc. Jpn.* **60**,1277 - 1280.
81. Tai, A., Tsukioka, K., Ozaki, H., Harada, H., and Izumi, Y. (1984) Preparation of durable modified Raney nickel catalyst (MRNi), *Chem. Lett.* 2083 - 2084.
82. Tai, A., Imachi, T., Harada, T., and Izumi, Y. (1981) Embedded nickel catalyst in polymer, *Chem. Lett.* 1651 - 1652.

83. Tai, A., Harada, T., Tsukioka, K., Osawa, T., and Sugimura, T. (1988) Enantioface-differentiating hydrogenation of prochiral ketones with asymmetrically modified nickel catalyst (MNi): studies of the durability of catalyst under repeated uses, *Proc. 9th Int. Congr. Catal. Calgary*, pp. 1082-1089.
84. Yokozeki, M., Shimokoshi, K., and nMiyazaki, E. (1985) Enantioselective hydrogenation of methyl acetoacetate over a modified nickel surface: MNDO calculation of stability for intermolecular interactions, *J. Phys. Chem.* **89,** 2397 -2400.
85. Tai, A., Watanabe, H., and Harada, T. (1979) Stereochemical studies of the hydrogenation with asymmetrically modified nickel catalysts: the hydrogenation of methyl 2-alkyl-3-oxobutyrate, *Bull. Chem. Soc. Jpn.* **52**, 1468 - 1472.
86. Harada, T., Tai, A., Ozaki, N., and Izumi, Y. (1980) Characterization of the surface of asymmetrically modified nickel catalyst: an approach for the development of highly effective enantioface-differentiating (asymmetric) hydrogenation catalyst, in *Proc. 7th Intern. Congr. on Catal.,* Tokyo.
87. Osawa, T.(1985) Enantioface-differentiating hydrogenation of 2-alkanones, *Chem. Lett.* 1609 - 1612.
88. Ito, K., Harada, T., and Tai, A. (1980) The stereo-differentiating (asymmetric) hydrogenation of the C=O double bond with a modified nickel catalyst, XXXV. A facile method for the preparation of optically pure -diols. *Bull. Chem. Soc. Jpn.* **53**, 3367 - 3368.
89. Hiraki, Y., Ito, K., Harada, T., and Tai, A. (1981) Enantioface-differentiating hydrogenation of ketosulfone, *Chem. Lett.* 131 - 132.
90. Tai, A., Ito, K., and Harada, T. (1981) Stereochemical studies of the hydrogenation with an asymmetrically modified Raney nickel catalyst. The hydrogenation of acetylacetone, *Bull. Chem. Soc. Jpn.* **54**, 223 - 227.
91. Tanabe, T. (1973) Asymmetric hydrogenation of the C=O double bond with modified Raney nickel XXVII. Asymmetric hydrogenation of acetylacetone, *Bull. Chem. Soc. Jpn.* **46**, 1482 - 1488.
92. Tanabe, T., and Izumi, Y. (1973) Asymmetric) hydrogenation of the C=O double bond with modified Raney nickel, XXVI. Asymmetric hydrogenation of keto esters, B*ull. Chem. Soc. Jpn.* **46**, 1550-1551.
93. Ito, K., Harada, T., and Tai, A. (1980) The stereo-differentiating asymmetric hydrogenation of the C=O double bond with a modified Raney nickel XXXV. A facile method for preparation of optically active pure diols. *Bull. Chem. Soc. Jpn.* **53**, 3367 - 3368; 1518 -1521.
94. Murakami, S., Harada, T., and Tai, A. (1980) The enantioface-differentiating hydrogenation of the C=O double bond with asymmetrically modified Raney nickel, XXXIII. The preparation of (R)- and (S)-1,3,-

butanediol from 4-hydroxy-2-butanone, *Bull. Chem. Soc. Jpn.* **53**, 1356 - 1360.

95. Tai, A., Imaida, M., Oda, T., and Watanabe, H. (1978) Synthesis of optically active common precursor of sex pheromone of pine sawflies: an application of enantioface-differentiating hydrogenation on modified nickel, *Chem. Lett.* 61 - 64.
96. Tai, A., Nakahata, M., Harada, T., and Izumi, Y. (1980) A facile method for preparation of the optically pure 3-hydroxytetradecanoic acid by an application of asymmetrically modified nickel catalyst, *Chem. Lett.* 1125 - 1126.
97. Ito, K., Harada, T., Tai, A., and Izumi, Y. (1979) A facile method for the preparation of optically pure 2,4-pentanediol. An application of asymmetrically modified nickel catalyst for the preparation of optically active compounds, *Chem. Lett.* 1049 - 1050.
98. Harada, T. (1980) The enantioface-differentiating (asymmetric) hydrogenation of the C=O double bond with modified Raney nickel XXXIV. The adsorption mode of 2-hydroxy acid on Raney nickel, *Bull. Chem. Soc. Jpn.* **53**, 1019 - 1022.
99. Ozaki, H., Tai, A., Kobatake, S., Watanabe, H., and Izumi, Y. (1978) Enantioface-differentiating (asymmetric) hydrogenation of ketoester with modified Raney nickel catalyst (MRNi). XXXI. A comparative study of reaction rates and optical yields, *Bull. Chem. Soc. Jpn.* **51**, 3559 - 3563.
100. Harada, T., Hiraki, Y., Izumi, Y., Muraoka, J., Ozaki, H., and Tai, A. (1976) Comparative study of the rate and optical yield on the enantioface-differentiating hydrogenation of acetoacetate with the Raney nickel modified with optically active substance (MRNi), in *'Progress in Catalysis ', Proc. 6th Intern. Congr. on Catal.*, London, Preprint B 41, p. 9.
101. Izumi, Y. (1977) A hypothesis on the mechanism of catalytic hydrogenation, *Proc. Japan Acad.* **53**, 38 - 41.
102. Ozaki, H. (1978) The enantioface-differentiating (asymmetric) hydrogenation of the C=O double bond with modified Raney nickel XXX. The temperature dependency of the optical yield, *Bull. Chem. Soc. Jpn.* **51**, 257 - 259.
103. Harada, T., Onaka, S., Tai, A., and Izumi, Y. (1977) Improvement of nickel catalyst for enantioface-differentiating (asymmetric) hydrogenation of methyl acetoacetate, *Chem. Lett.* 1131 - 1132.
104. Kishi, K., Ikeda, S., and Hirota, K. (1967) Infrared and ultraviolet studies of the adsorption of acetylacetone on evaporated iron and nickel films, *J. Phys .Chem.* **71**, 4384 - 4389.
105. Harada, T., and Izumi, Y. (1978) Improved modified Raney nickel catalyst for enantioface-differentiating (asymmetric) hydrogenation of methyl acetoacetate, *Chem. Lett.* 1195 - 1196

106. Hatta, A., Moriya, Y., and Suetaka, W. (1975) The infrared absorption spectra of nickel metal surface modified with optically active alanine from its aqueous solution, *Bull. Chem. Soc. Jpn.* **4 8**, 3441 - 3445.

107. Ninomiya, T., Kishi, K., and Ikeda, S. (1970) Ultraviolet study for the adsorpton of acetylacetone on D-tartaric acid preadsorbed nickel film, *Bull. Chem. Soc. Jpn.* 43, 2270.

108. Izumi, Y., Takizawa, H., Nakagawa, K.I., Mamura, R., Imaida, M., Ninomiya, T., and Yajima, S. (1970) Asymmetric hydrogenation of C=N double bond with modified Raney nickel. I. New determination method for the reaction mechanism using asymmetric catalyst, *Bull. Chem. Soc. Jpn.* **43**, 1792 - 1795.

109. Tanabe, T., Ninomiya, T., and Izumi, Y. (1970) Asymmetric hydrogenation of C=O double bond with modified Raney nickel. XV. Asymmetric hydrogenation of metal chelate compounds, *Bull. Chem. Soc. Jpn.* **43**, 2276.

110. Osawa, T., Mita, S., Iwai, A., Takayasu, O., Harada, T., and Matsuura, I. (1998) New preparation method of asymmetrically modified supported Nickel catalysts for the enantiodifferentiating hydrogenation of methyl aetoacetate, in *Preparations of catalysts VII., Proc. 7th Intern. Symp. Sci. Bases Preparation Heterog. Catalysts*, Delmon, B., Jacobs, P.A., Maggi, R., Martens, J.A., Grange P., and Poncelet, G. (eds.) Louvain-La Neuve, 1998, Elsevier, Amsterdam, p. 313-320.

111. Osawa, T., Mita, S., Iwai, A., Miyazaki, T., Takayasu, O., Harada, T., and Matsuura, I. (1997) Enantio-differentiating hydrogenation of methyl acetoacetate over asymmetrically modified supported nickel catalyst prepared from nickel acetoacetonate, *Chem. Lett.* 1131 - 1132.

112. Tanabe, K., and Uyeda, N. (1974) Oriented adsorption of amino acids on (001) nickel surface, *Bull. Inst. Chem. Res. KyotoUniv.* **52,** 616-647, *Chem. Abstr.* 1998, **128**, 90728g.

113. Petrov, Yu.I., Klabunovskii, E.I., and Balandin, A.A. (1967) Asymmetric hydrogenation over Raney Nickel modified with optically active compounds, *Kinetika i kataliz (Kinetics and Catalysis ,russ.)* **8**, 814-820, *Chem. Abstr*. 1968, **68**, 58786f.

114. Klabunovskii, E.I., Neupokoev, V.I., and Petrov, Yu.I. (1971) Dissymmetric skeletal cobalt catalyst, *Izv. AN. SSSR. Ser. khim.* 2067-2068, *Chem. Abstr*. 1972, **76**, 18286b.

115. Klabunovskii, E.I. (1972) Some general questions of mechanism of stereospecific catalysis on dissymmetric modified catalysts, *"Catalytic Ractions in Liquid Phase", Proc. III. All-Union Conf. on catalytical reactions in liquid phase,* 1971, Nauka Kazakh. SSR, Alma-Ata, p. 70-75, *Chem. Abstr.* 1973, **79**, 4824e.

116. Vedenyapin, A.A., Chankvetadze, B.G., and Klabunovskii, E.I. (1983) New method of preparation of Ni catalysts of asymmetric

hydrogenation, *Izv. AN. SSSR. Ser. khim.* 701, *Chem. Abstr.* 1983, **98**, 167691m.

117. Petrov, Yu.I., Klabunovskii, E.I., Kapitannikov, Yu.V., Tomina, L.D., Antonova, T.A., and Lipgart, E.N. (1974) Asymmetric hydrogenation of alkyl-derivatives acetacetic esters over modified Ni-catalysts, in "*Catalytic Reactions in Liquid Phase*", Proc. IV. All-Union Conf. on catalytic reactions in liquid phase, Kazakh SSR, Alma-Ata, 1974,V.1, p. 28-29, *Chem. Abstr.* 1975, **83**, 163519d.
118. Klabunovskii, E.I., Sokolova, N.P., Petrov, Yu.I., Patrikeev, V.V., Neupokoev, V.I., and Ribtsov, I.A. (1983) Asymmetric hydrogenation of acetylacetone on skeletal Ni catalyst, *Izv. AN SSSR. Ser. khim.* 838-841, *Chem. Abstr.* 1983, **99**, 104734c.
119. Zubareva, N.D., Chernysheva, V.V., Grigoryev, Yu.A., and Klabunovskii, E.I. (1987) Enantioselective hydrogenation of ethyl-acetoacetate on dissymmetric skeletal Ni catalysts, *Izv. AN SSSR. Ser. khim.* 525-527, *Chem. Abstr.* 1987, **107**, 236070b.
120. Lipgart, E.N., Petrov, Yu.I., and Klabunovskii, E.I. (1971) Asymmetric hydrogenation on skeletal Ni catalyst modified with optically active substances, Effect of reaction condition on conversion of acetacetic ester, *Kinetika i kataliz (Kinetics and catalysis, russ.)* **12**, 1491-1494, *Chem. Abstr.* 1972, **76**, 98985s.
121. Klabunovskii, E.I., Sokolova, N.P., Vedenyapin, A.A., Talanov, Yu.M., Zubareva, N.D., Polyakova, V.P., and Gorina, N.B. (1972) On skeletal Ru catalyst of asymmetric hydrogenation, *Izv. AN. SSSR Ser. khim.* 2361, *Chem. Abstr.* 1973, **78**, 71345p.
122. Klabunovskii, E.I., Sokolova, N.P., Vedenyapin, A.A., and Talanov, Yu.M. (1971) On asymmetric hydrogenation of carbonyl group in presence of modified Ru-catalyst, *Izv. AN. SSSR. Ser. khim.* 1803-1804, *Chem. Abstr.* 1972, **76**, 3317p.
123. Klabunovskii, E.I., Vedenyapin, A.A., Zubareva, N.D., Sokolova, N.P., and Talanov, Yu.M. (1975) On the asymmetrizing and catalytic activity of the Ru-catalysts, *React. Kinet. Catal. Letters* **2**, 291-296.
124. Klabunovskii, E.I., Sokolova, N.P., Neupokoev, V.I., Antonova, T.A., and Ternovskii, S.E. (1974) Dissymmetric Ru catalyst modified with optically active amino acids, *Izv. AN SSSR. Ser. khim.* 1361-1364, *Chem. Abstr.* 1974, **81**, 91016w.
125. Vedenyapin, A.A., Klabunovskii, E.I., Talanov, Yu.M., and Sokolova, N.P. (1975) Study of the catalytic and asymmetric activities of Ru-catalysts. Comm. I. Study of the effect of nature of carrier on dispersity and specific catalytic activity, *Kinet. Katal. (Kinetics and Catalysis, russ.)* **16**, 436-442, *Chem. Abstr.* 1975, **83**, 58047e, 1976, **84**, 4423x.
126. Vedenyapin, A.A., Klabunovskii, E.I., Talanov, Yu.M., and Sokolova, N.P. (1975) Comm.V. On mechanism of asymmetric hydrogenation,

Kin. i Kat. (Kinetics and Catalysis, russ.), 1975, **16**, 692-696, *Chem. Abstr.* 1975, **83**, 96278j.

127. Vedenyapin, A.A., Zubareva, N.D., Klabunovskii, E.I., Godyatsheva, I.P., and Rubtsov, I.A. (1975) Study of adsorption of hydrogen on Ru-catalyst, Comm.I. Effect of acidity of solution on strength of bond of hydrogen on Ru-black in solution, *Zh. Fiz .khim. (J.Phys.Chem., russ.)* **49**, 1480-1485.
128. Vedenyapin, A.A., Zubareva, N.D., Akimov, V.M., Klabunovskii, E.I., Giorgadze, N.G., and Barannikova, N.E. (1976) Study of Phase Composition of Skeletal Ru-catalysts, *Izv. AN SSSR. Ser. Khim.* 2340-2343, *Chem. Abstr.* 1977, **86**., 60926s.
129. Vedenyapin, A.A., Klabunovskii, E.I., Talanov, Yu.M., and Areshidze G.Kh. (1976) Effect of size of particles of Ru in catalyst, *Izv. AN SSSR. Ser. khim.* 2628-2630, *Chem. Abstr.* 1977, **86**, 89128u.
130. Zubareva, N.D.,Vedenyapin, A.A., and Klabunovskii, E.I. (1978) Comm. II. Adsorption of hydrogen on skeletal Ru catalyst in liquid phase, *Zh. Fiz. khim. (J. Phys. Chem., russ.)* 52, 407-410, *Chem. Abstr.* 1978, **89**, 3108t.
131. Dmitriev, R.V., Zubareva, N.D., Vedenyapin, A.A, Klabunovskii, E.I., and Minachev, Kh.M. (1977) Study of state and behaviour of hydrogen adsorbed on Ru-catalysts, *Izv. AN SSSR, Ser. khim.* 986-990, *Chem. Abstr.* 1977, **87**, 44662t.
132. Dmitriev, R.V., Zubareva, N.D.,Vedenyapin, A.A., Klabunovskii, E.I., and Minachev, Kh.M. (1988) Chemosorption and spillover of hydrogen on skeletal Ru, *Izv. AN SSSR. Ser. khim.* 1712-1715, *Chem. Abstr.* 1988, **109,** 55180h.
133. Vedenyapin, A.A., Klabunovskii, E.I., Sokolova, N.P., Zubareva, N.D., Areshidze, G.Kh., Leonova, E.V., Savitskii, E.M., Polyakova, V.P., and Gorina, N.B. (1974) Catalytic properties of skeletal Ru,Cu-Ru and Ni-Ru catalysts, in *"Alloys of noble metals and their substitutes". Proc. of All-union Meeting,* Moscow, Report N 65. p.11.
134. Vedenyapin, A.A., Klabunovskii, E.I., Sokolova, N.P., Zubareva, N.D., Areshidze, G.Kh., Leonova, E.V., Ivanov, A.K., Savitskii, E.M., Polyakova, V.P., and Gorina, N.B. (1977) Preparation of alloys Ru-Al, Ru-Cu-Al and Ru-Ni-Al and preparation based on its dissymmetric catalysts, in *" Alloys of Noble Metals"*, Moscow, Nauka Publ., p. 187-189, *Chem. Abstr.* 1977, **87**, 123241h.
135. Klabunovskii, E.I., Vedenyapin, A.A., and Areshidze, G.Kh. (1975) Asymmetric activity of skeletal Cu-Ru-Catalysts in hydrogenation of acetacetic ester. *Izv. AN SSSR. Ser. khim.* 2311-2313, *Chem. Abstr.* 1976, **84**, 43254n.
136. Kuznetsova, T.A, Vedenyapin, A.A., and Klabunovskii, E.I. (1984) Behavior of Cu-Ru catalysts in enantioselective hydrogenation, *Izv. AN. SSSR. Ser. Khim.* 1002-1005, *Chem. Abstr.* 1984, **101**, 130194f.

137. Klabunovskii, E.I., Vedenyapin, A.A., Areshidze, G.Kh., Polyakova, V.P., and Gorina, N.B. (1975) On asymmetric activity of skeletal Cu-Ru-catalysts, *Kin.i Kat. (Kinetics and Catalysis,russ.)* **16**, 1081-1082.
138. Klabunovskii, E.I., Vedenyapin, A.A., and Airapetov, Yu.S. (1978) Enantioselective hydrogenation on heterogeneous metal catalysts, *React. Kinet. Catal. Letters* **9**, 73-77, *Chem Abstr*. 1979, **90**, 137149h.
139. Vedenyapin, A.A., Zubareva, N.D., Akimov, V.M., Areshidze, G.Kh., Barannikova, N.E., and Klabunovskii, E.I. (1978) Study of phase composition of skeletal Ru-Cu catalysts and hydrogen absorption by them, *Izv. AN SSSR Ser. khim.* 1942-1945, *Chem. Abstr*. 1978, **89**, 204768d.
140. Petrov, Yu.I., Klabunovskii, E.I., Lipgart, E.N., and Neupokoev, V.I. (1972) Asymmetric hydrogenation of acetacetic ester on modified skeletal Ni,Co and Cu catalysts, *Catalytic reactions in liquid phase. Proc. 3rd All-union Conf. on catalytic reactions in liquid phase,* 1971, Nauka, Publishers Kazakh. SSR, Alma-Ata, p.254-263, *Chem. Abstr*. 1973, **79**, 17883u.
141. Neupokoev, V.I., Petrov, Yu.I., and Klabunovskii, E.I. (1976) Asymmetric hydrogenation of acetoacetic ester on skeletal Co and Cu catalysts modified with D-(+)-tartaric acid, *Izv. AN SSSR Ser. khim.* 113-117, *Chem. Abstr*. **84**, 150160a.
142. Zubareva, N.D. Klabunovskii, E.I., and Kheifits, L.A. (1989) Stereoselective hydrogenation of mixture of menthon-isomenthon on heterogeneous Ni, Ni-Co and Co-catalysts, *Izv. AN SSSR Ser. khim.*, 1920-1923, *Chem. Abstr*. 1990, **112**, 7725e.
143. Vedenyapin, A.A., Klabunovskii, E.I., and Areshidze, G.Kh. (1977) Kinetics and mechanism of hydrogenation of C=O group in acetoacetic ester on skeletal Cu catalyst, *Izv. AN SSSR Ser. khim.*, 1416-1418, *Chem. Abstr*. 1977, **87**, 101672x.
144. Vedenyapin, A.A., Airapetov, Yu.S. Klabunovskii, E.I., Fridman, Ya.D., and Mikhailyuk, L.Ya. (1977) On enantioselective hydrogenation of ethyl acetoacetate on skeletal Cu-catalyst modified with amino acids, *Izv. AN SSSR Ser. khim.*, 2385, *Chem. Abstr*. 1978, **88**, 37224u.
145. Fridman, Ya.D., Mikhailuk, L.Ya., Petrenko, S.V., Pavlov, V.A., Airapetov, Yu.S., and Klabunovskii, E.I. (1977) Formation of mixed compounds Cu with acetylacetone and amino acids in solution and in adsorbed layer, *Koordinats. khim. (Coordinative Chemistry, russ.)* 3, 1309-1315, *Chem. Abs.* 1978, **88**, 56573y.
146. Klabunovskii, E.I., Pavlov, V.A., and Fridman, Ya.D. (1984) On chirodiastaltic interaction in mixed complexes metal-amino acid-diketone (ketoester), *Izv. AN SSSR Ser. khim.,* 1005-1010, *Chem. Abstr*. 1984, **101**, 79240p.

147. Vedenyapin, A.A., Klabunovskii, E.I., Leonova, E.V., Areshidze, G.Kh., and Barannikova, N.E (1978) Mechanism of modification of skeletal Cu-catalyst of asymmetric hydrogenation, *Izv. AN SSSR Ser. khim.* 206-207, *Chem. Abstr*. 1978, **88**, 3592q.
148. Klabunovskii, E.I., Neupokoev, V.I., and Petrov, Yu.I. (1970) Asymmetric hydrogenation of acetoacetic ester on modified skeletal copper catalyst, *Izv. AN. SSSR Ser. khim.* 2839-2840, *Chem. Abstr.* 1971, **74**, 99415b.
149. Chernysheva, V.V., and Klabunovskii, E.I. (1983) Enantioselective hydrogenation of ethyl acetoacetate on Cu-Ni catalysts. Effect of pH modification and nature of modifier, *Izv. AN SSSR Ser. khim.* 2023-2026, *Chem. Abstr*. 1984, **100**, 5815n.
150. Vedenyapin, A.A., Klabunovskii, E.I., and Vlasenko, Yu.V. (1981) Structure of dissymmetric clusters leading the enantioselective heterogeneous catalysis, *Izv. AN SSSR, Ser. khim.* 2519-2523, *Chem. Abstr*. 1982, **96**, 41547j.
151. Klabunovskii, E.I., Vedenyapin, A.A., Areshidze, G.Kh., and Ivanov, A.K. (1976) On dissymmetric activity of modified Cu-Ni catalysts, *Izv. AN SSSR, Ser. khim.* 2642-2643, *Chem. Abstr*. 1977, **86**, 120756y.
152. Vlasenko, Yu.V., Klabunovskii, E.I., Vedenyapin, A.A., and Brezzhunova, L.N. (1977) Catalytic and asymmetric activity of mixed Cu-Ni catalysts supported on aerosil in hydrogenation of acetylacetone, *Izv. AN SSSR Ser. khim.*, 680-682, *Chem. Abstr.*, 1977, **87**, 84459p.
153. Vedenyapin, A.A., Chankvetadze, B.G., and Klabunovskii, E.I. (1984) Studies of Copper- Nickel catalysts for enantioselective hydrogenation, *React. Kinet. Catal. Letters* **24**, 77-80.
154. Zubareva, N.D., Chernysheva, V.V., Zhdan, P.A., Chukin, G.D., Akimov, V.M., Nefedov, B.K., and Klabunovskii, E.I. (1986) Studies of effect of genesis of modified powder Cu-Ni catalysts on enantioselectivity of hydrogenation, *Kin. i Kat. (Kinetics and Catalysis , russ.)* **27**, 1264-1267, *Chem. Abstr*. 1987, **106**, 155734y.
155. Vedenyapin, A.A., Chankvetadze, B.G., Tsivinskaya, L.K, Akimov, V.M., and Klabunovskii, E.I. (1987) Studies of Cu-Ni supported catalysts for enantioselective hydrogenation, *React. Kinet. Catal. Letters* **33**, 53-58, *Chem. Abstr.* 1988, **108**, 74488k.
156. Vedenyapin, A.A., Chankvetadze, B.G., Akimov, V.M., and Klabunovskii, E.I. (1988) Studies of formation of dissymmetric Cu-Ni catalysts, *Izv. AN SSSR Ser. khim.*, 2213-2217, *Chem. Abstr*. 1989, **110**, 211882a.
157. Klabunovskii, E.I., Vedenyapin, A.A., Karpeiskaya, E.I., and Pavlov, V.A. (1981) Enantioselective hydrogenation on dissymmetric Cu-Ni and Pd catalysts, in *"New Horizons in Catalysis*, Proc. 7th Int. Congr. Catal., 1980, Tokyo, Kodansha, Tokyo, 1980, pp. 390-399, *Chem. Abstr*. 1981, **95**, 226360r.

158. Vedenyapin, A.A., Klabunovskii, E.I., Vlasenko, Yu.V., Akimov, V.M., and Kharlamov, V.N. (1980) Study of adsorption and hydrogenation of acetylacetone over catalysts Cu-Ni-aerosil modified with (*RR*)-(+)-tartaric acid, *Izv. AN SSSR, Ser. khim.* 1750-1754, *Chem. Abstr.* 1981, **94,** 14836e.
159. Chernysheva, V.V., Zubareva, N.D., and Klabunovskii, E.I. (1984) Enantioselective hydrogenation of ethyl acetoacetate on Cu-Ni-catalysts. Effect of condition of preparation of catalyst, *Izv. AN SSSR Ser.khim.* 2443-2444.
160. Klabunovskii, E.I., Vedenyapin, A.A., Karpeiskaya, E.I., and Pavlov, V.A. (1981) Enantioselective hydrogenation on dissymmetric Cu-Ni and Pd catalysts, *Stud. Surf. Sci. Catal.* 7A, 390-401.
161. Neupokoev, V.I., Klabunovskii, E.I., and Petrov, Yu.I. (1973) Asymmetric hydrogenation of acetacetic ester on disymmetric Co, *Kin. i Kat. (Kinetics and Catalysis, russ.)* **14**, 447-451, *Chem. Abstr.* 1973, **79**, 18008z.
162. Klabunovskii, E.I., Neupokoev, V.I, Petrov, Yu.I., Fasman, A.B., and Areshidze, G.Kh. (1980) Enantioselective hydrogenation of C=O bond on skeletal Co catalyst modified with optically active amino acids, *Izv. AN SSSR Ser. khim.* 121-124, *Chem. Abstr.* 1980, **93**, 25559x.
163. Zubareva, N.D., Klabunovskii, E.I., and Dorokhin, G.V. (1991) Enantioselective properties of dissymmetric supported Ni and Ni-Co catalysts in hydrogenation ethyl acetoacetate, *Izv. AN SSSR Ser. khim.* 1769-1773.
164. Zubareva, N.D., and Klabunovskii, E.I. (1988) Heterogeneous Ni-Co catalysts in enantioselective hydrogenation of ethyl acetoacetate, *Izv. AN SSSR, Ser. khim.* 1172-1174, *Chem. Abstr.* 1989, **110**, 38604c.
165. Zubareva, N.D., Ryndakova, I.A., and Klabunovskii, E.I. (1988) Studies of hydrogenation of ethyl acetoacetate on dissymmetric powder Ni-Co catalysts, *Kin. i Kat. (Kinetics and Catalysis, russ.)* **29**, 1485-1488, *Chem. Abstr.* 1989, **111**, 23080a.
166. Vedenyapin, A,A., Kuznetsova, T.I., and Klabunovskii, E.I. (1983) On sinergism of asymmetric action of bimetallic hydrogenation catalysts, *Izv. AN SSSR, Ser. khim.* 1455, *Chem. Abstr.* 1983, **99**, 111419g.
167. Vedenyapin, A.A., Kuznetsova, T.I., Klabunovskii, E.I., Akimov, V.M., Gorina, N.B., and Polyakova, V.P. (1983) Behavior of skeletal Cu-Pd-catalysts in enantioselective hydrogenation, *Izv. AN SSSR, Ser. khim.* 2206-2210, *Chem. Abstr.* 1984, **100**, 22245u.
168. Vedenyapin, A.A., Kuznetsova, T.I., and Klabunovskii, E.I. (1987) On ligand effect in enanatioselective hydrogenation on Cu-Pd-catalysts, *Izv. AN SSSR, Ser. khim.* .277-280, *Chem. Abstr.* 1987, **107**, 235823n.
169. Vedenyapin, A.A., Zubareva, N.D., Areshidze, G.Kh., Kuznetsova, T.I., and Klabunovskii, E.I. (1988) On ligand effect in dissymmetric Cu-Ru

and Cu-Pd catalysts, *Izv. AN SSSR, Ser. Khim* .2729-2731, *Chem. Abstr.* 1989, **111**, 6719y.

170. Kuznetsova, T.I., Murina, I.P., Vedenyapin, A.A., Akimov, V.M., and Klabunovskii, E.I. (1988) On catalytic and enantioselective activity of supported Cu-Pd and Ni-Pd catalysts, *React. Kinet. Catal. Letters* **37**, 363-365.
171. Konenko, I.R., Gorshkova, L.S., Parfenova, N.M., and Klabunovskii, E.I. (1980) Enantioselective hydrogenation of ethyl acetoacetate on modified with (*RR*)-(+)-tartaric acid intermetallic hydrides of rare earth metals and Ni, *Izv. AN SSSR, Ser. khim.* 431-432, *Chem. Abstr.* 1980, **93**, 25872u.
172. Savitskii E.M., Klabunovskii E.I.,Konenko I.R., Parfenova N.M., and Mordovin V.P. (1983) Intermetallic compounds Ln Ni_5,La_{1-x} $PrNi_5$ as sorbents and hydrogenation catalysts (Ln = La,Ce,Pr,Nd,Sm,Gd), *J. Less-Common Metals,* **89**, 528.
173. Konenko, I.R., Parfenova, N.M., Klabunovskii, E.I., and Mordovin, V.P. (1983) Intermetallic compounds based on rare earth metals with Ni as a new class of catalysts of hydrogenation, in *"Alloys of rare metals with special physical properties"*, Nauka, Moscow, p. 106-111, *Chem. Abstr.* 1984, **100**, 91965r.
174. Starodubtseva, E.V., Konenko, I.R., Klabunovskii, E.I., Savitskii, E.M., and Mordovin, V.P. (1984) $LaNi_{5-x}Cu_x$-hydrides modified with (*RR*)-tartaric acid as catalysts of hydrogenation of ethyl acetoacetate, *Izv. AN SSSR Ser. khim.* 758-762, *Chem. Abstr.* 1984, **101**, 54152y.
175. Konenko, I.R., Starodubtseva, E.V., Fedorovskaya, E.A., Klabunovskii, E.I., Savitskii, E.M., and Mordovin, V.P. (1984) Sorption and catalytic properties of intermetallic compounds $LaNi_3$,$LaNi_{5-x}Cu_x$ and their hydrides, *Izv. AN SSSR, Ser. khim.* 754-758, *Chem. Abstr.* 1984, **101**, 12769z.
176. Konenko, I.R., Starodubtseva, E.V., Stepanov, Yu.P., Fedorovskaya, E.A. Slinkin, A.A, Klabunovskii, E.I., Savitskii, E.M., and Mordovin, V.P. (1985) Intermetallic compounds Ln $Ni_{5-x}Cu_x$ and their hydrides in processes of hydrogen sorption and olefin hydrogenation, *Kin. i Kat. (Kinetics and Catalysis, russ.)* **26**, 340-344, *Chem. Abstr.* 1985, **103**, 76809m.
177. Starodubtseva, E.V., Konenko, I.R., Fedorovskaya, E.A., Klabunovskii, E.I., and Mordovin, V.P. (1986) Dissymmetric properties of hydrides of intermetallic compounds based on rare earth metals, Ni and Co, modified with (*RR*)-(+)-tartaric acid, *Izv. AN SSSR, Ser. khim.* 1960-1965, *Chem. Abstr.* 1987, **106**, 175463n.
178. Konenko, I.R., Starodubtseva, E.V., Stepanov, Yu.P., Fedorovskaya, E.A., Slinkin, A.A., Klabunovskii, E.I., and Mordovin, V.P. (1986) Sorptional, catalytic and structural characteristics of intermetallic

compounds LaNi $_{5-x}Co_x$ and their hydrides, *Kin. i Kat. (Kinetics and Catalysis, russ.)* **27**, 456-462, *Chem. Abstr*. 1986, **104**, 231316q.

179. Konenko, I.R., Starodubtseva, E.V., Urazbaeva, K.A., Zubareva, N.D., Klabunovskii, E.I., Fedorovskaya, E.A., Slinkin, A.A., and Mordovin, V.P. (1987) Relation of catalytic sorption and dissymmetric properties to the structural features of intermetallic compounds, their hydrides and Ni-Co-powders, *Proc. 6th Int. Symp. Heterogen. Catal.*, Sofia, Part 1, p. 300-305, *Chem. Abstr*. 1988, **109,** 210500h.
180. Starodubtseva, E.V., Konenko, I.R., Fedorovskaya, E.A., Klabunovskii, E.I., and Mordovin, V.P. (1988) Effect of preparation conditions of dissymmetric $LaNi_3Co_2H$-catalysts on their hydrogenating and enantioselective properties, *Izv. AN SSSR, Ser. khim.* 2725-2728, *Chem. Abstr.* 1989, **111**, 6908j.
181. Konenko, I.R., Starodubtseva, E.V., Urazbaeva, K.A., Fedorovskaya, E.A., Klabunovskii, E.I., Slinkin, A.A., and Mordovin, V.P. (1988) Influence of the transition d-metals in the lattice of the intermetallic compounds and their catalytic,sorption and magnetic characteristics, in *" Hydrogen Energy Progress", Proc. 7th World Hydrogen Energy Conf.,* Pergamon Press, Moscow, V.2, p.1273-1288.
182. Konenko, I.R., Starodubtseva, E.V., Urazbaeva, K.A., Fedorovskaya, E.A., Klabunovskii, E.I., Slinkin, A.A., and Mordovin, V.P. (1988) Catalytic properties of intermetallides Ln $(NiM)_5$ and their hydrides in hydrogenation reactions, *Kin. i Kat. (Kinetics and Catalysis, russ.)* **29**, 993-996, *Chem. Abstr*. 1989, **110**, 29523y.
183. Klabunovskii, E.I., Konenko, I.R., Zubareva, N.D., Starodubtseva, E.V., Urazbaeva, K.A., Fedorovskaya, E.A., Slinkin, A.A., and Mordovin, V.P. (1988) Preparation of active and stable catalysts based on intermetallides of *f*- and *d*- metals, their hydrides and Ni-Co powder catalysts for hydrogenations of C=O and C=C bonds, 3rd *Czechoslovak Conf. on Preparation and properties of heterogeneous catalysts,* Bechine- near-Tabor, Summaries p. 67-68.
184. Klabunovskii, E.I., Petrov, Yu.I., and Vedenyapin, A.A. (1974) On regularities of stereospecific catalysis on modified with optically active ligands heterogeneous hydrogenation catalysts, *2nd Soviet-France Seminar on Catalysis,* Kiev, Preprint N7. p. 33-39.
185. Klabunovskii, E.I., Komarova, E.N., and Petrov, Yu.I. (1967) On asymmetric hydrogenation on modified Ni catalysts and parameters of optical activation, Referate, *Intern. Symp. Mechanism Heterogen. Catal.,* Sofia.
186. Chernysheva, V.V., Murina, I.P., and Klabunovskii, E.I. (1983) Study of inhibition of enantioselective hydrogenation of ethyl acetoacetate, *Izv. AN SSSR Ser. khim.* .757-760, *Chem. Abstr*. 1983, **99**, 21647a.

187. Klabunovskii, E.I., Vedenyapin, A.A., and Karpeiskaya, E.I. (1975) On some structural and energetic factors in enantioselective hydrogenation, *The first Research Conference on Catalysis,* Osaka, Preprints p.16.
188. Klabunovskii, E.I., Mordovin, V.P., Pisarev, Yu.N., and Artem'ev, V.A. (1997) Catalyst for conversion of methanol into synthesis gas, *Pat. Russian Federation,* 2087190, 20.08.1997, *Chem. Abstr.* 1998, **128**, 90727f.
189. Klabunovskii, E.I., Mordovin, V.P., Pisarev, Yu.N., and Artem'ev, V.A. (1997) Manufacture of catalyst for conversion of methanol into hydrogen-carbon monoxide mixture, *Pat. Russian Federation* 2087189, 20.08.1997, *Chem. Abstr*. 1998, **128**, 90728g.
190. Murina, P., Vedenyapin, A.A., and Klabunovskii, E.I. (1988) Catalytic hydrogenation of acetylamino-beta-ketobutyric ester on modified metallic catalysts, *Izv. AN SSSR, Ser. khim.* 999-1001, *Chem. Abstr*., 1989, **110**, 39332z.
191. Zubareva, N.D., Khatskevich, I.E., Kuznetsova, T.I., Bitiev, Sh.G., Karpeiskaya, E.I., Vedenyapin, A.A., and Klabunovskii, E.I. (1983) Preparation of *Z*- and *E*-methyl esters of benzoylaminomethyl-ethylacrylic acid and its enantioselective hydrogenation on heterogeneous catalysts, *Izv. AN Latv. SSR. Ser. khim.* 477-482, *Chem. Abstr*. 1983, **99**, 212892m.
192. [a] Heinen, A.W., Papadogianakis, G., Sheldon, R.A., Peters, J.A., and van Bekkum, H. (1999) Factors effecting the hydrogenation of fructose with a water soluble Ru-TPPTS complex. A comparison between homogeneous and heterogeneous catalysis, *J. Mol. Catal. A: Chem.* 142, 17-26.

[b]). Veksler, M.A., Petrov, Yu.I, Geling, N.G., and Klabunovskii, E.I. (1974) Study of stereospecific hydrogenation of fructose on skeletal Ni using the mathematic simulation, *Izv. AN SSSR, Ser. khim.* 53-57, *Chem. Abstr*. 1974, **80**, 108766v.
193. Orito, Y., Niwa, S., and Imai, S. (1976) Asymmetric hydrogenation of methyl acetoacetate using Nickel-Platinum metal-Kieselguhr catalysts modified with tartaric acid, *Yuki Gosei Kagaku Kiokaishi* (*J. Synth. Org. Chem. Jpn.)* **34**, 236 - 239.
194. Orito, Y., Niwa, S., and Imai, S. (1976) Asymmetric hydrogenation of methyl acetoacetate using Nickel-Palladium-Kieselguhr catalysts modified with tartaric acid, *Yuki Gosei Kagaku Kiokaishi*, (*J. Synth. Org. Chem. Jpn.)* **34**, 672 - 674.
195. Orito, Y., Niwa, S., and Imai, S. (1977) Agency of Industrial Science and Technology. Asymmetric catalyst, *Jpn. Kokai*, 77.05.692, *Chem. Abstr.* 1977, **86**, P 146521t.
196. Orito, Y., Niwa, S., and Imai, S. (1977) Effect of additives on the asymmetric hydrogenation of methyl acetoacetate using Nickel-

Palladium- Kieselguhr catalysts modified with tartaric acid, *Yuki Gosei Kagaku Kiokaishi*, (*J .Synth. Org. Chem. Jpn.)* **35**, 672-674.
197. Orito, Y., Niwa, S., and Imai, S. (1977) Highly selective asymmetric hydrogenation of methyl acetoacetate with Nickel-Palladium-Kieselguhr catalyst modified with tartaric acid, *Yuki Gosei Kagaku Kiokaishi, (J. Synth. Org. Chem. Jpn.),* **35**, 753 - 754.
198. Nitta, Y., Kawabe, M., and Imanaka, T. (1987) Enantioselectivity of tartaric acid modified Ni-Al_2O_3 catalysts, *Appl. Catal.* **30**, 141-149.
199. Nitta, Y., Sekine, F., Imanaka, T., and Teranishi, S. (1982) Effect of preparation variables on enantioselectivity of supported Nickel catalysts modified with tartaric acid, *J. Catal.* **74**, 382 -392.
200. Nitta, Y., Yamanishi, O., Sekine, F., Imanaka, T., and Teranishi, S.(1975) Effect of support on enantioselectivity of modified catalyst, *J. Catal.* **79**, 475 -480.
201. Nitta, Y., Sekine, F., Sasaki, J., Imanaka, T., and Teranishi, S (1983) Conversion dependence of enantioselective hydrogenation of methyl acetoacetate with modified Ni-silica catalyst, *J. Catal.* **79**, 211- 214.
202. Nitta, Y., Sekine, F., Sasaki, J., Imanaka, T., and Teranishi, S. (1981) Hydrogen pressure dependence of the asymmetric hydrogenation of methyl acetoacetate with modified Ni/SiO_2 catalyst, *Chem. Lett.* 541-542.
203. Nitta, Y., Imanaka, T., and Teranishi, S. (1985) Preparation chemistry of precipitated Ni-silica catalysts for enantioselective hydrogenation, *J. Catal.* **96**, 429 - 438.
204. Nitta, Y., Imanaka, T., and Teranishi, S. (1983) Kinetic studies of enantioselective hydrogenation of methyl acetoacetate on modified Ni/SiO_2 catalyst, *J. Catal.* **80**, 31 - 39.
205. Nitta, Y., Utsumi, T., Imanaka, T., and Teranishi, S. (1986) Effect of additives on morphological and catalytic properties of Ni-silica catalysts for enantioselective hydrogenation, *J. Catal.*, **101**, 376 - 388.
206. Nitta, Y., and Imanaka, T. (1988) Effect of pore diffusion in liquid-phase enantioselective hydrogenation with Ni-silica catalysts, *Bull. Chem. Soc. Jpn.,* **61**, 295 - 297.
207. Nitta, Y., Kawabe, M., Sun, L., Ohmachi, Y., and Imanaka, T. (1989) Preparation of uniformly dispersed Nickel-silica catalysts from synthetic Nickel-Chrysotile, *Appl. Catal.* **53**, 1 -28.
208. Fu L., Kung H.H., and Sachtler, W.M.H. (1987) Particle size effect on enantioselective hydrogenation of methyl acetoacetate over silica-supported nickel catalyst, *J. Mol. Catal.* **42**, 29-36.
209. Sachtler, W.M.H. (1985) Asymmetric sites in heterogeneous catalysis, *Catalysis of Organic Reactions* (Dekker) p.189 - 206.
210. Groenewegen, J.A., and Sachtler, W.M.H. (1972) The infrared spectrum of glycine chemisorbed by supported nickel, *J. Catal.* **27**, 360 -373.

211. Groenewegen, J.A., and Sachtler, W.M.H. (1974) Infrared spectra of some amino acids adsorbed on silica -supported nickel, *J. Catal.* **33**, 176 -183.
212. Groenewegen, J.A., and Sachtler, W.M.H. (1975) Mechanism of asymmetric hydrogenation on modified nickel catalyst, *J. Catal.* **38**, 501-504, 360 -373.
213. Groenewegen, J.A. (1974) Adsorption complexes in asymmetric catalysis (Mechanism of enantioselectivity derived from IR data of relevant bifunctional compounds chemisorbed on Ni/ SiO_2. Thesis. Promoter Sachtler, W.M.H., Leiden Univ., 75 pp.
214. Hoek, A., and Sachtler, W.H.M. (1979) Enantioselectivity of nickel catalysts modified with tartraic acid or nickel-tartrate complexes, *J. Catal.* **58**, 276 - 286.
215. Bostelaar, L.J., and Sachtler, W.M.H. (1984) Effect of catalyst reduction and modification conditions on the enantioselectivity of Ni/SiO_2 catalysts for liquid-phase hydrogenation of a prochiral keto compound, *J. Mol. Catal.,* **27**, 377-385.
216. Bostelaar, L.J., and Sachtler, W.M.H. (1984) The role of alkali halides in the enantioselective hydrogenation of a prochiral keto-compounds over modified nickel catalysts, *J. Mol. Catal.* **27**, 387 - 395.
217. Bostelaar, L.J. (1984) Enantioselective hydrogenation on supported nickel catalysts, Thesis. Promoter Sachtler W., Leiden Univ., 157 pp.
218 Smith, G.V., and Musoiu, M. (1979) Asymmetric hydrogenation over modified Raney Nickel, *J. Catal.* **60**, 184 -192.
219. Bartok, M., Wittmann, G.W., Gondos, G., and Smith, G.V. (1987) Homogeneous and heterogeneous catalytic asymmetric reactions. I. Asymmetric hydrogenation of the prochiral C=C bond on a modified Raney nickel catalyst, *J. Org. Chem.* **52**, 1139 -1141.
220. Solladie-Cavallo, A., Hoernel, F., Schmitt, M., and Garin, F. (2003) Heterogeneous asymmetric hydrogenation and deuteration of 2-methyl-2-pentenoic acid over Pd supported catalysts: proton/deuterium exchange, *J. Mol. Catal. A : Chem.* **195**, 181-188.
221. Smith, G.V., Tjandra, S., Musoiu, M., Wiltowski, T., Notheisz, F., Bartok, M., Hannus, I., Ostgard, D., and Malhotra, V. (1996) Modified activities and selectivities of silated oxidized-reduced Pd and Pt catalysts, *J. Catal.* **161**, 441-450.
222. Wittmann, G., Bartok, G.B., Bartok, M., and Smith, G.V. (1990) Homogeneous and heterogeneous catalytic asymmetric reactions. 3. New investigations concerning the preparation of reproducible MRNi catalyst, *J. Mol. Catal. A: Chem*., **60**, 1-10.
223. Gross, L.H., and Rys P (1974) Heterogeneous catalytic asymmetric hydrogenation, *J. Org. Chem.* **39**, 2429 - 2430.
224. Yasumori, I., Inoue, Y., and Okabe, K. (1975) Mechanism of asymmetric hydrogenation on nickel surface coordinated with an

optically active molecule, in *Catalysis: Heterogeneous and Homogeneous,* Proc. Int. Symp. on Relations between heterogeneous and homogeneous catalytic phenomena, Brussels, 1974, Elsevier, Amsterdam, p. 41-50.

225. Yasumori, I. (1978) Mechanism of enantioface-differentiating hydrogenation on modified nickel surfaces, *Pure Appl. Chem.*, **50**, 971-978.
226. Inoue, Y., Okabe, K., and Yasumori, I. (1981) X-ray photoelectron spectra of adsorbed methyl acetoacetate and coordinated tartaric acid, aspartic acid and alanine on the Nickel surface, *Bull. Chem. Soc. Jpn.* **54**, 613-614.
227. Hubbell, D.O., and Rys, P. (1970) Asymmetric hydrogenation of ketones. An example of rate-dependent specificity, *Chemie,* **24**, 142-144.
228. Fish, M.J., and Ollis, D.J. (1977) Characterization of enantioselective hydrogenation catalysts: transient electrochemical oxidation of D-(+)-tartaric acid on Nickel, *J. Catal.* **50**, 353-363.
229. Schwab, G.M., and Rudolph, L. (1932) Katalytische Spaltung von Racematen durch Rechts und Links Quarz, *Naturwissenschaften* **20,** 363 - 364.
230. Terent'ev, A.P., Klabunovskii, E.I., and Patrikeev, V.V. (1950) Asymmetric synthesis by means of catalysts supported on right and left quartz, *Repts. of Academy of Sci. USSR*, **74**, 947-950, *Chem. Abstr.* 1951, **45**,3798c.
231. Bennett, A., Cristie, S., Keane, M.A., Peacock, R.D., and Webb, G. (1991) Enantioselective hydrogenation of methyl acetoacetate over nickel catalysts modified with tartaric acid, *Catalysis Today,* **10**, 363 - 370.
232. Jackson, S.D., Allan, E., Webb, G., and Young, N.C., (1998) New heterogeneous Nickel catalysts for enantioselective hydrogenation, in *Preparations of catalysts VII.*, Proc. 7th Int. Symp. Sci. Bases Preparation Heterog. Catal., Delmon, B., Jacobs, P.A., Maggi, R., Martens, J.A., Grange, P., and Poncelet, G. (eds.) Elsevier, Amsterdam, p. 313-320.
233. [a] Jackson, S.D. (1998) New heterogeneous nickel catalysts for enantioselective hydrogenation, *Stud. Surf. Sci. Catal.*, **118**, 305-312.
    [b] Laine, R.M., Num, G., Wood, B.J. and Dawson, M. (1980) Asymmetric hydrogenation of 2-methypyridine to 2-methylpiperidine. Raney Nickel modified with a molecular template, *The 7th International Congress on Catalysis,* June 30, 1980.
234. Balandin, A.A. (1970) Multiplet theory of catalysis. Part III. Theory of hydrogenation. Classification of organic reactions. Theory of complicate reactions. Structural algebra in chemistry. (russ., Klabunovskii, E.I., ed.), Publ. Moscow University, 475 pp.

235. Vedenyapin, A.A., Klabunovskii, E.I., Sokolova, N.P., Zubareva, N.D., Areshidze, G.Kh., Leonova, E.V., Ivanov, A.K., Savitskii, E.M., Polyakova, V.P., and Gorina, N.B. (1977) Preparation of alloys of Ru-Al, Ru-Cu and Ru-Ni-Al and formation from its the dissymmetric catalysts, in *Alloys of noble metals* (russ.), Savitskii, E.M. (ed.), Moscow, Nauka, p.187-189, *Chem. Abstr.* 1977, **87**, 123241.
236. Klabunovskii, E.I. (1951) Asymmetric synthesis by means of catalysts based on optically active quartz, Cand. Diss. Thesis, M.V. Lomonosov Moscow State Univ. (supervisor: Terent'ev, A.P.), Moscow.
237. Klabunovskii, E.I. (1966) Studies in the field of stereospecific catalysis, Doctor Diss. Thesis, N.D. Zelinskii Institute of organic Chemistry, Academy of Sciences of the USSR, Moscow.
238. Vedenyapin, A.A. (1979) Studies in the field of asymmetric catalysis on metals, Doctor. Diss. Thesis, N.D. Zelinskii Institute of organic Chemistry, Academy of Sciences of the USSR, Moscow.
239. Pavlov, V.A. (2000) Asymmetric induction in metal-complex catalysts and hydride reductions, Doct. Diss. Thesis, N.D. Zelinskii Institute of organic Chemistry, Academy of Sciences of the USSR, Moscow, (supervisor Klabunovskii, E.I.).
240. Petrov, Yu.I. (1966) Studies of catalytic asymmetric hydrogenation Cand. Diss. Thesis, N.D. Zelinskii Institute of Organic Chemistry, Academy of Sciences of the USSR, Moscow.
241. Godunova, L.F. (1970) Studies of chemistry of catalytic transformations of optically active terpene compounds, Cand. Diss. Thesis, N.D. Zelinskii Institute of Organic Chemistry, Academy of Sciences of the USSR, Moscow (supervisors Klabunovskii, E.I and Balandin, A.A.).
242. Neupokoev, V.I. (1972) Asymmetric hydrogenation of C=O bond on modified cobalt and copper catalysts, Cand. Diss. Thesis, N.D.Zelinskii Institute of Organic Chemistry, Academy of Sciences of the USSR, Moscow (supervisor Klabunovskii, E.I.).
243. Talanov, Yu.M. (1974) Studies of catalytic and asymmetric activities of Ru catalysts in reaction of liquid phase hydrogenation of ethyl acetoacetate, Cand. Diss. Thesis, N.D. Zelinskii Institute of Organic Chemistry, Academy of Sciences of the USSR, Moscow (supervisors Klabunovskii, E.I. and Vedenyapin, A.A.).
244. Zubareva, N.D. (1976) Studies of asymmetric and adsorption properties of Ru catalysts, Cand. Diss. Thesis, N.D. Zelinskii Institute of Organic Chemistry, Academy of Sciences of the USSR, Moscow (supervisors E.I.Klabunovskii E.I. and Vedenyapin A.A.).
245. Areshidze, G.Kh. (1976) Studies of phase composition, adsorption, catalytic and asymmetric properties of copper-ruthenium and copper-nickel catalysts, Cand. Diss.Thesis, N.D. Zelinskii Institute of Organic Chemistry, Academy of Sciences of the USSR, Moscow (supervisors Klabunovskii, E.I and Vedenyapin, A.A.).

246. Airapetov, Yu.S. (1978) Regularities of asymmetric hydrogenation of carbonyl-compounds on sceletal copper catalysts modified with optically active amino acids, Cand. Diss. Thesis, N.D. Zelinskii Institute of Organic Chemistry, Academy of Sciences of the USSR, Moscow (supervisors Klabunovskii, E.I. and Vedenyapin, A.A.).
247. Piloyan, S.R. (1979) Studies of heteroligands chelates of copper,nickel and cobalt as models of intermediate complexes in enantioselective catalysis, Cand. Diss. Thesis, N.D. Zelinskii Institute of Organic Chemistry, Academy of Sciences of the USSR, Moscow (supervisors Klabunovskii, E.I and Pavlov, V.A.).
248. Chernysheva, V.V. (1984) Studies of effects of genesis of modified skeletal nickel and powder copper-nickel catalysts and effect of additives of organic compounds for their properties in enantioselective hydrogenation, Cand. Diss. Thesis, N.D. Zelinskii Institute of Organic Chemistry, Academy of Sciences of the USSR Moscow (supervisor Klabunovskii, E.I.).
249. Chankvetadze, B.G. (1985) Behavior of dissymmetric nickel and copper -nickel catalysts in enantioselective hydrogenation, Cand. Diss. Thesis, N.D. Zelinskii Institute of Organic Chemistry, Academy of Sciences of the USSR, Moscow (supervisors Vedenyapin, A.A. and Klabunovskii, E.I.).
250. Kuznetsova, T.I. (1985) Behavior of dissymmetric palladium- copper, ruthenium-copper and palladium-nickel catalysts in enantioselective hydrogenation, Cand. Diss. Thesis, N.D. Zelinskii Institute of Organic Chemistry, Academy of Sciences of the USSR, Moscow (supervisors Vedenyapin, A.A. and Klabunovskii, E.I.).
251. Urazbaeva, K.A.(1990) Modified hydrides of intermetallic compounds as catalysts of asymmetric hydrogenation, Cand. Diss. Thesis, N.D. Zelinskii Institute of Organic Chemistry, Academy of Sciences of the USSR, Moscow (supervisors Klabunovskii, E.I. and Konenko, I.R.).
252. Starodubtseva, E.V. (1990) Sorption and catalytic properties of intermetallic compounds in reaction hydrogenation of propylene and enantioselective hydrogenation of ethyl acetoacetate, Cand. Diss. Thesis, N.D. Zelinskii Institute of Organic Chemistry, Academy of Sciences of the USSR, Moscow (supervisors Klabunovskii, E.I. and Konenko, I.R.).
253. Giorgadze, N.G. (1990) Adsorption, catalytic and electrochemical properties of supported catalysts based on palladium and nickel, Cand. Diss. Thesis, Tbilisi University, Gruz. SSR (supervisors Vedenyapin, A.A. and Chivadze, G.O.).
254. Lopez-Martinez, A., and Keane, M.A.(2000) The gas phase hydrogenation of 2-butanone over supported nickel catalysts: introduction of enantioselectivity, *J. Mol. Catal. A : Chem.* **153**, 257-266.

255. Osawa, T., Ozawa, A., Harada, T., and Takayasu, O. (2000) Enantio-differentiating hydrogenation of methyl acetoacetate over fine nickel powder with in situ modification, *J. Mol. Catal. A: Chem.* **154**, 271-275.
256. Osawa, T., Mita, S., Iwai, A., Takayasu, O., Hashiba, H., Hashimoto, S., Harada, T., and Matsuura, I. (2000) Enantio-differentiating hydrogenation of methyl acetoacetate over tartaric acid-NaBr-modified supported nickel catalyst prepared from nickel acetylacetonate, *J. Mol. Catal. A: Chem.* **157**, 207-216.
257. Osawa, T., Hayashi, Y., Ozawa, A., Harada, T., and Takayasu, O. (2001) Application of an in situ modification of nickel catalysts to the enantio-differentiating hydrogenation of methyl acetoacetate, *J. Mol. Catal. A: Chem.* **169**, 289-293.
258. Osawa, T., Sakai, S., Deguchi, K., Harada, T., and Takayasu, O. (2002) High durability of asymmetrically modified nickel catalysts prepared by in situ modification, *J. Mol. Catal. A : Chem.* **185**, 65-69.
259. Balandin, A.A. (1969) Modern state of the multiplet theory of heterogeneous catalysis, *Advances in Catalysis*, V.19., pp 1-210, Acad. Press, N.Y., London.
260. Pavlov, V.A., Piloyan, S.R., and Klabunovskii, E.I. (1980) Studies of optically active heteroligand chelate complexes. Comm. 7. Heteroligand complexes as models of intermediate complexes in enantioselective hydrogenation catalysis, *Izv. Acad. Nauk. SSSR, Ser. khim.* 545 - 551.
261. Sakata, Y., Domen, K., Maruya, K., and Onishi, T. (1990) The structure of tartaric acid adsorbed over nickel catalyst observed by FT-IRAS, *Catal Lett.*, 4, N8.
262. Osawa, T., Amaya, Y., Harada, T., and Takayasu, O. (2003) Enantio-differentiating hydrogenation of methyl acetoacetate over asymmetrically modified reduced nickel catalysts. The effect of the nickel sources on the enantio-differentiating ability, *J. Mol.Catal. A: Chem.* **211**, 93-96.
263. Smith, G.V., and Notheisz, F. (1999) Enantioselective hydrogenations, in *Heterogeneous Catalysis in Organic Chemistry,* Academic Press, NY., Chapter 1, 1-28.
264. Musoiu, M. (1975) *Ph.D. Dissertation,* Southern Illinois University, Carbondale, Il.
265. Boyd, W.A. (1972) *Ph.D. Dissertation*, Southern Illinois University, Carbondale, Il.

Chapter 5

ASYMMETRIC HYDROGENATION OF 2-OXOCARBOXYLIC ACID ESTERS AND UNSATURATED CARBOXYLIC ACIDS ON MODIFIED PT AND PD CATALYSTS

Abstract

Although the "Orito reaction", a highly effective enantioselective heterogeneous hydrogenation of 2-oxocarboxylic acid esters on Pt-alumina catalysts modified with cinchona alkaloids, was elaborated long ago, only during the 90's did a rapid growth of studies occur in this area. The reaction received strong development in experimental and theoretical aspects. Now about 300 papers have been published on the elaboration of the new catalytic systems with natural and synthetic chiral modifiers. Based on a number of *alpha,beta*-dicarbonyl-groups, a number of types of substrates and a number of new catalysts supported on zeolites were found. Based on detailed mechanistic studies, practical processes, such as the syntheses of the pharmaceutical, (*R*)-(-)-pantolactone, were accomplished with enantioselectivities above 95%.

5.1. General

The most efficient enantioselective heterogeneous hydrogenation catalysts to date are the Raney nickel catalysts, modified with tartaric acid, for the hydrogenation of the C=O group of 3-oxocarboxylic acid esters; the alumina supported platinum catalysts, modified with cinchonas, for the hydrogenation of the C=O group of 2-oxocarboxylic acid esters (and *alpha*-keto lactones); and supported palladium catalysts, modified with cinchona and other compounds, the for hydrogenation of C=C bond in prochiral unsaturated carboxylic acids. The latter two catalytic systems have been studied intensively, especially recently, and produced extremely high enantio-selectivities, quite comparable with results attained with the chiral metal-complex catalysts. As for Ni catalysts, it took about fifty years to improve the first results on asymmetric hydrogenation of C=O and C=C bonds in prochiral compounds (ethyl methyl ketone, 2-phenylcinnamic acid) on metal-quartz catalysts (see, Chapters 1 and 2, and Schwab et al. [1], Terent'ev et al. [2], Klabunovskii et al. [3], and others [4-12]), and increase from the first *ee* of 0.1% to the *ee* of 98.6%, from the hydrogenation of ethyl 3-cyclopropyl-3-oxopropionate on Raney Nickel, modified with *(2R,3R)*-tartaric acid (Nakagawa et al. [13]).

Enantioselective heterogeneous hydrogenations on platinum-alumina catalysts modified with cinchonidine (Cnd) is an asymmetric organic reaction that has become a very intensively studied heterogeneous process. That can be explained by high enantioselectivities and enhanced reaction rates leading to important chiral compounds with high enantiomeric purity, up to 98%. The catalyzing metal, platinum, may be supported on a diverse range of materials including silica, titania, and charcoal, but only with alumina are the best results obtained for the enantioselective hydrogenation of the oxo-group in 2-oxocarboxylic acid esters, mainly pyruvates. High enantiomeric excesses are comparable to the effectiveness of chiral homogeneous complex catalysts when 5% Pt-alumina catalyst is modified with cinchona alkaloids having a particular stereostructure, like cinchonidine, or better 10,11-dihidrocinchonidine (DHCnd), or 9-methoxy-10,11-dihydrocinchonidine (MeODHCnd). There are many reviews summarizing the recent studies of this catalytic system and the mechanism of the enantioselective heterogeneous hydrogenation [14-28].

The first attempts to use the asymmetric hydrogenation of prochiral compounds with C=C and C=N bonds over chiral modified heterogeneous Pt and Pd catalysts were not effective. Lipkin and Stewart [29] found that the (+)-10,11-dihidrocinchonidine salt of 3-methyl-3-phenylacrylic acid (*beta*-methylcinnamic acid), **1**, can be hydrogenated over Adams PtO_2 at 3 bar hydrogen in ethanol solution into 3-phenylbutanoic acid, **2**, with an *ee* of 8-9%. A lower enantioselectivity was observed in the hydrogenation of the (+)-10,11-dihidrocinchonidine salt of the 3-(1-naphthyl)-3-phenylacrylic acid [*beta*-(*alpha*-naphtyl)cinnamic acid], **3**. Similarly, in the hydrogenation of the C=N bond in acetophenoxime, **4**, on Pt-black at room temperature (RT) in the presence of menthoxyacetic acid modifier, (*R*)-(1-phenylethyl)amine, **5**, was obtained with an *ee* of 9.22%, and the hexahydrofluorenon oxime, **6**, was hydrogenated into amine **7**, with an optical rotation of $[\alpha]_{589} = + 6.8^{o}$ [30] (Scheme 5.1.).

Smith et al. [31-33] elaborated a new type of the enantioselective modified heterogeneous catalysts by attaching chiral silyl ethers to Pt and Pd surfaces. They found an effect of such "siliconation" on the catalytic activities and selectivities of highly dispersed Pd and Pt catalysts. Chiral alkoxytrimethylsilanes, $R^{*}OSiMe_3$, appeared to partly decompose on Pd and form residues which impart enantioselectivity to the catalytic surfaces. Three different types of 1% Pd-silica catalysts were prepared with modification by the silyl ethers of (1*R*)-(-)-nopol, (1*R*)-(-)-myrtenol, and (1*S*)-*endo*-(-)-borneol and were assessed by the liquid phase hydrogenations of the prochiral acids, 2-methyl-3-phenylacrylic acid (*alpha*-methylcinnamic acid), **8**, and (*E*)-2-methylpent-2-enoic acid, **9** (Scheme 5.2.). The resulting catalysts possed enantioselective activity in hydrogenation of C=C bonds with an *ee* of 22.5%. Asymmetric hydrogenations of prochiral compounds containing C=C and C=O bonds also were successful using immobilized chiral complexes [34,35].

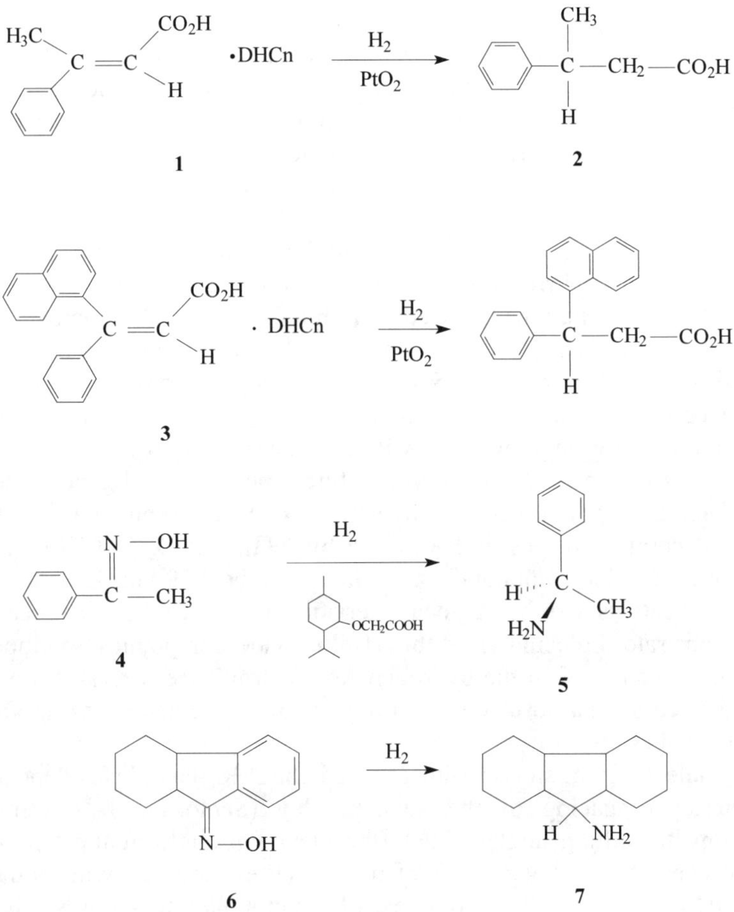

Scheme 5.1.

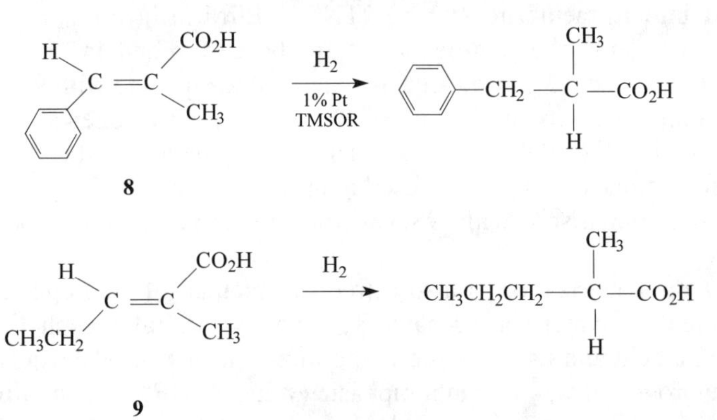

Scheme 5.2.

Immobilization via entrapment is a frequently used method of heterogenization. Mesoporous solids like silicas with pore diameters in the range of 2-25 nm were used by Raja et al.[36] for heterogenization of cationic Rh and Pd complexes with chiral ligands: $[Rh(COD)X]^+CF_3SO_3^-$ and $[Pd(allyl)X]^+CF_3SO_3^-$ in the enantioselective hydrogenation of methyl benzoylformate into methyl mandelate in methanol solution at 40°C and 20 bar hydrogen. Chiral ligands, X, used in Rh complexes are: (*S*)-(-)-2-aminomethyl-1-ethylpyrrolidine (1); (1*R*,2*R*)-(+)-1,2-diphenyl-ethylenediamine (2); and (*S*)-(+)-1-(2-pyrrolidinylmethyl)pyrrolidine (3). For Pd complexes only ligand (3) was used. The examinations showed that homogeneous Rh complexes with ligands (1) and (2) revealed no asymmetric effects. Only the more constrained ligand (3) gave *ee*'s of 53-55% for Rh and Pd complexes. Rh and Pd complexes with ligand (3) immobilized on silicas showed high enantioselectivities (*ee* of 94%) only on silicas with pores size 38-60 A°.

A rhodium complex with chiral phosphane ligand was also intercalated into sodium hectorite by cation exchange (Sento et al.[37]). The intercalated compound was characterized by FTIR, XRD, and TEM and the basal spacing of the compound was estimated to be 2.29 nm. This novel heterogenized catalyst exhibited a characteristic chiral as well as size recognition of the substrate molecule (like the "Tailor-made compounds" method used earlier by Balandin [38] in the hydrogenation of tripticene derivatives over Ni) and was used in enantioselective hydrogenation of the itaconates (methylene-succinic acid esters).

Selectivity of smectite-intercalated chiral Rh-complexes depended on the interlayer spacing of the swollen clay (Sento et al.[39]). Smectites possessing the tilting angle of 90° (the elevation angle of the longitudinal straight chain to the layer plane of the smectite) modified with both chiral $[Rh(COD)(S,S\text{-DIOP})]ClO_4$ complex (A) and bulky quaternary dimethyl-dioctadecylammonium bromide (B) as structural tuning guests were novel host-guest catalysts. With a mixture of catalysts {18.3 (A) + 42.8 (B)-LiTN} included into Li-taeniorite {where "TN" is $Li[Mg_2Li](Si_4O_{10})F_2 \times H_2O$} at 20°C in methanol, the hydrogenation of the C=C bond in (1*R*)-(+)-alpha-pinene, **1** (Scheme 5.3.), produced an excess of *exo*-pinane with 92% selectivity in comparison to an *ee* of 87.5% from the hydrogenation with the homogeneous [Rh(DIOP)] catalyst. The hydrogenation of the C=O bond in 4-*tert*-Bu-cyclohexanone, **2**, revealed a diastereoselectivity (*cis/trans* ratio) of 99.6% in comparison with 98.6% for the corresponding homogeneous catalyst.

Copolymerization is also a frequent method of heterogenization. A water-soluble polymer was prepared by coupling a chiral phosphane, PPM, to polyacrylic acid and used for the immobilization of $[Rh(NBD)_2]OTf$. At 22 bar of hydrogen in water or in a biphasic (water-AcOEt) system, this catalyst accomplished the enantioselective hydrogenation of the precursor of *N*-AcPhe with an *ee* of 89 % (Malmstrom et al. [40]).

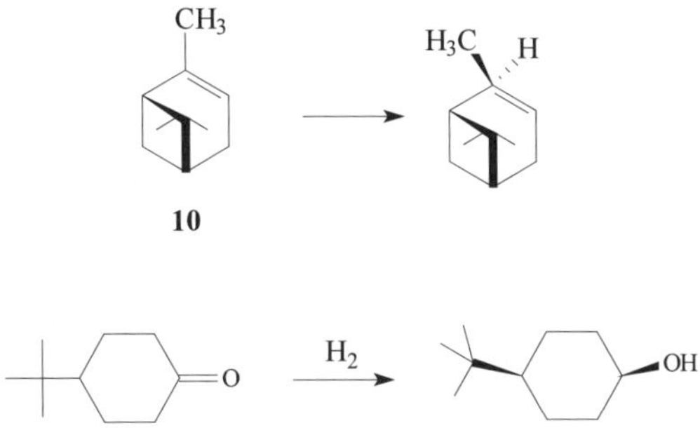

10

11

Scheme 5.3.

Poly[(*S*)-glycidylmetacrylate]co-ethyleneglycol dimethacrylate, an enantiopure copolymer, was transformed into optically active polyamino alcohols. This copolymer was used as a ligand of a Ru complex catalyst in the asymmetric hydrogenation of acetophenone (Herault et al. [41]).

A polymerizable rhodium complex, [Rh{(+)-DIOP}(A)], {where (A) is the deprotonated form of the ligand ethyl 2-acetoacetoximethacrylate}, was obtained by the reaction of [Rh(COD)(A)] with (+)-DIOP at -80^{0}C. Supported chiral complexes have been obtained by the copolymerization of [Rh{(+)-DIOP}(A)] with *N,N*-dimethylacrylamide. The enantioselective hydrogenation of the precursor of (*S*)-phenylalanine methyl ester produced the amino acid with an *ee* of 67%. (Mastrorilli et al. [42]).

Another approach is to build up a polymer network around a complex. Pavlov et al.[43] studied a chiral liquid crystal matrix, in which the Wilkinson catalyst, [$RhCl(PPh_3)_3$] was embedded in cholesteryltridecanoate and catalyzed the enantioselective hydrogenation of 2-acetamidocinnamic acid into *N*-acetylphenylalanine with an *ee* of 60% (see details in Chapter 3).

The heterogenized [Rh(BPPM)] complex embedded in micelles in a membrane reactor (Dwars et al.[44]) proved to be enantioselective in the hydrogenation of precursors of *alpha*-amino acids.

Recently important results were obtained using catalysts supported on chiral natural materials containing polypeptides and polysaccharides. According to Yin et al. [45] a wool-Pd complex at 30^{0}C and 1 bar hydrogen was found to catalyze the asymmetric hydrogenation of 4-hydroxy-4-methylpentan-2-on (diacetone alcohol), 1, into (*R*)-2-methylpentane-2,4-diol, 2, with an *ee* of 73% and 3-methylbutan-2-one, 3, into (*R)*-3-methylbutan-2-ol, 4, with an *ee* of 100% (Scheme 5.4.).

The optical yields were greatly affected by the Pd content in Pd-wool complexes. The catalyst could be used several times without appreciable change in *ee*.

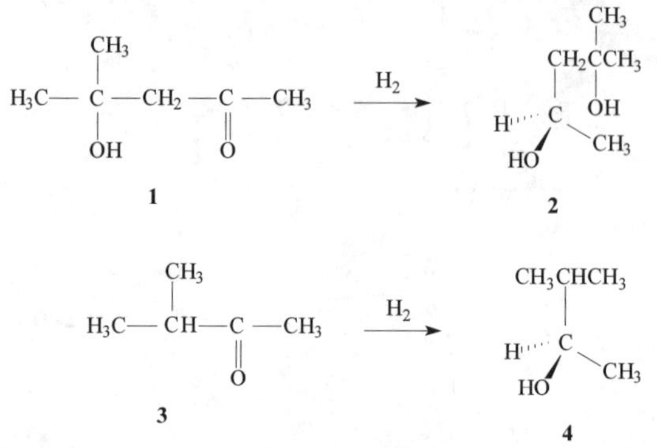

Scheme 5.4.

Yuan et al. [46] found that a Pt-silica catalyst bound with chitin (poly-*N*-acetylglucosamine) is active at 30°C and 1 bar hydrogen in the kinetic resolution of racemic 1-phenylethanol, (1*R* and 1*S*), producing (*R*)-(+)-1-cyclohexylethanol, (2*R*) (Scheme 5.5).

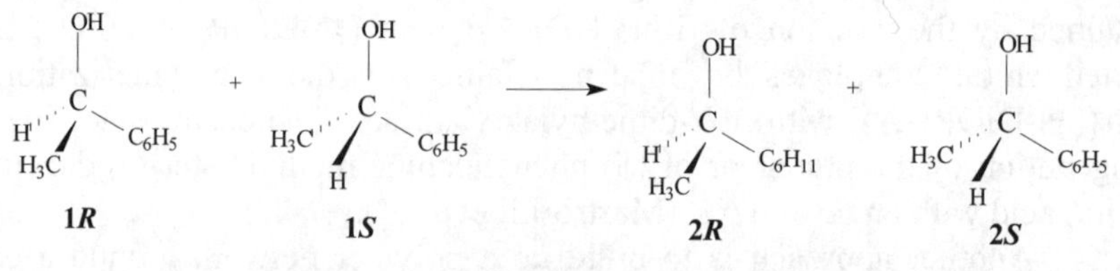

Scheme 5.5.

The optical selectivity of the product amounted to 100% under optimal conditions (unfortunately, the *ee* was calculated only from the optical rotation of the product mixture), but was greatly affected by the circumstances (Pt content of the silica-chitin complex, reaction temperature, solvent, conversion and the amount of added HCl). The catalyst proved to be very stable and the optical selectivity was maintained at 100% upon reuse (see also Blaser [47,48]).

Yin et al.[49] have found a similar catalyst based on a Pd-silica complex bound with chitosan (polyglucosamine) that is highly enantioselective at ambient conditions in the hydrogenation of acetophenone, propiophenone, 3-methylbutan-2-one and 4-methylpentan-2-one into their corresponding alcohols with *ee*'s of 99-100 %.

Padgett, Beamer et al. [50,51] prepared a novel type of catalysts composed of Pd supported on specifically prepared silica gels which had been precipitated from Na silicate with HCl in the presence of sulfates of cinchonidine (Cnd), cinchonine (Cn), quinidine (Qnd), and quinine (Qn). These catalysts proved to be enantioselective in the hydrogenation of 2-methylcinnamic acid with *ee*'s of approximately 3% (see the structure of cinchona alkaloids used for the enantioselective hydrogenation of *alpha*-keto esters in Scheme 1.6. in Chapter 1).

Later, several groups studied another type of catalysts composed of Pt (and Pd) supported on alumina and modified with cinchona alkaloids. Striking results were obtained using this new type of modification of Pt catalysts with cinchona alkaloids. Platinum supported on carriers, mainly alumina, and modified with alkaloids, exhibited very high enantioselectivities in the hydrogenation of 2-oxocarboxylic esters into 2-hydroxycarboxylic esters. It is interesting to note that before the discovery of this reaction, Orito's group [52-56] in 1976-1977 studied a supported Ni-Kieselguhr (1:1) catalyst modified with (2*R*,3*R*)-tartaric acid and promoted with 1% noble metals (Pt, Pd, Rh, or Ru) in the hydrogenation of methyl acetoacetate (MAA) to MHB and found small increases in *ee*'s (for catalysts promoted with above mentioned metals the *ee*'s were 61.9%, 55.3%, 55.6%, and 49.5%, respectively) in comparison with the catalyst without these promoters, from which an *ee* of 53.2% was obtained [52] (in this book *ee* values are corrected in comparison with data given in papers).

Later, in 1979, Orito et al.[57-61] studied novel catalysts containing noble metals (Pt-alumina and Pt-charcoal modified with cinchonidine) in the hydrogenation of MePy, MeBf, and EtBf. Optical yields in the hydrogenation of ethyl benzoylformate, EtBf, to (*R*)-(-)-ethyl mandelate reached 89.5% [57,58].

The chemical yield of methyl mandelate reached 87-92% and the *ee* changed, depending on the nature of the solvent, from 61.5% in $MeCO_2Me$ up to 81.9% in $EtCO_2Me$ [57,61]. EtBf was hydrogenated on Pt-C-Cnd with a somewhat larger *ee* in comparison to MeBf [58]. Enantioselective ability of the catalysts strongly depend on the condition of preparation, such as the conditions of reduction of charcoal supported H_2PtCl_6 and the nature of the carrier. Thus, maximum optical yield was observed only when using the commercial charcoal "Norit Extra" of Japanese origin and when the reduction of the Pt-salt supported on the charcoal was performed in sodium formate solution. If after reduction the catalyst was treated with acetic acid and heated in flowing hydrogen, the optical yield was raised 4-5%. The role of AcOH as a solvent will be considered later (see Part 5.5.).

Using alumina instead of charcoal as a carrier of Pt yielded no advantages in the hydrogenation of EtBf, but in the hydrogenation of MePy with addition of alkaloids, Pt-alumina proved to be more effective then Pt-C with the best *ee* being 86.6% [57-61]. Modification of 5% Pt-alumina catalysts with Cn and Qnd were less effective than modification with Cnd and Qn and resulted in configurational enantiomer (*S*)-(+)-MeLa instead of (*R*)-(-)-MeLa. This fact indicated that the configuration of the resulting hydroxy ester is determined by the configuration of the C-8 and C-9 centers in the alkaloid molecules.

Procedures for the preparation of such catalysts for asymmetric hydrogenations are very simple. For example, 5% Pt on carbon catalyst is simply treated with a 1% alcoholic solution of Cnd at 25^{0}C. The product is washed and used in the hydrogenation of MeBf in benzene solution at 60 bar hydrogen and 25^{0}C.

Orito's works [57-61] did not attract any attention for a long time; only ten years later did great interest arise in this system now called "Orito reaction", because it was developed by Orito in the National Chemical Laboratory of Industry (Tokyo). Since then detailed studies of these chiral modified Pt and Pd catalytic systems have been performed by several groups: Blaser (Novartis, at present Solvias Basel)[22,26-28,62-78], Baiker (ETH, Zurich)[17-21,79-118], Perez (Coll. France, Paris)[119], Nitta (Techn. College, Niihama)[120-140], Wells and Hutchings (Univ. Hull and Univ. Cardiff)[23,24,141-154,276], Margitfalvi (Centr.Res.Inst.Acad.Sci., Budapest)[155-162], Tungler and Sheldon (Techn.Univ, Budapest and Univ. Delft)[163-167,230], Bartok (Univ. Szeged)[168-195], Reschetilowski (Techn. Univ., Dresden)[196-201], Smith (Southern Illinois Univ., Carbondale)[31,33,171,173,202-204], Augustine (Seton Hall Univ., South Orange)[205-206], and Blackmond (Univ. Hull)[207-212]. Other metal catalysts modified with tartaric acid were studied by the groups of Izumi and Tai (Inst. Protein Res., Univ. Osaka)[213-219], Klabunovskii (Inst. Org. Chem, Acad. Sci., Moscow)[2,7,9,43,167,220-223], and Yasumori (Inst. Tech., Tokyo)[224-226].

The research works mentioned above were aimed mainly in the following directions:

- stereochemical control of the reaction determining the absolute configuration of the products and chirodiastaltic interactions[182]
- effect of the modifier structure
- conditions of the reactions and the properties of the catalysts: the nature and structure of metal catalysts and carriers
- detailed mechanism of the reaction.

The exclusive stereospecificity of this reaction was examined using the system of Pt-alkaloid-ethyl pyruvate. It was found that the optical yield and the rate of the reaction strongly depend on many parameters:

a) properties of catalysts
   - size of catalyst crystallites
   - nature of precursors and the preparation mode of the catalyst
   - preliminary treatment of catalyst

b) procedure of modification of catalysts
   - the structures of alkaloid molecules
   - condition and mode of modification
   - quantities of the alkaloid added

c) structure and purity of substrate molecules

d) effect of solvents
e) conditions of reaction
- temperature and hydrogen pressure
- hydrogen concentration in the solution and the extent of conversion.

Large enantioselective effects were observed only in the cases of Pt-alumina-*alpha*-keto ester systems. *Beta*-keto esters or *beta*-diketones can not be hydrogenated enantioselectively by this catalytic system. The best conditions of the reaction were the following: a solution at 25°C, hydrogen pressure up to 10 bar, Pt crytallites larger than 2 nm and supported on *gamma*-alumina, 10,11-dihydrocinchonidine (DHCnd) or 9-methoxy-10,11-dihydrocinchonidine (MeODHCnd) as best modifiers. Under optimum conditions the reaction rate was high and the optical yield of *alpha*-hydroxy esters reached above 95%.

The mechanism of hydrogenation of pyruvates consists of the addition of dihydrogen to the adsorbed '*trans*' form of pyruvate leading to (*R*)-(-)- or (*S*)-(+)-lactates, depending on the stereostructure of the alkaloid-modifier. The great peculiarity of this reaction is the sharp increase in the rate of hydrogenation in the presence of modified catalyst, both when the catalysts was preliminarily treated with the alkaloid or when the modifier was added simply to the reaction mixture. This alkaloid *in situ* effect is explained by the phenomenon of "ligand acceleration" (Jacobsen, Sharpless et al. [227]).

Recently great interest arose in the process of preparation of optically active 2-hydroxycarboxylic acids and their esters. Especially concerning the preparation of lactic acid (and lactates) and mandelic acid (and mandelates) using heterogeneous asymmetric hydrogenation of the corresponding 2-oxocarboxylic acids and esters over Pt-alumina catalysts in the presence of cinchona alkaloids as chiral modifying agents. Enantiomeric excesses of the products in this process reached above 95% in the best cases.

Using cinchonidine as a modifier of Pt-alumina catalysts, the reaction resulted in the formation of products with preponderances of one enantiomer of the resulting 2-hydroxycarboxylic acid or ester of (*R*)- configuration. But in many papers devoted to this process there is some confusion in the assignment of configuration to the optically active acid (or ester) [167]. Even the same authors (Blaser, Orito, Baiker) in many of their own papers gave either correct or incorrect assignments of the sign of optical rotation to the configuration of lactate or mandelate.

It is well known from monographs (for example, see [228]) that (*S*)-lactic acid and its lactates (MeLa, EtLa), and (*R*)-mandelic acid and its mandelates (MeMn, EtMn) have (+)-rotations (at 589 nm), while the (*R*)-enantiomers have the (-)-rotation. Thus, (*S*)-(+)-lactates correlate configurationally with (*S*)-(+)-mandelates and (*S*)-(+)-butan-2-ol. Similar correlations apply to enantiomers (*R*)-(-)-mandelic and (*R)*-(-)-lactic acids. These correct correlations were accepted e.g. in the papers by Minder, Baiker et al.[92], and Orito et al. [57,58,60]. On the other hand the incorrect (*R*)-(+) and (*S*)-(-) correlations were published by Orito et al.[59], Sutherland, Wells et al. [141], Reschetilowski et al. [197,200], Blaser et al.[62,65], Wehrli, Baiker et al.[78,80], Zuo [229], and Tungler et al.[230]. To avoid these confusions and to emphasize that (-)-rotating 2-hydroxycarboxylic acids and their esters have (*R*)-configuration, while (+)-rotating enantiomers have (*S*)-configuration, in this book the corrections will be inserted into all cited papers (interesting to note that most salts of (*S*)-(+)-lactic acid are levorotatory, therefore $[\alpha]_{589}$ of L-(+)-lactic acid is -13.5°, if it is measured in an 1.5 N NaOH solution).

5.2. Properties of catalysts

5.2.1. Nature of metal

The use of catalysts modified with cinchonidine (Cnd) based on Rh-alumina or Ir-alumina yields moderate *ee*'s (around 30%) in the enantioselective hydrogenation of pyruvates. Catalysts composed of Pd-C and Ru-alumina were found to produce lower enantioselectivities and Raney Ni modified with (2*R*,3*R*)-tartaric acid was inactive for this reaction (Blaser et al. [62]). Platinum catalysts were found to be most suitable. The enantioselectivity strongly depends on the conditions of the preparation, even on the conditions of the reduction of H_2PtCl_6. Thus a high optical yield (*ee* of 89.5%) was reached using commercial "Norit Extra" carbon and preparing the catalyst by reduction of the Pt-salt with sodium formate (Orito et al. [58]). If, after reduction, the catalyst was treated with acetic acid and heated in flowing hydrogen, the optical yield from hydrogenation of EtBf increased by 4-5%.

5.2.2. Nature of carrier

Using alumina instead of carbon as the carrier of Pt has no advantage in the hydrogenation of EtBf, but in the hydrogenation of MePy, with addition of alkaloids, Pt-alumina was more effective than Pt-C producing the best *ee* of 86.6%.

Activated carbons were oxidized different ways for the preparation of Pt-C catalysts. Such catalyst were modified with (-)-Cnd and used in the hydrogenation of MePy. After activation, their catalytic activities depend on the concentration of remaining surface oxygen measured by XPS (Fraga et al.[231]).

Farkas and Tungler et al. [232] studied the support effects in the enantioselective hydrogenation of isophorone over Pd catalysts prepared on different carbon supports with different specific surface areas and on activated carbons with different surface chemistries. The Pd catalysts, obtained by different reduction methods of the catalyst precursors had different dispersions, and the low dispersions turned out to be advantageous for high enantioselectivity in the hydrogenation of the C=C bond in isophorone.

Various Pd-black catalysts modified with DHVin differing in their preparation method showed different *ee's* from the hydrogenation of isophorone (Farkas and Tungler et al. [233]).

Solvent	5% Pt-C *ee* (%)	5% Pt-alumina *ee* (%)	Addition of 0.1 g alkaloid
None	74.0	-	None
MeOH	64.6	32.6	None
MeOH	71.2	86.6	Cnd
EtOH	78.7	83.2	Cnd
iPrOH	72.2	80.6	Cnd
EtOH	76.6	-	Qn
MeOH	-	82.4	Qn

Table 5.1. Asymmetric hydrogenation of methyl pyruvate (MePy) to (*S*)-(+)-methyl lactate (MeLa) on 5% Pt-C and 5% Pt-Alumina catalysts modified with supplemental addition of different alkaloids in different solvents (according to Orito et al. [59]).

The hydrogenation of MePy to (*R*)-(-)MeLa proceeded somewhat less effectively; the *ee* values increased from 64% to 73.1% in the series of solvents, THF, MeOH, AcOEt, and iPrOiBu, But it was found that hydrogenation without solvent was even somewhat more effective yielding an *ee* of 74%. The addition of 0.1 g of Cnd to that reaction mixture did not affect the *ee* value, whereas addition of Cnd or Qn in the presence of solvent increased *ee* (Table 5.1.).

5.2.3. Zeolites as carriers

The steric structure of the matrix-carrier can make an additional contribution to the asymmetric reaction, especially if the carrier contains chiral characteristics. Thus, Smith et al. [202] showed that *alpha*-cyclodextrin as the host compound can induce asymmetric reactions in the reduction of acetophenone and Tanaka et al. [234] showed that the pyridine-borane-cyclodextrin system induces chirality with an *ee* of 91%. Pd supported on a co-polymer of *beta*-cyclodextrin-epichlorhydrin hydrogenates 2-methylpent-2-enoic acid with an *ee* of 9.3% whereas Pt supported on *beta*-cyclodextrin polymer produced an optical yield of up to 15.7% (Smith et al. [202]).

Zeolites can be used as supports for the preparation of chiral heterogeneous catalysts in different ways [235]:

a) anchoring chiral complexes by covalent bonds

b) encapsulating large chiral complexes
c) modifying zeolites with chiral compounds

Method a) was used by Corma et al. [236-239] to prepare heterogenized hydrogenation catalysts. Proline based ligands were anchored through their silanol groups to the walls of modified USY zeolite producing a "super-micropore" system with 1.2-3 nm pore sizes. Rh complexes of these immobilized ligands were active catalysts in the enantioselective hydrogenation of 2-acetamidocinnamic acid to *N*-acetylphenylalanine.

Method b) consists of the encapsulation of voluminous complexes within the micropores of zeolite. It was used in the enantioselective epoxidation on a Mn(III)Salen complex-zeolite catalytic system [240-242].

Method c) is exemplified in the chiral modification of surfaces of zeolites with chiral compounds. Sundarababu et al. [243] modified NaX and NaY zeolites with (-)-ephedrine and used them in the asymmetric photolysis of ketones with *ee*'s above 10%. A strange behavior of the nature of carriers for the configuration of products in the latter reaction was observed: the NaY-(-)-ephedrine zeolite system gave (+)-rotating products, while the NaX-(-)-ephedrine zeolite system gave (-)-rotating products. Zeolite H-Y modified with (*R*)- or (*S*)-dithiane-1-oxide showed catalytic activity in the asymmetric decomposition of racemic 2-butanol (Hutchings) [244].

This result raised hopes of modifying zeolite supported metals by chiral compounds and using these catalysts in asymmetric hydrogenations. Nitta et al. [132] used Ni catalysts supported on Zeolite Y (specific area 550 m^2/g) in the asymmetric hydrogenation of the *beta*-keto ester MAA to MHB. The catalysts (Ni:Zeolite Y = 1:1) were prepared by precipitation of Ni on the Zeolite, followed by calcination and reduction of the resulting catalyst in a hydrogen stream. Modification of the catalyst was carried out in an aqueous solution of tartaric acid at pH 5.1. This catalyst produced an *ee* of 45.6% in the hydrogenation of MAA to MHB at 10 bar hydrogen and 60^0C. For comparison, under the same conditions, a catalyst composed of Ni-silica modified with (2*R*,3*R*)-tartaric acid gave MHB with an *ee* of 56.2%. Also modified Ru ion-exchanged zeolite catalysts were used in the enantioselective hydrogenation and diastereoselective hydrogenations of cyclic *beta*-keto esters [245-247].

The first example of an enantioselective hydrogenation reaction in the gas phase was published byHutchings's group [153,154,244]. They accomplished the asymmetric decomposition of racemic 2-butanol with rather good diastereoselectivity using a Y zeolite modified with chiral dithiane oxide.

Reschetilowski et al. [196-201] elaborated a novel Pt catalysts supported on zeolites, which proved to be rather enantioselective, effective, and stable in the liquid phase hydrogenation of EtPy. Pt-zeolite (5%) was prepared by impregnating the zeolite with an aqueous solution of $H_2Pt(OH)_6$ + HNO_3. The resulting material was calcinated and reduced in flowing hydrogen at

250^{0}C [199,201] or at 200^{0}C [200]. However, more effective catalysts resulted from Pt on zeolite Y or ZSM-35, with ratios of Si:Al = 2.5-5, when 1-5 mg of Cnd were added to the reaction solution during hydrogenation of EtPy. At 70 bar hydrogen and 20^{0}C (*R*)-(-)-EtLa was obtained with an *ee* of 86.1% in AcOH solution. It was found that other zeolites (mordenite and erionite) were also effective supports for Pt over which (*R*)-(-)-EtLa was obtained with an *ee* of 85-87%. At the same time catalysts based on zeolite Beta and ZSM-5 were less effective (see Table 5.2.) in the catalytic production of (*R*)-(-) EtLa. In Table 5.2. the Si:Al ratio changes from 2.5 to 15 and ZSM-35 showed the highest *ee* (68%) in the hydrogenation of EtPy to (*R*)-(-)-EtLa.

	zeolite Y	ZSM-35	zeolite Beta	ZSM-5
Si/Al ratio	2.5	5	12.5	15
Pore size (nm)	0.71	0.35-0.54	0.55-0.76	0.51-0.56
Particle size (nm)	2.9	9.6	12.0	8.7
ee (% of (*R*)-(-)EtLa)	62	68	53	47

Table 5.2. Characteristics of zeolites used for the preparation of 5% Pt-zeolite catalysts.

Reschetilowski et al. [200] also gives a comparison of the efficacy of Pt-zeolite-Cnd catalysts in the hydrogenation of EtPy based on alumina enriched with zeolite Y or ZSM-35 and silica enriched Zeolite Beta or ZSM-5 as carriers (Fig.5.1.).

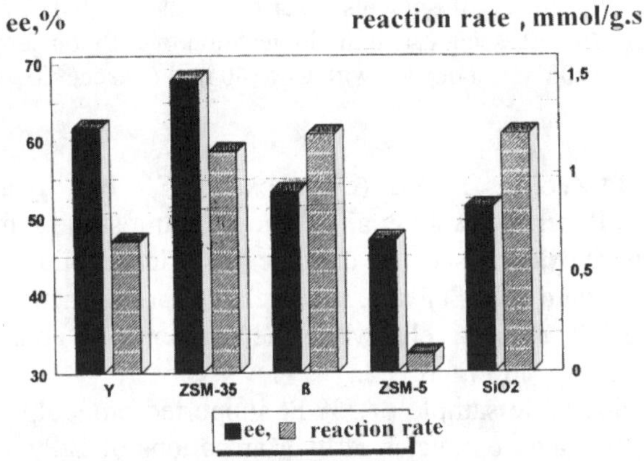

Figure 5.1. A diagram for comparison of enantioselectivity and reaction rate in the hydrogenation of EtPy on Pt-containing catalysts (according to Reschetilowski et al.[200])

Figure 5.1. shows that the best *ee* values were obtained on catalysts based on alumina enriched zeolites like ZSM-35 and zeolite Y. Optical yields did not depend on the amounts of catalysts but increased sharply with diminishing amounts of added Cnd reaching an optimal value with the addition of 1 mg Cnd per 100 mg of catalyst.

Repeated hydrogenations of up to 20 cycles were achieved without diminishing *ee's* (about 75%) on a sample of 5% Pt-ZSM-35 contingent upon the addition of a new portion of Cnd (1.2 mg) in each cycle, even though the amount of catalyst was reduced by one-fourth after each filtration (Reschetilowski [199-201]).

Table 5.3. shows dependence of the *ee* values on the polarities of solvents in the hydrogenation of EtPy on Pt-zeolite catalysts in comparison with Pt-alumina-Cnd catalysts.

Solvent	dielectric constant	5% Pt/Y		5% Pt/ZSM-35		5% Pt/γ-Al_2O_3	
		r_o	*ee*	r_o	*ee*	r_o	*ee*
methanol	32.63	0.37	61.6	1.1	69.8	1.08	75.1
ethanol	24.3	0.42	67.7	1.2	64.5	2.28	75.0
acetone	20.7	0.57	69.8	1.17	73.3		
propan-2-ol	20.1	0.88	64.3	2.01	71.3	3.05	76.7
butan-2-ol	15.5	0.84	68.7	1.69	72.3		
dichloromethane	9.08	0.69	76.3	1.09	75.0	2.20	81.4
acetic acid	6.15	0.56	83.9	0.69	86.1		
cyclohexane	2.02	0.80	74.5	1.79	77.5	2.94	79.9
hexane	1.89	0.74	71.3	2.07	74.7		

Table 5.3. Influence of solvents polarity on the *ee* (%) and the reaction rate (mmol/ sec·g cat) in hydrogenation of EtPy on supported Pt catalysts (200 mg) modified with Cnd (40 mg) (Reschetilowski et al. [200]).

Using 5% Pt-Zeolite Y-Cnd (cyclohexane, 30 bar hydrogen, 20-30°C) Boehmer and Reschetilowski et al. [198] showed almost the same *ee* values (82-84%) on Pt catalysts based tge cationic modified zeolites, HNaY, CaNaY, NaY, MgNaY, and (Nd-Pr)NaY, with relative proton activities of 10.2, 5.2, 0.0, 5.6, and 7.0, respectively, while the reaction rate constant increased in this series as shown in Figure 5.2.

A very stable sample of 5% Pt supported on zeolite ($H_{0.7}Na_{0.3}Y$) in AcOH solution gave ten cycles of hydrogenations of EtPy with an *ee* of 85-87% and gave 15 cycles in cyclohexane with an *ee* of 70-78% (Reschetilowski [199,201], Bartok [311]).

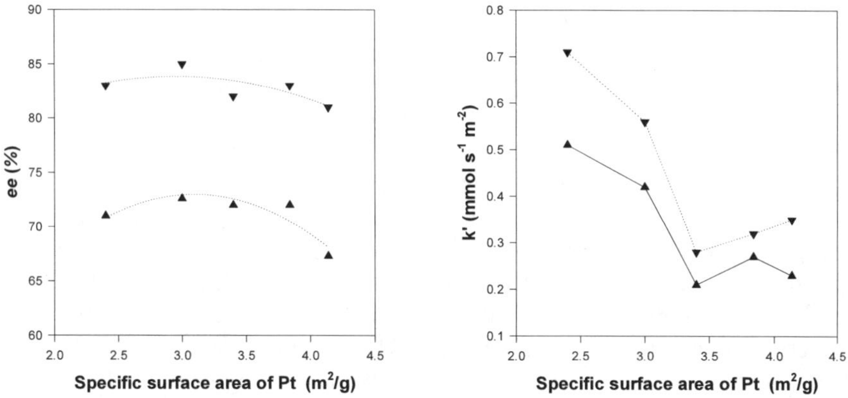

Figure 5.2. Dependence of the enantiomeric excess, *ee*%, (left) and the initial reaction rate constant, k', (right) of ethyl lactate on the specific surface area of platinum and the relative proton activity of the modified zeolites during the hydrogenation of ethyl pyruvate on 5% Pt-zeolite catalysts modified with Cnd in acetic acid (▼) and in cyclohexane (▲) [198]. The catalysts from left to right were: Pt/HNaY (10.2), Pt/CaNaY (5.2), Pt/NaY (0), Pt/MgNaY (5.6), and Pt/(Nd-Pr)NaY (7.0) (proton activities are given in parentheses).

5.2.4. Effect of crystallite size on enantioselectivity

Both the metal particle size and the crystallite size of supported metal catalysts have decisive effects on enantioselectivity, but other parameters of the reaction often become important as well; for example, the nature of the metal, the nature of the support, the method of preparation, the salt used for preparation, the mode of catalyst reduction, and the nature of catalyst pretreatment, such as high temperature heating or sintering.

Bartok et al. [174] recommended that the pretreatment of Pt-alumina (Engelhard 4759) in hydrogen at 400°C must be carried out before the modification with Cnd or DHCnd to avoid hydrogenation of the quinoline skeleton, which will lead to a decrease of optical yield and, in fact, caused the hydrogenation of EtPy to stop at approximately 70% conversion.

Mastalir et al. [248] prepared highly dispersed Pt nanoparticles immobilized in swelling clay minerals modified with DHCnd. The modified smectites were impregnated with H_2PtCl_6 and reduced with $NaBH_4$. TEM images revealed the formation of ultrafine, monodispersed Pt nanoparticles with mean crystallite sizes of 0.8 nm (montmorillonite) or 1.2 nm (synthetic hectorite). Both catalysts proved to be mildly enantioselective in the asymmetric hydrogenation of ethyl pyruvate to (*R*)-ethyl lactate. The low catalytic performance was ascribed to steric reasons. The constrained environment between the layers prevents the formation of the proper chiral active sites for enantioselective hydrogenation.

For the preparation of Pt-C catalysts activated carbon was oxidized either in the presence or absence of a liquid phase. The catalyst was modified with (-)-Cnd and was used in the hydrogenation of MePy. The catalytic activity depended on the concentration of surface oxygen remaining after activation as measured by XPS (Fraga et al. [231]).

Farkas et al.[232] studied the support effects in the enantioselective hydrogenation of isophorone over Pd catalysts prepared on different carbon supports with different specific surface areas and with different surface chemistries. In the cases of Pd catalysts different precursors resulted in different dispersions after reduction. The low dispersion proved to be advantageous for high enantioselectivity in the hydrogenation of the C=C bond in isophorone.

Various Pd-black catalysts modified with DHVin and differing in their preparation method showed different *ee*'s in the hydrogenation of isophorone (Farkas et al.[233]).

Enantioselective efficacy of heterogeneous metal catalysts depends on many factors, one of which is crystallinity. The effect of metal crystallinity was studied by several groups: Gross and Rys [249], Sachtler [250], and Nitta [131] used Ni catalysts, Klabunovskii [3,25,220,222] studied Ru catalysts, Blaser and Baiker [63] used Pt, while Nitta [128] and Zuo [229] studied Pd catalysts. These investigations revealed very important regularities in the hydrogenations of C=O and C=C bonds in prochiral keto esters and olefinic acids.

It is well known that increasing crystallinity of the metals is beneficial for obtaining higher *ee*'s in C=O hydrogenations (Klabunovskii [3,8,9,220] and Sachtler [250,251]).

The effect of crystallite size on the optical yield was first observed by Vedenyapin et al. [222] in the hydrogenation of ethyl acetoacetate into ethyl 3-hydroxybutyrate during progressive loading of Ru in Ru-silica catalysts that had been modified with (2*R*,3*R*)-tartaric acid. The reaction proved to be structure sensitive. The most effective catalyst proved to be the one with a 4.5 nm crystallite size, while catalysts with crystallite sizes 1.6 and 8.0 nm revealed lower asymmetric abilities. The sizes of the Ru crystallites were increased by increasing reduction times of the catalysts during their preparations.

Similar structure sensitivities was found by Gross and Rys [249]. They found that the enantioselectivities of Raney nickel catalysts were increased by increasing the temperature and time of NaOH etching of Ni-Al. Nitta et al. studied the Pd-TiO_2-Cnd catalyst system in the hydrogenation of 2-phenylcinnamic acid [120-130] and the Ni-silica-Tart catalyst system in the hydrogenation of MAA [131-140].

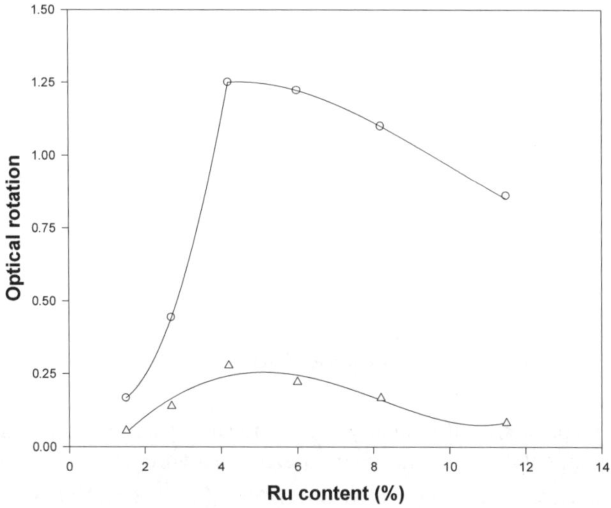

Figure 5.3. Effect of loading of Ru on enantioselectivity in the hydrogenation of ethyl acetoacetate into ethyl 3-hydroxybutyrate over Ru-silica catalysts modified with (2*R*,3*R*)-tartaric acid (reduction time: 0.5 (upper curve) or 5 h (lower curve); crystallite sizes after 0.5h reduction time were: 1.6 nm (at 1.5% Ru), 4.5 nm (at 4.2% Ru) or 8 nm (at 11.5% Ru) (according to Vedenyapin et al. [222]).

Also, they compared [131] enantioselectivities in hydrogenation of MAA over Raney Ni-Tart at 60^{0}C and 10 bar hydrogen in AcOEt solutions to enantioselectivities of amorphous Ni-boride (Ni-B) and Ni-phosphide (Ni-P) catalysts (prepared by reduction of Ni-salts with $NaBH_4$ or NaH_2PO_2, respectively) and found that a mean crystallite size of 6-8 nm produced *ee* values of 55%, which are comparable to Raney Ni prepared from Ni-Al alloy or from decomposition of Ni-formate. Ni-B and Ni-P catalysts have very small crystallite sizes and exhibit lower enantioselectivities (Figure 5.4.). The Figure also contains data (as curve 4) from the paper by Gross and Rys [249].

According to Molvinger et al. [252] Ni-boride modified with (1*R*,2*S*)-(-)-norephedrine produced catalysts that can hydrogenate acetophenone to (*R*)-(+)-1-phenylethanol with an *ee* of 90%. This catalyst can be recycled with no loss of performance. Such excellent enantioselectivity is connected with the formation of 1,2,3-oxazaborilidene anchored to the surface of Ni-boride. In addition to acetophenone, 4-methylpentan-2-one and isophorone were also hydrogenated (with lower *ee*'s) to products with (*S*)-configurations.

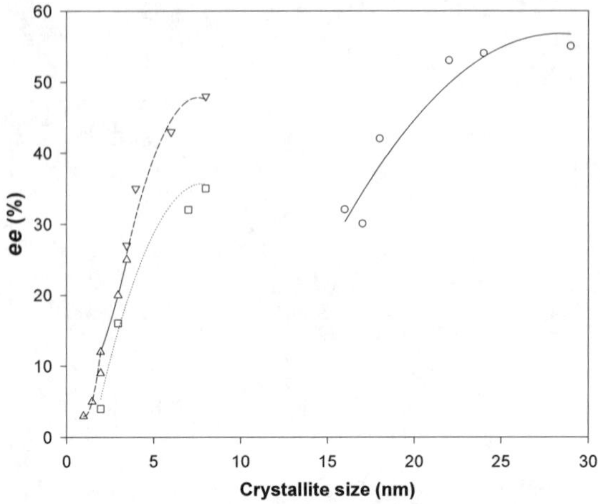

Figure 5.4. Dependence of optical yields on mean crystallite size of nickel in various catalysts in the enantioselective hydrogenation of methyl acetoacetate over RNi (□), Ni-B (Δ), Ni-P (∇) [131] or RNi (o) [249] catalysts.

As Nitta et al. [136,137] have found, amorphous Ni catalysts revealed low enantioselectivities and only catalysts with crystallite sizes around 10.0 nm revealed rather high *ee*'s. Increasing Ni loading on the support increases the *ee* associated with Ni crystallite diameter on carriers [131]. In general, as noted by Fish and Ollis [10,11] through electrochemical studies of this reaction on chirally modified Ni-catalysts, reactions of molecules adsorbed on an optically selective metal surface site require more area than one adsorbed on an optically non-selective site. Therefore optically selective centers of Ni catalyst must consist of an ensemble of Ni atoms large enough to adsorb both modifier and reactant on one nickel crystallite. A similar stereochemical model of orientation of adsorbed modifier molecules of (2*R*,3*R*)-tartaric acid and ethyl acetoacetate on the edge of nickel crystallite catalyst was proposed by Klabunovskii and Vedenyapin [220] and discussed by Izumi [215,218] and Tai [214].

The conditions of preparation of catalysts from its precursor Ni-salts affect the crystallite size and crystallite size distribution. Therefore small amounts of the additives, such as, Pt or Pd salts, have a favorable effect during reduction of the catalysts and the process of its formation [138]. Orito et al. [52-56] were first to show that enantioselectivity in the hydrogenation of MAA into MHB over Ni catalysts supported on Kieselguhr and modified with (2*R*,3*R*)-tartaric acid can be increased from 53% to 62% after including in the composition of the catalyst 1% of a noble metal, the best being Pt or Pd. Loading of the metal on Kieselguhr is of importance; only the

Ni:Pt:Kieselguhr ratio of 1:0.01:1 produced an an *ee* above 90% (see Chapter 4, part 4.3.).

Although the method of purification of Kieselguhr used for preparation of these catalysts was not indicated [52-56], likely the carrier contained some additives favorable for stereoselective reactions because pure silica gel used as the carrier was found to be less effective than Kieselguhr. In the case of Ni-silica-Tart, the increased loading of Ni from 20% to 75% gave increased mean crystallite sizes from 6 to 9 nm that resulted in hydrogenation of MAA to *ee*'s of 47.3% and 57.2%, respectively. Moreover, according to Nitta [138], the addition of $PdCl_2$ during preparation of the catalyst from a Ni-salt-precursor improved the reduction of the Ni-salt and subsequent enantioselectivity.

Optimal crystal sizes can be arranged by changing the temperature of catalyst treatment. Thus, the reduction by Orito et al. [52,54] of Ni-silica-Tart catalysts at 300°C proved to be mild enough to provide high *ee*'s but reduction made by Sachtler [250,251] at 400°C was too severe for Ni-alumina catalysts. Therefore crystallite size distribution seems to have a very important role in determining enantioselectivity of supported catalysts. Catalysts with larger crystallite size and with narrower crystallite size distribution produce higher *ee*'s, because smaller crystallites have higher catalytic activities but lower enantioselectivities (according to Nitta et al. [136]). On the other hand, catalysts with large crystallites produce only low catalytic activity; therefore in the case of supports such as silica, which give moderate crystallites of nickel around 6.0 to 10.0 nm, higher reduction temperatures (400°C) seem to be suitable for the preparation of catalysts with high enantioselectivities.

Another problem hindering true evaluation of the role of crystallinity of metal catalyst consists of the inhibition effect of the product. Namely, the predominant product of the reaction, (*R*)-(-)-3-hydroxybutyrate, can decrease enantioselectivity of the reaction owing to the fact that the relative adsorption coefficient of the (-)-enantiomer on modified chiral Ni centers is somewhat larger than those of the racemic product, (found by Chernyshova et al. [253] and by Neupokoev et al. [254]). Therefore, introducing (*R*)-(-)-ethyl 3-hydroxybutyrate (EHB) [253,254] into the reaction mixture of ethyl acetoacetate (EAA) or (-)-methyl 3-hydroxybutirate (MHB) [134] into the reaction mixture of methyl acetoacetate (MAA) will diminish optical activity of the resulting 3-hydroxybutyrates. So, in order to make an accurate evaluation of the effect of crystallite size, the extent of asymmetric reaction needs to be controlled (for details see later in Part 5.5.).

To produce supported Ni catalysts with constant loading, Nitta [137] used as a carrier the layered silicate, chrysotile, and prepared the compound, Ni-chrysotile, $Ni_3(OH)_4Si_2O_5$. The reduction of this compound gave homogeneously dispersed *Ni-on-silica* catalyst. The crystallite size of these catalysts were controlled by the changing of the reduction temperature, while the general loading of Ni remained constant. In the hydrogenation of MAA these catalysts gave *ee*'s from 22% to 35%, when the crystallite sizes increased from 5 to 6.5 nm, and *ee*'s from 30% to 40% at crystallite sizes 8 and 10.5 nm.

5.2.5. Enantioselective hydrogenation of C=C bond in unsaturated carboxylic acids

Until now there were not many examples of asymmetric hydrogenation of C=C bonds on Ni-Tart and Pt-Cnd catalytic systems. Bartok et al.[171-173,183] found that the Na-salt of (*E*)-2-phenylcinnamic acid gave sodium (*S*)-(+)-2,3-diphenylpropionate with an *ee* of 17%, whereas the hydrogenation of the acid gave only 0.21 %. Although Perez et al.[119] showed earlier that the latter chiral acid can be prepared with an *ee* of 30.5% using the Cnd modified Pd-C catalyst instead of Ni. This finding indicated that Pd-catalysts modified with Cnd are favored for asymmetric hydrogenations of C=C bonds in prochiral unsaturated acids.

Indeed, as it was mentioned above, Smith [31-33] accomplished the enantioselective hydrogenation of unsaturated carboxylic acids using Pt or Pd catalysts modified with chiral silyl ethers.

Quite recently in the asymmetric hydrogenation of **(*E*)**-2-methylpent-2-enoic acid in hexane over Engelhard Pd-alumina catalysts (E 40692 and E 5220), modified with Cnd, were used. Deuteriumation experiments showed that double bond migration and *E-Z* isomerization occurs during deuteriumation, but these processes are less important at the higher pressure of hydrogen (40 bar) and in the presence of modifier Cnd (Salladie-Cavallo et al.[255]).

Aminocinnamic acid derivatives on catalysts modified with Cnd or DHVin proceeded with low *ee's*. Only in the diastereoselective hydrogenation of *N*-acetyl-dehydrophenylalanyl-(*S*)-prolinanilide into dipeptide was a higher *ee* (68%) achieved (Tungler et al. [166]).

The mechanism of enantioselective hydrogenation of unsaturated carboxylic acids over cinchona modified Pd catalysts was studied in the hydrogenation of tiglic acid (2-methylbut-2-enoic acid) into 2-methylbutanoic acid. Applications of the derivatives of Cnd, modified at the C_9-OH group and the quinuclidine *N*-atom, proved that both functional groups are involved in the enantiodiscriminating step. Addition of the strong base DBU to tiglic acid prior to hydrogenation revealed that one Cnd molecule interacts with the dimer of tiglic acid on the metal surface. Calculation showed that the acid dimer-Cnd intermediate is stabilized by hydrogen bonding revealing both the OH group and the quinuclidine *N*-atom of Cnd (Borszeky et al. [84]).

Nitta et al. [120-130] have found that Pd-titania modified with Cnd is the most suitable catalyst for hydrogenation of 2-phenylcinnamic acid because titania is a nonporous carrier and gave metal crystallites of the proper size.

This catalyst was reduced in flowing hydrogen at 200^{0}C in order to increase the crystallite size of Pd. In this case, the first results on hydrogenation of (*E*)-2-phenylcinnamic acid reached an of *ee* 44%. With the addition of 7% water to the solvent mixture of EtOAc and EtOH the *ee* reached 58% (Nitta et al. [121,122]).

In optimal conditions the hydrogenation of (*E*)-2-phenylcinnamic acid proceeded with an *ee* of 72% (Nitta et al. [128]), whereas the hydrogenation of the C=C bond in isophorone on Pd modified with (-)-dihydroapovin-caminic acid ethyl ester, (-)-DHVin, reached an *ee* of 55% (Tungler, Nitta et al.[129]). The differences in *ee* values were explained by the interactions between modifier and the reactant and by their different basicities.

Thus, there are some important differences between the asymmetric hydrogenations of C=O and C=C bonds :

1) For C=O hydrogenations solvents should be dry; on the other hand, for C=C hydrogenations the addition of water is favorable and increases *ee* values.

2) For C=C hydrogenations the reactions with unmodified catalyst were much faster then those with modified catalyst, such as with Cnd. This is quite different from results reported for the hydrogenation of *alpha*-keto esters on Pt-alumina, where the addition of the modifier alkaloid leads to chiral products and strongly accelerates the reaction (Bartok et al. [171,172], Tungler, Nitta et al. [129]).

The structure of Pd catalyst proved to be of decisive influence for accomplishing good enantioselectivity (Nitta [128]). Pd metal particles located in micropores of the support proved to be inaccessible to the bulk molecules of the modifier and reactant and to the intermediate complex [Modifier-Reactant] resulting in low *ee*'s. Therefore, using nonporous carriers are preferable for enantioselective reactions. But high metal dispersion on the surface of the catalyst is also detrimental to enantioselectivity (Tai [214]).

Nitta et al. [131-140] showed in a number of papers that *ee* values in the hydrogenation of MAA increased with increasing crystallite size of Ni. This effect confirmed the data of Klabunovskii et al. [220,222] who obtained results for a Ru-silica catalyst modified with (2*R*,3*R*)-tartaric acid in the hydrogenation of EAA into ethyl 3-hydroxybutyrate that could be explained by the increasing fraction of the metal surface atoms occupying regular crystal faces large enough (Fu et al. [251], Klabunovskii [220]) to have enough area to accommodate the large molecules of reactants and modifiers. The same suggestions were made later to explain hydrogenation of EtPy on Pt-alumina-Cnd catalysts (Wehrli et al. [63,69,78,80]). They found that the dispersion of Pt should be lower than 0.2 for good *ee*'s, and that the modifier cinchona alkaloid is anchored with the quinoline ring on the Pt surface (see below, on "Template model" of Wells et al.[141-152]).

In contrast, Zuo et al. [229,256] prepared highly efficient Pt nanocluster catalysts stabilized with polyvinylpyrrolidone polymer (PVP) modified with Cnd and noted that achievement of high *ee*'s in the formation of MeLaPd requires the existence of very small crystallite sizes. On finely dispersed catalysts MePy and EtPy were hydrogenated into (*R*)-lactates with *ee*'s of 97.6% and 92.2% , respectively. Zuo et al.demonstrated that the reaction is structure insensitive on clusters of about 1.4 nm, which is quite different from conventional supported chiral catalysts.

However, according to LeBlond et al. [257], MePy and EtPy were hydrogenated into lactates with the same *ee*'s, which is different from Zuo et al. [229]. Moreover, Zuo et al. [229] erroneously reported (*R*)-(+)-lactates instead of the correct correlations: (*R*)-(-) and (*S*)-(+) lactates[167].

The pore diameter limitation also was found for Pd-Cnd catalysts active in the C=C hydrogenation of (*E*)-2-phenylcinnamic acid. On a 5% Pd-silica catalyst the *ee*'s increased with increasing average pore diameter of silica, and the conclusion was that Pd metal particles should be located in pores large enough to accommodate the bulky alkaloid-modifier molecules and the substrate forming intermediate complex {modifier-reactant} that have been identified on the surface of Pd-titania catalyst (Nitta et al. [126]). Palladium metal particles in smaller pores are difficult to modify and they behave therefore as non-selective centers.

The comparison of a number of Pd-supported catalysts on non-porous and porous materials confirmed that non-porous ultrafine carriers like titania, seem to be most suitable for hydrogenation of (*E*)-2-phenylcinnamic acid. Thus, for porous and non-porous supports results of hydrogenation were as follows: on SiO_2 49.1% and 30.7%, on TiO_2 62.0% and 29.4%, respectively (Nitta et al. [128]).

The effect of Pd loading was also very important for accomplishing high *ee*'s. It was found that only a 5% Pd-titania catalyst (non-porous titania) gave the best *ee* (58.6%) with the highest rate of reaction. Catalysts with 20% Pd on titania or Pd-black gave only *ee*'s of 17.3% and 8.0%, respectively; thus the most effective catalyst proved to be a 5% Pd-titania-Cnd with a Pd dispersion of 0.2.

In general, the reactions of enantioselective hydrogenations of *beta*-keto esters on modified Ni or of *alpha*-keto esters on modified Pt are structure sensitive reactions. In both reactions, enantioselectivities increase with increasing crystallite size of Ni or Pt.

In the case of Pd catalysts supported on non-porous carriers, like SiO_2 or TiO_2, *ee's* depend on dispersion as found in the case of Pd-TiO_2-Cnd in the hydrogenation of (*E*)-2-phenylcinnamic acid (Nitta et al. [128]). These results showed that the reaction is mildly structure sensitive in the dispersion range higher than 0.2, but at lower Pd dispersions, it is similar to Pt and Ni catalysts in the enantioselective hydrogenation of C=O bonds in *alpha*- and *beta*-keto esters (Figure 5.5.).

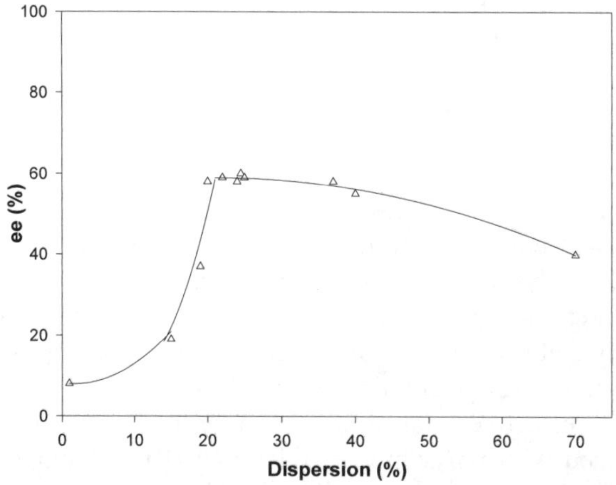

Figure 5.5. Correlation between the dispersion of palladium and the enantioselectivity in the hydrogenation of (*E*)-2-phenylcinnamic acid on Pd-TiO_2 catalyst modified with cinchonidine (according to Nitta et al. [128]).

On the other hand, the sharp decrease of *ee* at dispersions lower than 0.2 is similar to that observed in enantioselective hydrogenation on Pd modified with a bulky vinca alkaloid, according to Tungler and Nitta et al. [129].

The behavior of Pd catalyst in hydrogenation of (*E*)-2-phenylcinnamic acid can be explained in terms of the adsorption mode of the intermediate complex [modifier-substrate] on the Pd surface. The adsorption of the bulky Cnd molecule on large Pd crystallites hinders the C=C bond's approach to the Pd active center and it leads to a diminishing of the *ee* and the rate. If the molecule to be hydrogenated is small enough, as in the case of the hydrogenation of isophorone on Pd modified with vinca alkaloid (Tungler [164,165,230,258-260]), the *ee* increased with decreasing Pd crystallite sizes, and Pd-black (crystallite sizes 0.1 nm) proved to be a more effective catalyst than Pd-C with crystallite sizes around 0.5 nm [258]. The same holds true for the comparison of the strong hindrances in the hydrogenation of the C=O bond in pyruvate and the C=C bond in (*E*)-2-phenylcinnamic acid. Borszeky et al. [84] proposed an arrangement of an adduct {Cnd-acid dimer, 1:1} on the flat Pd crystallite surface for the hydrogenation of tiglic acid (Nitta et al. [128]).

Thus, enantioselective hydrogenation of pyruvates on Pt clusters supported on alumina modified with Cnd as well as finely dispersed colloidal Pt-PVP (according to Boennemann et al. [261], Koehler et al. [262,263], and Zuo et al. [229,256]) exhibits less structure sensitivity with the best *ee*'s occurring over very small clusters, which contradicts results on the common Pt-alumina catalysts with particle sizes of 3-4 nm.

Solvent-stabilized Pt and Pd nanoparticles, of sizes 2.3-2.8 and 2.7-3.8 nm, respectively, have been prepared by metal vapor synthesis routes and modified with Cnd. These catalysts were used in the hydrogenation of EtPy, and the Pd catalysts produced EtLa of inverse configuration, that is, instead of the expected (*R*)-EtLa, the (*S*)-EtLa was obtained. The Pt particle size distribution showed a higher degree of monodispersity after use in catalysis (Collier et al. [264]).

Vorlop et al. [265,266] suggested a novel strategy for heterogenizing homogeneous and colloidal chiral catalysts based on the three-dimensional entrapment of catalyst by electrostatic attraction between a polyelectrolyte and an active metal in the polymeric matrix. Based on this method, they entrapped chirally stabilized Pt-colloid in alginate (polymannuronate) and conducted twenty-five hydrogenation cycles of EtPy.

A rather effective catalyst proved to be the system of rhodium nanoclusters stabilized with polyvinylpirrolidone (PVP) and supported on finely dispersed oxides (gamma-alumina, silica, or titania) and modified with Cnd (Huang et al. [267]). With this system EtPy was hydrogenated at 25^{0}C and 70 bar hydrogen in THF with a TOF of 58.6 min^{-1} and an *ee* of 65.4%.

Rhodium nanoclusters stabilized by PVP, supported on TS-1, and modified with Cnd, Cn, or Qn (4 mmol/L) were used for the hydrogenation of EtPy at 5^{0}C and 70 bar hydrogen with an *ee* of 63.1% (Ma et al. [268]).

A similar catalytic system based on finely dispersed PVP-stabilized Pt-nanoclusters supported on conventional carriers and modified with Cnd accomplished the enantioselective hydrogenation of 2,2,2-trifluoroacetophenone into the (*R*)-alcohol at 20 bar hydrogen in a mixture of *o*-dichlorobenzene and ethanol (Zhang et al. [269]). The same catalyst was not very effective in the asymmetric decomposition-hydrogenation of racemic 3-hydroxybutan-2-one. The enantiomeric excess reached a maximum *ee* > 40% at a molar ratio of the modifier to reactant of about 1:650 in dichloromethane-ethanol solvent (Zuo et al. [270]).

Finely dispersed PVP-stabilized iridium, supported on alumina and modified with Cnd produced only an *ee* of 34.1% in the hydrogenation of MePy at 10^{0}C and 40 bar hydrogen in alcohol or AcOH (Zuo et al. [271]).

A rhodium nanocluster stabilized with PVP and modified with Cnd or Qn (4.3 mmol/L in THF) was active in the enantioselective hydrogenation of EtPy at 25^{0}C and 50 bar hydrogen with a TOF of 941 h^{-1} and an *ee* of 42.2% (Huang et al. [272]).

According to Boennemann [261], colloidal Pt-alumina catalyst, was prepared by reducing an aqueous solution of Pt salt in the presence of protonated DHCnd with the addition of the stabilizing colloid PVP (MW 10.000 [262,263]). Particle sizes of the resulting catalysts ranged from 1.5 to 4 nm and the hydrogenation of EtPy at room temperature and 1 bar hydrogen resulted in an *ee* of 76%. The enantioselectivity increased with diminishing dispersion. A catalyst of 1% Pt-alumina-Cnd with a particle size of 3.5 nm was compared with a Pt-PVP-Cnd catalyst system in the hydrogenation of EtPy and the *ee*'s were 34% and 42%, respectively.

The statement of Baiker [21,63] that good enantioselectivity can be achieved with Pt particle sizes above 3-4 nm is based on obtaining large ensembles where modifier and reactant can adsorb best on the large crystallites. Indeed, in the hydrogenation of EtPy on Pt-alumina-Cnd (room temperature and 70 bar hydrogen) Blaser et al. [69] found a maximal *ee* of 80% at 0.2 dispersion which diminished to 30% with dispersion increasing to 0.8 simultaneously with the rate of reaction.

With this conclusion (according to Zuo et al. [229,256]) the structure sensitivity originated from the structure of the surface of the catalysts, but stereochemical hindrances of reacting molecules probably confused this effect. On other samples of Pt-alumina cluster catalyst with crystallite sizes of 1.1-3.3 nm, prepared by adsorption of clusters [Pt-PVP] on Al_2O_3 with further rinsing of PVP from surface of catalyst, *ee*'s were found to be lower, at about 8%, for particles of 3-3.5 nm in comparison with *ee*'s of 88-90% for particles of 1.5 nm. In this case, in the hydrogenations of MePy, a very high *ee* was found: (*R*)-methyl lactate was obtained with an *ee* of 97.6%.

The role of crystalline size of Ni-Tar modified catalysts was studied also by Fu, Sachtler et al [251]. Increasing the loading of metal on the surface of the carriers resulted in increasing particle sizes; however, Ni catalysts changed their exact composition of large and small particles during modification in solutions of tartaric acid due to the leaching of small particles (this process was absent in the modification of Ru-silica catalysts [220,222]). It was suggested [251] that the "true" method for studying this effect is at constant metal loading for Ni on silica, modified with (2*R*,3*R*)-tartaric acid and active in the hydrogenation of MAA. Sachtler [250,251] proposed that the effect of loading on *ee* consists of pore diffusion and therefore the studies of crystallite size can be investigated only at constant loading. The change in nickel particle size was accomplished using the method in which larger particles grow at the expense of smaller ones during the formation and decomposition of Ni-tetracarbonyl at temperatures above 50°C.

$$Ni + 4CO \rightleftharpoons Ni(CO)_4$$

Hydrogenations of MAA were carried out both in gas-phase and liquid phase with flowing hydrogen. Optical yields were 8-15% in the former and 30-32% in the latter cases, and no significant structure sensivities were observed. In the range of 4-10 nm no influence of particle size was seen in the *ee*'s of the methyl 3-hydroxybutyrate product [251]. The results were considered confirmation of the “dual site model" (Sachtler [250]), which states that one sort of Ni atoms (or ions) on the surface coordinates the two molecules of modifier (tartaric acid) formatting two five membered chelate rings. The role of the other Ni atoms consists of adsorption of a dihydrogen molecule and dissociating it into hydrogen atoms.

According to Wehrli et al. [63], dispersion of crystallites in Pt-silica-Cnd catalysts from 0.3 to 0.8 (characterized by CO adsorption) produced *ee*'s in range of 65% to 35%. in the enantioselective hydrogenation of EtPy. Increasing the particle size was achieved by increasing the temperature of the reduction of the catalyst. Wehrli et al. [80] also found highest *ee*'s from catalysts prepared from Pt salts deposited on alumina by reducing them with formate or with HCHO. Catalysts reduced in flowing hydrogen gave the highest dispersion, 0.05, and the lowest enantioselectivity. Thus they concluded that attainment of highest *ee's* (80%) in the hydrogenation of EtPy required the Pt-alumina-Cnd catalyst to have crystallite sizes of 3-4 nm [80].

5.3. The structure of the modifiers

Although cinchona alkaloids and especially cinchonidine, Cnd, proved to be the most effective chiral modifier for the catalytic system of Pt-alumina, in the liquid phase enantioselective hydrogenations of the carbonyl group in pyruvic acid esters, efforts to understand the mechanism of action of this catalyst system has continued to the present. The efforts may be divided into two categories: finding natural modifiers other than cinchona alkaloids and examining new effective amino alcohols, which are modeled after the structure of known cinchona modifiers.

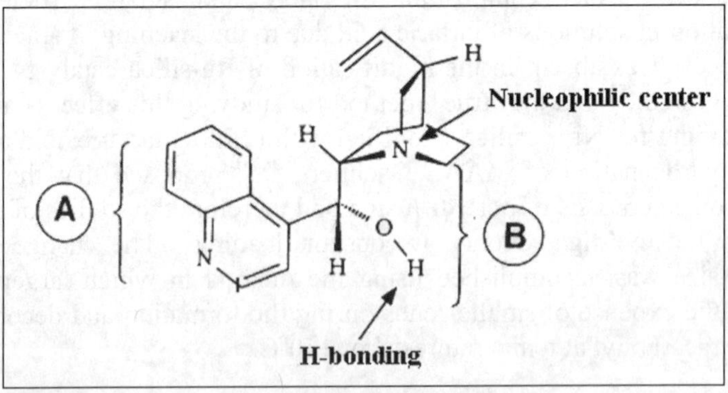

Scheme 5.6. The different parts of the cinchonidine (Cnd) molecules.
Part A: adsorbs on the surface of Pt.
Part B: asymmetric region of the molecule.

Elucidating the reasons for the efficacy of alkaloids as modifiers of Pt catalysts and searching for new modifiers with structures similar to the alkaloids of the cinchona alkaloid group is of great interest from the theoretical and practical points of view. The strategy for the search of structures of new modifiers must correspond to the basic requirements: to adsorb strongly on the surface of the metal (Part A of the cinchonidine molecule in Scheme 5.6.) and to match the stereochemical interaction through H-bonding of the substrate (Part B) with the nucleophilic center of the N-atom of the quinuclidine moiety of Cnd.

To these requirements the alkaloids of the cinchona group correspond moderately well owing to the their peculiar structure. The quinoline group of the Cnd molecule provides strong adsorption of the modifier on the surface of Pt while the stereochemically important part of molecule provides energetic and stereochemical interactions with the pyruvate molecule through the *N*-1 atom of quinuclidine ring.

Modifier structure			Solvent	*ee*	Configuration
C8	C9	R_1		(%)	
Cinchonidines					
H	OH	CH=CH2	EtOH	76	*R*
H	OH	CH_2CH_3	EtOH	79	*R*
H	OH	CH_2CH_3	AcOEt	81	*R*
H	OH	CH_2CH_3	PhMe	83	*R*
H	OMe	CH_2CH_3	EtOH	78	*R*
H	OH	CH_2CH_3	AcOEt	80	*R*
H	OH	CH_2CH_3	PhMe	81	*R*
H	OAc	CH_2CH_3	EtOH	20	*R*
H	H	CH_2CH_3	EtOH	44	*R*
H	OH	CH_2OH	EtOH	≈80	*R*
Cinchonines					
H	OH	$CH=CH_2$	EtOH	56	*S*
H	OH	$CH=CH_2$	MeOH	58	*S*
H	OH	CH_2CH_3	MeOH	54	*S*

Table 5.4. The effect of the structure of modifier alkaloids, cinchonidine (Cnd) and cinchonine (Cn), on the enantioselective hydrogenation of ethyl pyruvate into ethyl lactate over 5% Pt-alumina.
Reaction conditions: EtPy (10.9 g), 5% Pt-alumina (100 mg), modifier alkaloids (10 mg), solvent (20 mL), temperature 25-30^{0}C, 70-100 bar hydrogen (mainly according to Blaser [69]).

Recently a number of studies by the groups of Orito [57-60], Blaser [62,64,68,69,71,72], Baiker [17-19,79,88,90,92,104,105,111], Wells [23,141,147,148], Tungler and Sheldon [164,230,258, 259], and Bartok [169] were devoted to searching for principal methods of selection of new effective modifiers. Most detailed studies were made using the model reaction of the asymmetric hydrogenation of ethyl pyruvate over Pt-alumina catalysts modified with cinchona alkaloids. Structural alterations at various parts of the cinchonidine molecule led to significant changes in enantioselectivity. Table 5.4. summarizes the data (mainly according to Blaser [69]) on the enantioselective hydrogenation of EtPy over 5% Pt-alumina catalysts modified with alkaloid derivatives of Cnd or Cn.

For hydrogenation of both pyruvates and MeBf (methyl benzoylformate) Modification of Pt-alumina catalysts with Cn, Qn, or Qnd was less effective than modification with Cnd. Modifying with Cnd and Qn resulted in the (*S*)-(+) isomer of MeLa rather then the (*R*)-(-) isomer as in the case of modifying with Cn. This indicated that the absolute configuration of the resulting hydroxy ester is determined by the configurations at the C_8 and C_9 centers in the alkaloid molecules (see Scheme 5.6.).

Detailed investigations of the effects of structural changes in the Cnd and Cn modifier molecules on the enantioselectivity of hydrogenations of EtPy revealed that derivatives of Cnd yielded (*S*)-(+)-EtLa, while derivatives of Cn yielded (*R*)-(-)-EtLa.

The alkaloids Cnd and Cn are diastereomers that differ only in their configurations at C_8 and C_9, and this difference results in the formation of lactate products with opposite configurations. This was shown by the interaction of the pyruvate molecule with just the C_8 and C_9 centers of the alkaloids.

The next important conclusion concerned the role of the OH-group at the C_9 center in Cnd. Changing the OH for OAc or for H (annihilation of the chiral C_9 center) strongly diminished the *ee* values of EtLa [69]. However, Blaser et al.[68] found that methylation of the OH group in DHCnd resulted in the most efficient modifier of Pt-alumina catalysts reaching an *ee* value of 94%, which was the highest value ever reported for an enantioselective hydrogenation of EtPy over these catalytic systems. Table 5.5. summarizes the data from the hydrogenation of EtPy in different solvents.

In the hydrogenations of ethyl pyruvate, maximal optical yields of 91-94 % were achieved, especially if the reaction was carried out in AcOH or propionic acid instead of the more commonly used toluene or ethanol. It was a very important observation that the *ee* depended on the acidity of acids used as solvents (conditions of reaction are the same as in Table 5.4.).

The nature of groups at C_3 do not affect *ee* values, because the vinyl-group at C_3 in Cnd is hydrogenated in a few minutes after the beginning of the reaction into an ethyl-group forming 10,11-dihydrocinchonidine (DHCnd), which proved to be an even more effective modifier than Cnd itself. Therefore, most investigators preferred using the Pt catalysts modified with DHCnd.

Alkylation of the N-atom at the N-1 center in the quinuclidine part of Cnd completely annihilates enantioselectivity of the catalyst, which indicates the crucial role of the N-1 atom in the formation of an active intermediate complex [69].

Modifier	Solvent	ee (%)	rate (mol g_{cat}^{-1} min^{-1})
DHCnd	ethanol	82	< 0.1
DHCn	toluene	87	0.11
DHCn	acetic acid	92	0.12
MeODHCnd	ethanol	82	-
MeODHCnd	toluene	88	0.08
MeODHCnd	acetic acid	94+1	0.16
MeODHCnd	propionic acid	91	0.11
MeODHCnd	butyric acid	89	0.09
MeODHCnd	isobutyric acid	86	0.07
MeODHCnd	nonanoic acid	82	0.07
MeODHCnd	triethylamine	3	0.02
MeODHCnd	piperidine	4	0.007
MeODHCnd	morpholine	7	0.004
MeODHCnd	quinoline	30	<0.001
DHCnD	acetonitrile	28	0.007

Table 5.5. Hydrogenation of ethyl pyruvate on 5% Pt-alumina modified with 10,11-dihydrocinchonidine (DHCnd) or 10,11-dihydro-O-methylcinchonidine (MeODHCnd) in various solvents at room temperature and 100 bar hydrogen (according to Blaser et al.[68,69]).

The aromatic system of the quinoline nucleus provides strong adsorption of the alkaloid molecule on the Pt surface. Even partial hydrogenation of the quinoline nucleus diminishes *ee* values by as much as 50% during the formation EtLa [21,69].

From these data it is clear that high enantioselectivity in the hydrogenation of pyruvate over modified Pt-alumina catalysts results from modifier molecules that have the structure of heteroaromatic alcohols, like R_1-$CH(OH)$-CH_2-R_2, where R_1 is the quinolyl group and R_2 is the quinuclidyl group.

5.3.1. Modifiers with structures similar to cinchona alkaloids

In accordance with the general stereochemical requirements, Baiker et al.[17-19,21,79,88,90,92,104,105,111] prepared a number of optically active modifiers, such as (*R*)-2-(1-pyrrolidyl)-1-(1-naphthyl)ethanol (PNE, 2a in Scheme 5.7.) with structures similar to the structures of cinchonidine alkaloids.

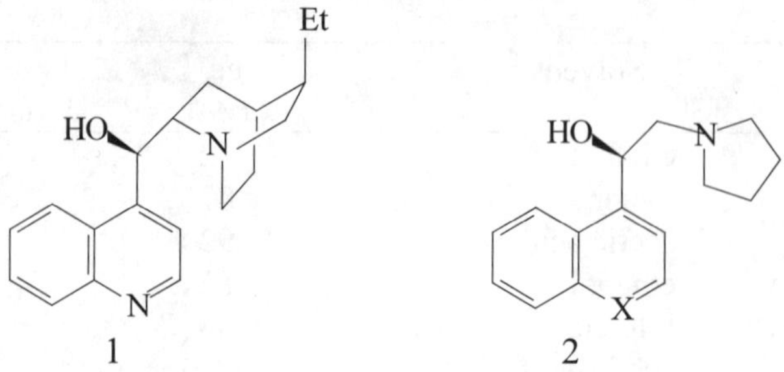

Scheme 5.7. The structures of DHCnd (1) and similar amino alcohols with naphthyl (2a, X = CH) and quinolyl (2b, X = N) aromatic groups.

Table 5.6. shows that the enantioselectivity in the hydrogenation of EtPy into (*S*-)-(+)-EtLa on 5% Pt-alumina modified with PNE, was 75%, which is similar to the 73% measured at lower pressure on a Cnd modified catalyst.

Modifier	H_2 pressure (bar)	Reaction time (h)	Conversion (%)	*ee* (%)
1	1	1.0	100	73
1	75	0.5	100	87
2a	1	1.0	100	68
2a	10	0.2	100	75
2a	25	1.0	99	47
2a	75	0.5	100	46
2b	1	1.0	100	48
2b	25	0.5	98	55
2b	75	0.5	100	66

Table 5.6. Comparison of the effect of modifiers DHCnd (**1** in Scheme 5.7.) and amino alcohols with naphthyl (**2a**) and quinolyl (**2b**) aromatic groups in the enantioselective hydrogenation of ethyl pyruvate on 5% Pt-alumina at various hydrogen pressures adapted data from Wang et al.[90]

The quinoline group, however, proved to be less effective than the naphthyl group (*ee* = 48-66%). The modifier PNE could be used as an additive to the reaction mixture without the usual preliminary procedure of modifying the catalyst. At optimal concentrations the relationship of EtPy to PNE was found to be equal to 30,000. Suggesting flat adsorption of EtPy and PNE on the Pt (111) facet, in which the complex molecule [Platinum-EtPy-PNE] requires an area of about 20 Pt surface atoms, results in 10% PNE being sufficient for full coverage of the surface.

The effect of solvent on the use of this model modifier is similar to that found for the Cnd modifier and the most effective solvent proved to be nonpolar ones (toluene, heptane) and also AcOH, which caused protonation of the *N*-1 center in PNE. With increasing temperatures, the *ee* reached a maximal value at 0^{0}C. Increasing the hydrogen pressure from 10 bar to 25 bar, decreased *ee's* from 72% to 60%, and 54% was measured at 40-70 bar. This decrease was a consequence of the partial hydrogenation of the aromatic quinoline ring (Figure 5.6.).

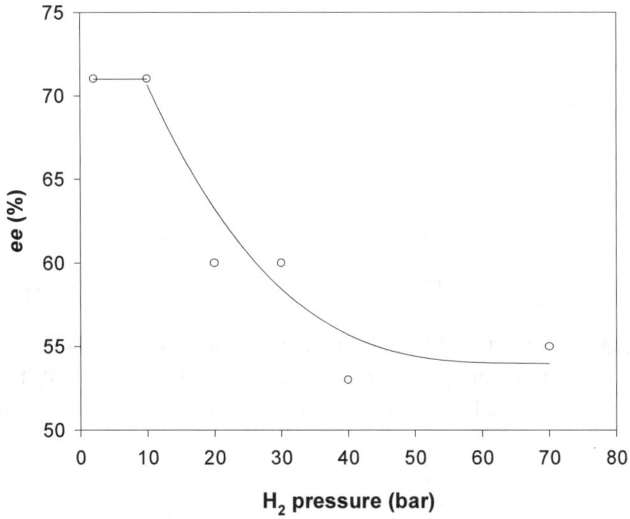

Figure 5.6. Enantiomeric excess as a function of hydrogen pressure in the hydrogenation of EtPy over 5% Pt-alumina-PNE in acetic acid at 25^{0}C (according to Minder et al.[92]).

This was a result of the weakening of the adsorption of the modifier on Pt due to breaking the definite orientation of the intermediate complex [Platinum-Modifier-Substrate]. Another picture was observed in the case of the Cnd modifier (Figure 5.7.).

At the beginning of the reaction the additional adsorption of the vinyl group of the alkaloid takes place, weakening the orientation of the modifier. However, during the course of the reaction this group is hydrogenated and the resulting DHCnd proved to be a more effective modifier than the original Cnd and increased *ee*. On the other hand, the partial hydrogenation of the quinoline ring of the Cnd weakened the adsorption of the modifier and diminished the *ee* of the resulting EtLa.

Indeed, the importance of this anchored group was revealed by an increase of *ee* from 48 to 66% when the naphthyl group in PNE was replaced with a quinoline group [90] (compare reactions at 75 bar in Table 5.6.)

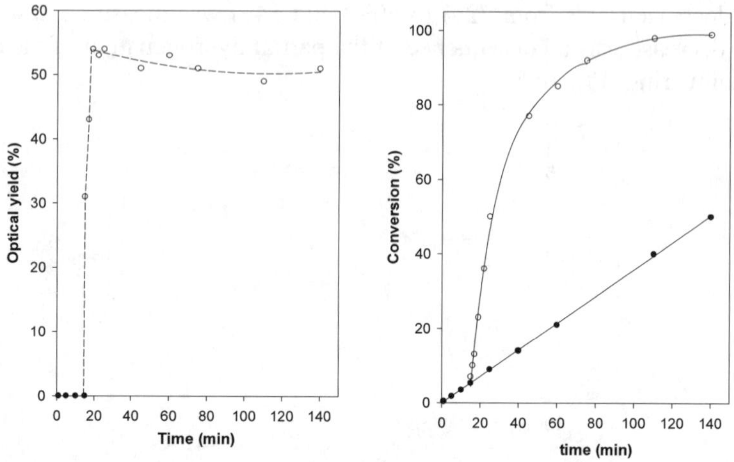

Figure 5.7. Effect of the addition of the modifier Cnd to the hydrogenation of ethyl pyruvate on 5 % Pt-alumina catalyst (mainly according to Baiker [21]). Optical yield (left) and conversion (right) with (o) and without (•) Cnd modifier.

To elucidate the stereochemical action of the PNE modifier the enantioselective hydrogenation of EtPy on 5% Pt-alumina in the presence of modifier-fragments of the PNE molecule were studied, and it was found that other structures close to the structure of Cnd can yield high catalytic and enantioselective activities during the hydrogenation of pyruvates. For example, Blaser et al.[62] found that some chiral amines, like 1-phenylethylamine and some amino alcohols, like ephedrine can reveal low to moderate enantioselectivities in the hydrogenation of pyruvates. But the most effective modifiers proved to be structures similar to PNE, which give good adsorption on the Pt surface and contain the aromatic part of the molecule as well as a chiral center with OH- and amino-groups that can react with the keto esters as shown in Table 5.7.

Table 5.7. demonstrates the crucial role of the large flat aromatic part of the modifier molecules, especially in the case of the anthracenyl-group (*ee* = 83%). But in the case of the tripticenyl moiety the enantioselectivity falls almost to zero, probably due to steric hindrances and the difficulties of the pi-bonding interactions with the surface Pt atoms [88]. However, according to Balandin and Klabunovskii (see review by Skvarchenko et al. [273]) tripticene derivatives, despite their steric molecular hindrances, could be hydrogenated on Ni catalysts, which likely have more homogeneous crystallite surfaces than platinum catalysts.

		H_2 pressure (bar)	*ee* (%)
10,11-Dihydrocinchonidine		75	87
(*R*)-5-(1-naphtyl)-(*N*-methyloxazolidine)		75	49
Ar	R		
1-Naphthyl	pyrrolidin-1-yl	1	68
4-Quinolyl	pyrrolidin-1-yl	75	48-66
Phenyl (1*S*)	pyrrolidin-1-yl	75	4
4-Pyridyl	pyrrolidin-1-yl	75	0
2-Naphthyl	pyrrolidin-1-yl	1	42
9-Anthracenyl	pyrrolidin-1-yl	10-75	83
1,9-Tripticenyl	pyrrolidin-1-yl	-	< 5
1-Naphthyl	dimethylamino	1	62
1-Naphthyl	di-isobutilamino	1	49
1-Naphthyl	morpholino	75	32
1-Naphthyl	methylamino	1	48
1-Naphthyl	amino	75	8
1-Naphthyl	(1*S*)-(1-phenylethyl)amino	25	32

Table 5.7. Effect of the aromatic part of the (1*R*)-1-aryl-2-(pyrrolidin-1-yl)ethanol type modifier molecules, (Ar-CH(OH)-CH_2R), on enantioselectivities in the hydrogenation of EtPy on 5% Pt-alumina catalyst (summarized data [88,111]).

Results received by Blaser, Baiker, et al. [69,90] confirmed that the most effective modifiers could be the compounds that contain an aromatic part of the molecule as in PNE. Indeed, molecules similar to PNE, containing the quinoline group instead of the naphthyl group, gave results close to those received with PNE (see Table 5.6.). These results showed that the center *N*-1 in the pyrrolidine ring did not play a crucial role in revealing enantioselectivity. Therefore the point of view of Augustine et al.[205,206] that the molecule of Cnd adsorbs in a perpendicular mode on the Pt surface seems not born out, rather the aromatic ring adsorbs parallel to the surface. Thus the use of (*R*)-(+)-1-(2-naphthyl)ethanol did not give a chiral product [97].

In the molecule PNE, which is simpler than the Cnd molecule, there is only a C7 center (see Scheme 5.7.), which is similar to the C9 center in Cnd, and proves to be enough for the creation of an effective modifier of Pt-alumina catalysts yielding EtLa with an *ee* of 70%. The function of the *N*-1 center is to give close bonding of the modifier to the pyruvate molecule involving the rigid quinuclidine ring in Cnd or the more flexible cycle in PNE. But in both cases the required sequences of centers, Ar-CH(OH)-CH-N-, is fulfilled. Protonation of *N*-1 in PNE during reaction in AcOH solution gave increased enantioselectivity (Schwalm et al. [97]).

Minder [104] used the commercially available (*R*)-(+)- or (*S*)-(-)-1-(1-naphthyl)ethylamines (see Scheme 5.8., enantiomers **1** and 2), which proved to be rather effective modifiers if the reaction is carried out at lower hydrogen pressures.

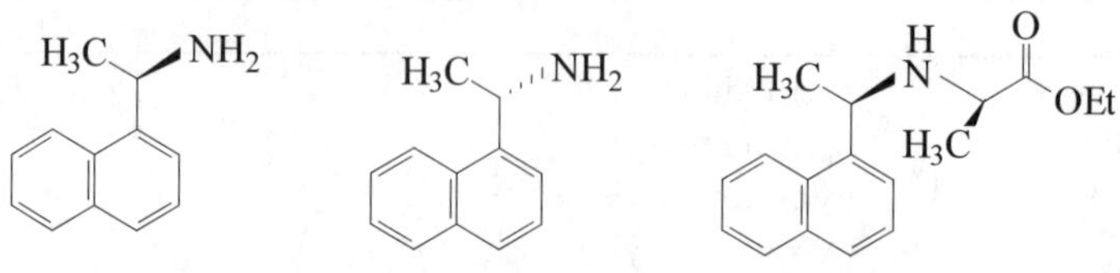

1 **2** **3**

Scheme 5.8.

With these modifiers an interesting interaction was observed in the hydrogenation of EtPy with the formation of compound 3 (Scheme 5.8.). That compound can itself act as a modifier of Pt catalyst in enantioselective hydrogenations (Table 5.8.) and can serve as a model by revealing the nonlinear effect of an autocatalytic process (see Avalos et al.[4], Girard et al. [5], Kagan et al. [6], Blackmond [211,212], and Soai [274]).

Modifier	pressure (bar)	conversion (%)	*ee* (%)
1	72-75	48	51 (*R*)
	22-25	54	55 (*R*)
2	72-75	52	53 (*S*)
	22-25	59	58 (*S*)
3	21-25	83	55 (*R*)

Table 5 8. A comparison of the chiral modifiers, (*R*)-(+)- and (*S*)-(-)-1-(1-naphthyl)ethylamines, in the enantioselective hydrogenation of ethyl pyruvate in AcOH at room temperature (reaction time is 1 h, mol reactant to mol modifier ratio is 1500). Adapted data from Minder et al. [104]

The structures of other modifiers similar to Cnd confirmed once more that modifier requirements for inducing enantioselectivity are an aromatic system and an amino-group. Results show that chiral catalytic systems can be prepared not only with cinchona alkaloids as modifiers but also with simpler amines and amino alcohols allowing the hydrogenation of the *alpha*-keto esters with *ee*'s up to 80% and, importantly, at lower hydrogen pressures [21].

5.3.2. Modifiers with structures different from cinchona alkaloids

Torey et al.[275] showed that in *alpha, beta*-unsaturated ketones (see Scheme 5.9.) the C=C bond can be asymmetrically hydrogenated over 10% Pd-C catalyst in MeCN solution with the addition of 0.5 equivalent of (-)-ephedrine at RT with an *ee* of 19-30%.

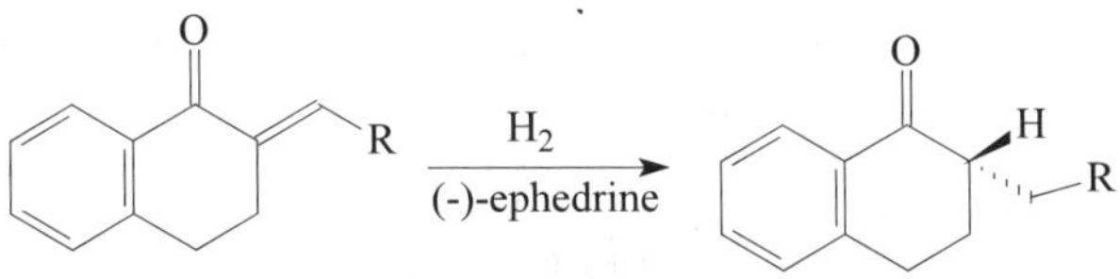

Scheme 5.9.

The hydrogenation of the EtPy on Pt-alumina modified with tartaric acid, 1-phenylethylamine or menthol was not enantioselective [22,26,62] and only ephedrine was an effective modifier with ee's of 5-25%.

Griffiths et al.[147] described attempts to use alkaloids (Scheme 5.10.) as modifiers like codeine (**1**), 7,8-dihydrocodeine (**2**) (*ee* = 3%), brucine (**3**), and strychnine (**4**). With brucine (but not with strychnine) *ee*'s of 10-12% were produced over 6.3% Pt-silica.

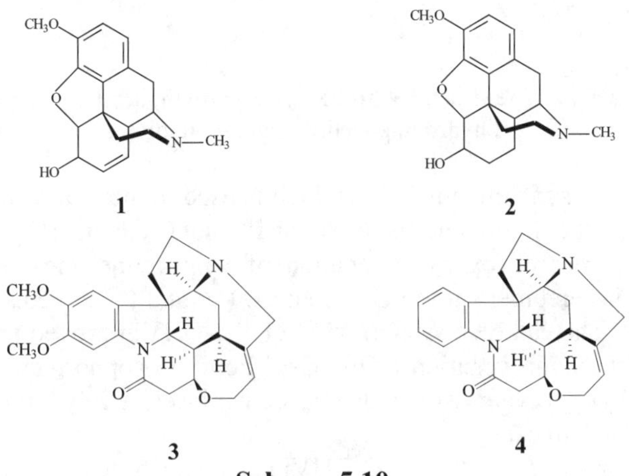

Scheme 5.10.

Enantioselective hydrogenation of MePy at RT and 10 bar of hydrogen over 6.3% Pt-silica modified with brucine resulted in (*S*)-MeLa with an *ee* of 20%, but no enantioselectivity was found in the hydrogenation of butane-2,3-dione. According to Wells et al.[276] brucine adsorbs on a Pt surface to form cavities for the enantioselective adsorption of MePy but not for the diketone. On the other hand, Pt catalyst modified with the morphine alkaloid, oxycodone, works as an enantioselective catalyst in the hydrogenations of both MePy and the diketone at 10 bar and 20°C with *ee*'s of 15% and produces (*R*)-enantiomers in each cases.

A new modifier prepared from L-triptophane proved to be effective in the enantioselective hydrogenation of EtPy over Pt catalyst supported on commercial alumina. The greatest enantioselective action was found by the modifier (*S*)-3-(1-methylindol-3-yl)-2-(methylamino)propan-1-ol. At ambient conditions an *ee* value of 43% was found. A slight increase of pressure led to a dramatic drop in *ee*. An interesting inversion of the sense of enantioselectivity was observed with this modifier when the reaction was carried out in AcOH instead of toluene (Szollosi et al. [175]).

Tungler, Sheldon et al. [164,230,258] found a novel modifier for Pd-carbon catalysts that proved to be active in the enantioselective hydrogenation of C=C and C=O bonds. This modifier, the alkaloid dihydrovinpocetine (the ethyl ester of (-)-dihydroapovincaminic acid, DHVin) possesses quite a different structure from the cinchona alkaloids (Scheme 5.11.).

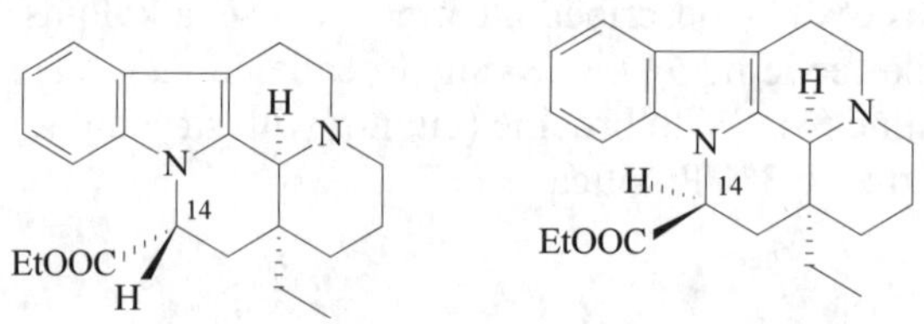

Scheme 5.11. The structures of the (-)-dihydrovinpocetine (left) and (+)-dihydrovinpocetine (right) alkaloids.

Vinpocetine is a synthetic medicine which is used in the treatment of oxygen-deficiency of the brain and is quite available (trade mark "Cavintone"). DHVin was prepared by hydrogenation of vinpocetine and consisted of a mixture of diastereomers at the C14 centers: (-)- and (+)-stereoisomers in the ratio of 91:8. The catalytic system of 5% Pd-C-DHVin proved to be enantioselective in the hydrogenation of the C=C bond in isophorone and the C=O bond in methyl pyruvate, which also was enantioselectively hydrogenated on Pt-alumina-DHVin catalyst.

The properties of the Pd-C-DHVin catalytic system are quite different from the cinchona alkaloid modified Pt catalysts. Thus, in the hydrogenation of the C=C bond in isophorone into the saturated ketone on 5% Pd-Carbon-DHVin, the *ee* values and rates of reaction changed in opposite directions and the effect of ligand acceleration, that was observed in the Pt-alumina-Cnd catalytic system, was absent. Also a very strange dependence of *ee* on the nature of supports was observed: *ee*'s increased in the following order: Al_2O_3 (5%), SiO_2 (6%), carbon (10%), $BaSO_4$ (16%), and TiO_2 (22%). Pd-black without support was the most efficient catalyst (*ee* 38%); enantioselectivity increased with dispersion of Pd and its maximal value was observed at 0.2-0.4 dispersion in the case of Pd-black-DHVin catalyst.

From the hydrogenation of isophorone different catalyst supported on carbon and modified with DHVin produced an unusual set of changes of *ee* values and absolute configurations of product depending on the nature of the catalytic metal and diminishing in the following order: 10% and (*R*) on Pd, 2.1% and (*R*) on Ir, 1.9% and (*R*) on Pt, 0.83% and (*S*) on Rh, and 0.3% and (*S*) on Ru.

Very interesting results were observed in studies of competing influence of modifiers DHVin and Cnd upon their simultaneous addition to one sample of catalyst in the enantioselective hydrogenation of EtPy at RT and 50 bar. Thus when increasing amounts of DHCnd were added to a Pt-alumina catalyst that was originally modified with DHVin, *ee* values of the resulting EtLa changed from 25% (*R*) (no DHCnd added) to 50% (*S*) (2 mmol Cnd added), which indicated that DHCnd is more strongly adsorbed and displaced DHVin from the active Pd centers.

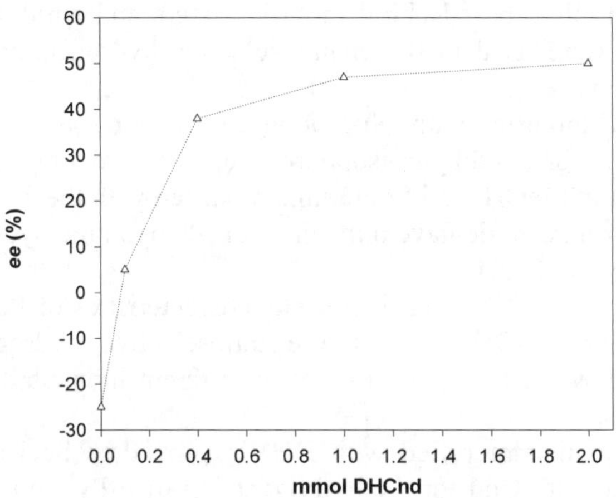

Figure 5.8. Enantiomeric excess values in the enantioselective hydrogenation of ethyl pyruvate into (*R*)-(-)- and (*S*)-(+)-EtLa on Pt-alumina catalyst modified with DHVin with increasing addition of DHCnd at RT and 50 bar (according to Tungler, Sheldon [230]).

The asymmetric inductions of (-)-DHVin and Cnd as chiral modifiers were compared in the Pd catalyzed hydrogenation of the C=C double bonds of 2-phenylcinnamic acid and isophorone. The differences in their effects and behaviors were attributed to the differences in the interactions between the modifier and reactant and their different basicities (Tungler et al. [129]).

Prominent nonlinear effects in enantioselectivity were observed when EtPy was hydrogenated over Pt-alumina in the presence of two cinchona alkaloid modifiers, Cnd and Qnd, which alone afford excesses of opposite enantiomers of EtLa. The changes in reaction rate and *ee* varied strongly with the type and amount of the alkaloid and with the order of their addition to the reaction mixture. Under ambient conditions in AcOH as solvent, Cnd afforded an *ee* of 90% to (*R*)-EtLa, but addition of an equimolar amount of Qnd to the Cnd modified catalyst reduced the *ee* to only 88%, even though Qnd alone provided 94% *ee* to (*S*)-EtLa. This results can be explained by preferential adsorption of Cnd on the surface of Pt via the quinoline rings, which lie approximately parallel to the Pt surface. In this position the Cnd molecule can interact with EtPy during H-uptake and control enantioselectivity. On the other hand, the weaker adsorbing Qnd adopts mainly a position with the quinoline plane tilted relative to the Pt surface, and this species cannot be involved in the enantioselective reaction (Huck et al. [277]).

Farkas et al.[278] compared the esters (-)-DHVin and (+)-DHVin with (-)-dihydro-apovincaminic acid as chiral modifiers in the enantioselective hydrogenation of EtPy and isophorone.

(*S*)-Proline based chiral modifier esters and amides containing aromatic rings were used in the enantioselective hydrogenation of isophorone and EtPy (Sipos et al.[279]).

Pd-C modified with (*S*)-proline[2-(2-naphthyl)ethyl] ester resulted in an *ee* of 23% for (*S*)-dihydroisophorone upon the hydrogenation of its C=C double bond in MeOH, and Pt-alumina modified with the 3-ethylindol derivative of (*S*)-prolineamide gave only an *ee* of 5% upon the hydrogenation of the C=O of EtPy in MeOH.

Sipos et al.[280,281] considered the characteristics of Pd-titania catalysts modified with (-)-DHVin in the enantioselective hydrogenation of isophorone. They used supports that were different in crystalline forms and in surface areas.

Pt-alumina modified with DHVin proved to be less effective than that modified with Cnd for the hydrogenation of EtPy; *ee* reached only 30% (Tungler, Sheldon [230]).

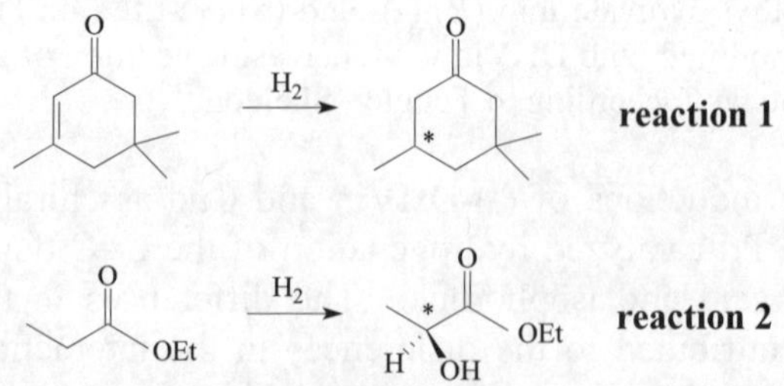

Scheme 5.12.

The hydrogenation of isophorone (reaction 1) and EtPy (reaction 2) were the studied reactions (Scheme 5.12.) using different Pd- and Pt-catalysts, modified with (-)-DHVin and (+)-DHVin (Table 5.9.) [258].

Reactions	Catalysts	Modifiers		
		(-)-DHVin	(+)-DHVin	DHCnd
		ee (%)	*ee* (%)	*ee* (%)
reaction 1	0.3 g Pd black	40	10	20
	0.5 g Pd powder	19	6	10
	0.1 g Pd/C	10	4	3
reaction 2	0.1 g Pt/Al_2O_3	30	14	72
	0.1 g Pt/C	27	13	55

Table 5.9. A comparison of enantioselective hydrogenation of isophorone (reaction 1) on Pd and ethyl pyruvate (reaction 2) on Pt catalysts modified with DHVin isomers and DHCnd (according to Tungler, Sheldon [230]).
Conditions: 100 cm^3 methanol, 0.1 g modifier, 0.5 g acetic acid, 0.05 mol isophorone or 0.1 mol EtPy.

Interaction of DHVin with EtPy to form and association was confirmed by circular dichroism in which it was found that the *N*-atom in DHVin reacted with the C=O group in pyruvate; the dichroic effect was increased when the *N*-atom was protonated by the addition of AcOH, analogous to the same effect in the case of Cnd. Adsorption on Pt through the heteroaromatic group and by the indole group of DHVin proceeds similarly to the adsorption of Cnd. The much stronger adsorption of the quinoline group in Cnd than the indol group in DHVin explains the greater effectiveness of catalysts modified with Cnd than with DHVin.

Comparison of the active parts in DHVin (left in Scheme 5.13) and Cnd (right) show their similarities.

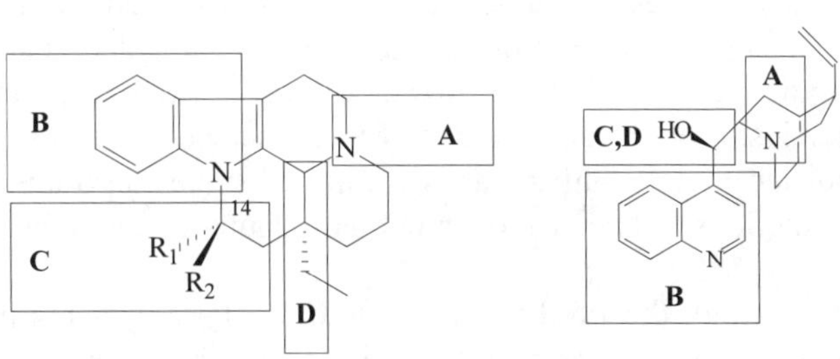

Scheme 5.13.

In DHVin it is possible to mark the four groups responsible for asymmetric reaction as A, B, C, and D, shown in Scheme 5.13. according to Tungler [164].

The effect of addition of AcOH in the hydrogenation mixture with Pt-catalyst modified with both DHCnd and DHVin can be explained by interaction of the basic center *N*-1 (region A) with the substrate molecule. The indole ring (in DHVin) and quinoline ring (in Cnd) (region B) are responsible for the coordination of the modifiers with metal centers on the surface of the catalyst.

The roles of regions C and D are less definite. (-)-DHVin and (+)-DHVin stereoisomers have different configurations at C-14 and different optical rotations, but as modifiers they lead to the same configuration of products.

Structures close to DHVin and DHCnd with regions A, B, C, and D can be identified in yohimbine and reserpine, too (Scheme 5.14.).

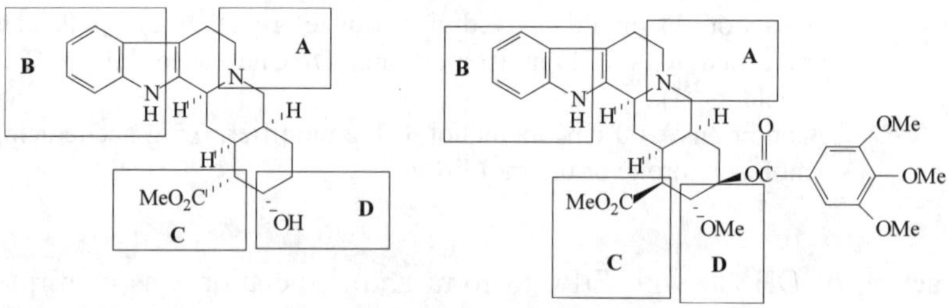

Scheme 5.14.

The effectiveness of these compounds as modifiers should be evaluated in future experiments. It is of interest that electrochemical reduction of the C=C bond in cumarin in the presence of yohimbine leads to the (*R*)-(+)-ketone with an *ee* of 13 % (see Chapter 6).

5.4. Structure of substrate

There are many classes of substrate structures involved in heterogeneous enantioselective hydrogenation on modified metal catalysts. They consist of 2-oxocarboxylic acids and their esters, ketones, diketones, keto lactones, unsaturated acids, oximes, and amides. Enantioselectivities of heterogeneous chirally modified metal catalysts are determined in an important way by the matched interactions between the functional groups of the substrate and the modifier.

For example, the Pt-alumina-alkaloid catalytic systems proved to be the best for enantioselective hydrogenations of 2-oxocarboxylic acids and their esters, especially the pyruvates (Blaser [22], Baiker [17-20]), and the formates (Orito et al.[57,58]), originally studied in the "Orito Reaction" (see Bartok et al.[169]). And recently very high *ee*'s were obtained in hydrogenations of pyruvaldehyde dimethyl acetal (Torok et al.[282] and Studer et al.[76]), ketopantolactone (Schuerch et al.[20]), substituted pyrrolidine-trione (Kunzle et al. [87]), and ketoglutarates (Balazsik et al.[283] and Felfoldi et al.[177]). The highest *ee*'s, 96-98.6%, were obtained with 2-oxoesters over 5% Pt-alumina catalyst modified with HCnd or MeO-DHCnd (Blaser [68]). And also, the hydrogenation of formates gave high *ee*'s; EtBf was hydrogenated with a somewhat larger *ee* than MeBf, 89.5% [58] versus 81.9% [57].

Recently reported have been several syntheses of chiral esters that have practical applications for the preparation of natural products and chiral synthones. Many good results were achieved due to special conditioning of the catalysts and adjustments of the solvents. Further, the reaction has been extended to include the hydrogenation of functionalized compounds; for example, carboxylic acids containing a C=C bond have been hydrogenated with chirally modified Pd catalysts supported on a new type of carriers (Smith et al.[202] and Nitta et al.[124]). And in the hydrogenation of the C=C bond in 2-phenylcinnamic acid recent amazing results were obtained over a Pd-TiO_2 modified catalysts that considerably extended the possibility of the given catalytic systems.

5.4.1. Hydrogenation of ketones

Hydrogenation of diketones and ketones proved to be not very effective (Vermeer et al.[150]); hydrogenations of the diketones, butane-2,3-dione and hexane-3,4-dione, on 6.5% Pt-silica-Cnd at 10 bar and 20^0C in CH_2Cl_2 solution produce *ee*'s of only 33% and 38%, respectively, although trifluoroacetophenone gave 56% (Bodmer et al. [99,100]). Likewise, Torey et al.[275] found that hydrogenation of the C=C side chain in an unsaturated cyclic ketone over Pd-charcoal catalyst modified with (-)-ephedrine at 0^0C in MeCN resulted in a product with an *ee* of only 36%.

The hydrogenation of a nonactivated ketone, acetophenone, over Pt supported on carbon and modified with Cnd was not very effective, producing an *ee* of only 20% (Perosa et al.[284]). However, electron-withdrawing groups in the aromatic ring increased the reaction rate and enantiomeric excess. Derivatives of acetophenone in the *o*-, *m*- and *p*-positions gave higher *ee*'s; for example, hydrogenation of 3,5-bis(trifluoroacetyl)acetophenone under ambient conditions over Pt-alumina-Cnd catalyst yielded a product with an *ee* of 60%. An interesting unprecedented behavior in this reaction was observed: addition of Cnd into the reaction media slowed down all hydrogenation reactions (Hess et al.[285]).

Von Arx et al.[286] studied the hydrogenation of 4-oxoisophorone (2,6,6-trimethylcyclohex-2-en-1,4-dione) over Pt-alumina and Pd-alumina catalysts modified with Cnd and discovered (over Pt-alumina-Cnd) the first example of an unprecedented selectivity in hydrogenation of a sterically hindered C=O bond in an *alpha, beta*-unsaturated diketone to an unsaturated alcohol (3,5,5-trimethylcyclohex-2-en-4-ol-1-one) with an *ee* of 14%.

Pyrone derivatives, 3,6-dimethyl-4-hydroxy-2-pyrone and 4-methoxy-6-methyl-2-pyrone, were hydrogenated over 5% Pd-titania modified with Cnd and Cn. During the reaction the quinuclidine *N*-atom of the alkaloid modifier can interact with the acidic OH group of 3,6-dimethyl-4-hydroxy-2-pyrone [113].

Solvent effects were studied in the enantioselective hydrogenation of 1-phenylpropane-1,2-dione over Pt-alumina-Cnd catalyst. The *ee* values and reaction rates are dependent on the hydrogen solubilities and dielectric constants in different solvents. The highest *ee*, 65%, of (*R*)-1-hydroxy-1-phenylpropanone was obtained in toluene. It decreased non-linearly with increasing solvent dielectric constant becoming close to zero in MeOH (Toukonitty et al.[287])

In the second hydrogenation step during the asymmetric hydrogenation of cyclohexane-1,2-dione over Pt-alumina-Cnd an enantiomeric excess of over 80% of (1*R*,2*R*)-*trans*-cyclohexane-1,2-diol was obtained due to kinetic resolution (Sonderegger et al.[288]).

Ethyl methyl ketone and methoxyacetone were transaminated with benzylamine over Pd-C catalyst in the presence of chiral modifiers like L-Ala, L-alaninol, or L-Ala-OR with *ee*'s of 20% in cyclohexane solvent. When the (*S*)-(-)-(1-phenylethyl)amine was the chiral transamination agent, the hydrogenation resulted in (*S*)-1-methoxypropan-2-amine with an *ee* of 70% (Gobolos et al.[289]).

Ethyl nipecotate (ethyl piperidine-3-carboxylate) was prepared with an *ee* of 24% by enantioselective hydrogenation of ethyl 1,4,5,6-tetrahydronicotinate over DHCnd modified Pd-C catalyst in solution (DMF, AcOH + water) and with an *ee* of 19% over Pd-TiO_2 in DMF solution. As Blaser et al.[290] noted, this example is the first asymmetric hydrogenation of a heterocyclic *alpha, beta*-unsaturated ester with significant *ee*'s using a chirally modified heterogeneous catalyst.

Compared to homogeneous chiral transition metal complexes, enantioselective hydrogenation of aromatic and heteroaromatic compounds over modified metal catalysts has the greater potential.

Maris et al. [112] hydrogenated furan-2-carboxylic acid over a Cnd modified 5% Pd-alumina catalyst at RT and 30 bar hydrogen in a solvent. (*S*)-tetrahydrofuran-2-carboxylic acid was obtained with an *ee* of 32%. When benzofuran-2-carboxylic acid was hydrogenated slowly, the *ee* increased to 50% at 29% chemical yield. The potential application of the method is limited by the competing hydrogenation of the quinoline rings of Cnd, which necessitate the presence of only small amounts of Cnd during the reaction. The reaction mechanism is analogous to that applied for *alpha,beta*-unsaturated carboxylic acids (see below, Part 5.7.). The *ee* values were lower in the hydrogenation of methylfuran carboxylic acids but 100% *de* values were achieved.

Diastereoselective hydrogenations of the bornyl and menthyl esters of furan-2-carboxylic acid in EtOH over 10% $Pd(OH)_2$-C catalyst at 20°C and 50 bar hydrogen were studied [291,292]. The optically active esters were reduced with $LiAlH_4$ and the *de* values of the resulting tetrahydrofuryl-alcohols were measured. The *de* value of the bornyl ester was 24%, and of the menthyl ester was 20%.

Exocyclic *alpha,beta*-unsaturated ketones can be enantioselectively hydrogenated. For example, (*E*)-2-benzilidene-1-benzosuberone was hydrogenated to (*S*)-2-benzyl-1-benzosuberone with an *ee* of 20% in AcOEt or acetonitrile at ambient conditions over Pd-C catalysts in the presence of (*S*)-proline (Fogassy et al.[293]) or with an *ee* of 53.7% in toluene over Pd black modified with Cnd (Fogassy et al.[294]).

On chiral modified metal catalysts very important results were obtained from the hydrogenation of ketopantolactone to pantolactone (see also, Chapter 7) with *ee*'s up to 79% (Baiker [17-20,82]). A very important role has been shown for the application of new type of modifiers other then the cinchona alkaloids by modeling their molecular structures (Schuerch [20]). Of interest is the elaboration of the effective preparation of intermediates and synthones for the syntheses of novel drugs via asymmetric hydrogenation of the keto ester, ethyl 2-oxo-4-phenylbutyrate to ethyl (*R*)-2-hydroxy-4-phenylbutyrate. An example is the preparation of benazepril (Blaser et al.[71-73]) (see also Chapter 7, reaction 7.31 and 7.32) using modified Pt-alumina-Cnd catalyst or the $[\{Rh(NBD)Cl\}_2(Norphos)]$ complex.

Unfortunately, the best *ee* (96%) obtained with the use of latter homogeneous complex was obtained only at a low substrate/catalyst ratio of 50. The heterogeneous catalyst, on the other hand, revealed an *ee* of 95%, high activity, and high productivity. The main problem accompanying the high rate and *ee* value was the substrate quality.

The main results from enantioselective hydrogenations on Pt-alumina were received from hydrogenations of pyruvates (the most promising model substrates), which procduced *ee*'s above 95%. Recently, however, some new examples of substrates were described revealing excellent enantioselectivities.

Bartok, Balazsik et al.[176,177,283] studied the enantioselective hydrogenation of 2-oxoglutaric acid and dialkyl 2-oxoglutarates over cinchona modified Pt/Al_2O_3 catalysts. They found that the hydrogenation of the keto-group in alkyl 2-oxoglutarates at 20°C and 10 bar hydrogen in AcOH solution over 5% Pt-alumina modified with MeO-DHCnd gave, with subsequent cyclization, the furane derivatives, (*R*)-alkyl-5-oxotetrahydrofuran-2-carboxylates (chiral building blocks for further syntheses), with *ee*'s of 93%. Increasing the hydrogen pressure to 20 bar gave products with somewhat higher *ee*'s, 94-96%. Modifying the catalysts with Cnd or Cn instead of

MeO-DHCnd produced products with *ee*'s of 83% (*R*), and 66% (*S*), respectively. Similar to the hydrogenation of EtPy on alkaloid-modified Pt-alumina, the general rate increased from 0.12 to 1.41 mmol/min·g catalyst after addition of the alkaloid modifier to the reaction mixture.

Unfortunately the report of Bartok, Balazsik et al. [283] did not indicate the *ee* of the hydroxyglutarate produced in initial catalytic hydrogenation step; therefore, it can not be excluded that the high *ee* of the cyclized product was obtained as a result of the diastereomeric cyclisation of hydroxyglutarate rather than as a result of the catalytic hydrogenation of the keto-group in 2-oxoglutarate. Felfoldi et al.[176] studied the modifier concentration dependence of the rate and enantioselectivity of hydrogenation of the diethyl 2-oxoglutarate and pyruvaldehyde dimethyl acetal over Pt-alumina modified with cinchona alkaloids. Using the Engelhard catalyst (E4759) in AcOH at ambient conditions an optical yield of 95-97% was achieved.

Bartok's group [282] described the first application of the heterogeneous cinchona modified Pt catalysts for the highly enantioselective synthesis of a chiral building block of an acetal with an *ee* of 96.5% using the hydrogenation of pyruvic aldehyde dimethyl acetal to lactaldehyde dimethyl acetal (Scheme 5.15.).

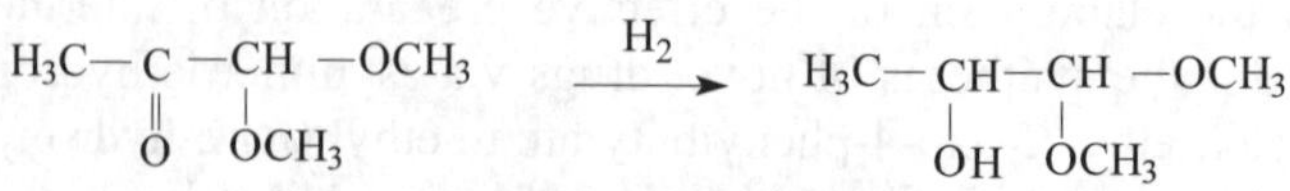

Scheme 5.15.

The reaction was carried out at 20°C and 1 bar hydrogen in AcOH solution over 5% Pt-alumina catalyst modified by Cnd or Cn. In the presence of Cnd-modifier, the (*R*)-product with an *ee* of 96.5% and a rate of 0.81 mol/g·catalyst·min was obtained, while with Cn as the modifier, the (*S*)-acetal with 88% *ee* and 0.72 mol/g·catalyst·min was formed. Increasing the hydrogen pressure decrease the values of *ee's* and increases the rates of reactions indicating a ligand acceleration mechanism; thus, the highest optical yields were obtained at the highest reaction rate.

Another example of high enantioselectivity (*ee* 97%) in the hydrogenation of pyruvaldehyde dimethyl acetal was found by Studer, Blaser et al.[76] carrying out the reaction on 5% Pt-alumina modified with MeO-DHCnd in AcOH as solvent at 60 bar and 25°C. This reaction was also studied by Torok, Bartok et al.[282], under mainly the same conditions but at 10 bar hydrogen and with Cnd as the modifier and resulted in products with *ee's* of 96.5 %(*R*)- and 88% (*S*)-isomers.

Very interesting results were obtained by Borszeky, Mallat, Baiker et al. [108-110] in the enantioselective hydrogenation of nitrogen containing substrates. It was found [108] that on Pd-alumina catalyst modified with Cnd, Cn, or ephedrine, the C=N bond in pyruvic acid oxime can be hydrogenated into alanine (Scheme 5.16.) at 0-45^{0}C and 10 bar hydrogen in polar solvents with *ee*'s of 14.1% (*S*)-isomer (Cnd), 12% (*R*)-isomer (Cn), or 26% (*S*)-isomer (ephedrine), respectively.

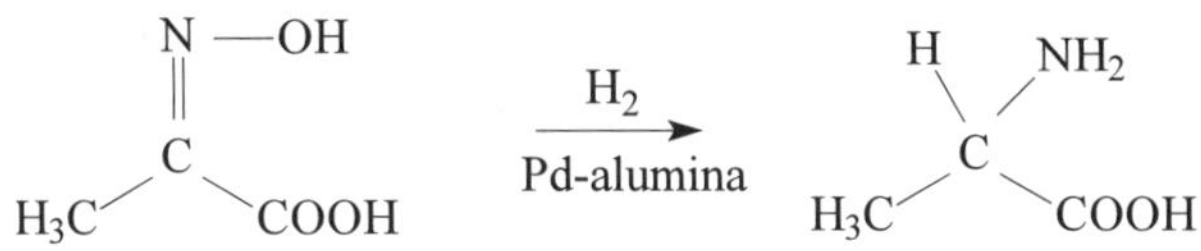

Scheme 5.16.

Enantioselectivity increased with higher temperatures and higher alkaloid to oxime ratios. The presence of the alkaloid modifier resulted in a decrease of the reaction rate by a factor of as much as 140, compared to the "racemic" hydrogenation without modifier. On 5% Pt-alumina-Cnd catalyst, at 60 bar, 20^{0}C, in AcOH, they showed [109,110] that pyruvamides with different amido groups gave modest *ee*'s with the best *ee* of 60% being observed only for trifluoroethyl pyruvamide.

In a similar way the hydrogenation of the cyclic imidoketone (Scheme 5.17.) proved to be enantioselective resulting in a product with an *ee* of 47%.

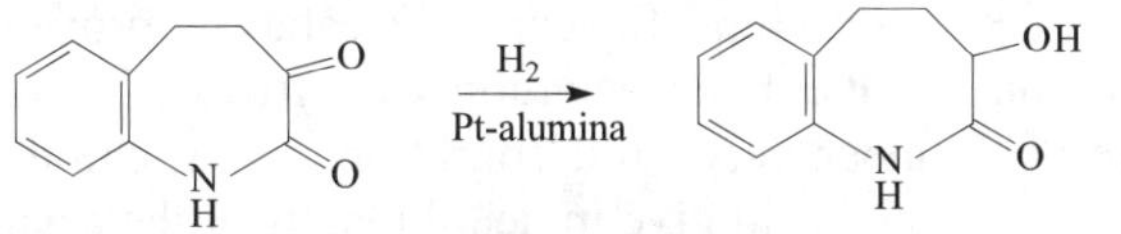

Scheme 5.17.

Kunzle, Baiker et al. [87] synthesized a cyclic imidoketone, 1-ethyl-4,4-dimethylpyrrolidine-2,3,5-trione (a structural analog of ketopantolactone), and hydrogenated it over 5% Pt-alumina-Cnd at 15^{0}C in toluene solution into (*R*)-1-ethyl-3-hydroxy-4,4-dimethylpyrrolidine-2,5-dione (Scheme 5.18.) with an *ee* of 91%.

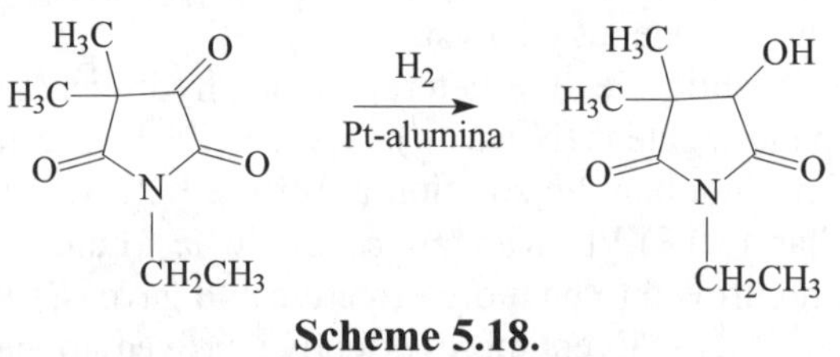

Scheme 5.18.

In polar solvents (EtOH or DMF), its *ee* dropped to 20%, but in AcOH its *ee* reached 70%, similar to the hydrogenation of ketopantolactone. The high concentration of hydrogen on the Pt surface favored increasing the *ee*;

therefore, a hydrogen pressures of 30 bar was chosen as the optimal value. With the Cnd concentration of 3 mg/l and the substrate to modifier ratio of 70,000, the optical yield reached 91%.

This process is used in the industrially important reaction, the hydrogenation of ketopantolactone (KPL) (4,4-dimethyltetrahydrofuran-2,3-dione) into pantolactone (PL) [(*R*)-(-)-3-hydroxy-4,4-dimethyltetrahydofuran-2-one] (Scheme 5.19.).

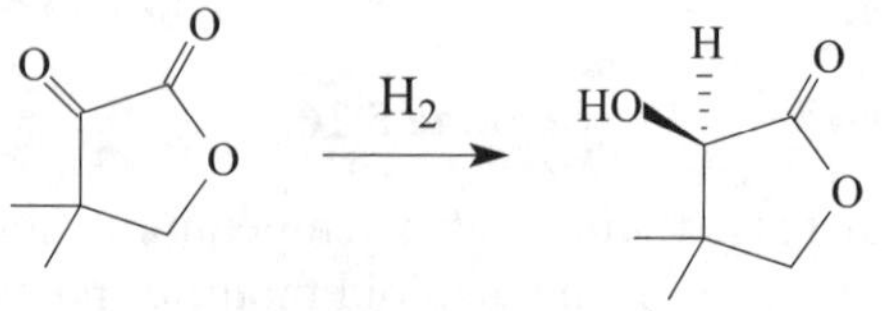

Scheme 5.19.

A very effective process for hydrogenating KPL using Rh and Ru chiral complex catalysts will be considered in Chapter 7. Here the enantioselective heterogeneous hydrogenation of KPL will be examined in connection with the question of mechanism of heterogeneous hydrogenation of the C=O bond on chirally modified Pt catalysts. In KPL the keto-carbonyl group is activated like the *alpha*-keto group in *alpha*-keto esters. The homogeneous hydrogenation of KPL on chiral Rh-complex catalysts proved to be very effective and similar to the hydrogenation of pyruvate in which *ee* values reached above 90% under very mild conditions using chiral metal complex catalysts [295-298]. Chapter 7 will give in detail results of the asymmetric synthesis of PL via hydrogenation of KPL using metal complexes with chiral phosphane ligands and producing *ee*'s up to 98%, which are of practical interest. Here we can note that Ru(II)- and Rh(I)-arene complexes were very effective in the enantioselective hydrogenation of activated carbonyl groups [108]. The use of Rh-complexes with a new chiral ligand (*S*)-dicyclopentylphosphane-dicyclopentylphosphinite-pyrrolidine allowed the hydrogenation of KPL into PL with an *ee* of 96% at 20°C and 96.9% at 70°C. The Rh-complex supported on silica gave PL with an *ee* of 91%, which is the same value obtained using the unsupported homogeneous catalyst.

The first enantioselective heterogeneous hydrogenation of KPL was reported in a Japanese patent (Niwa [299]). Over a 5% Pt/C catalyst previously modified with Cnd in ethanolic solution at reflux, KPL was hydrogenated in benzene at 60 bar to (*R*)-PL with an *ee* of 36%. Later, a Pd/C catalyst modified with Cnd proved to be more effective and gave (*R*)-PL with an *ee* of 52% (see Baiker [17-20]). Even more effective proved to be a Pt-alumina-DHCnd catalysts that produced (*R*)-(-)-PL with an *ee* of 79% under optimal conditions (22^{0}C and 70 bar) in toluene at low concentrations of alkaloid (Schuerch et al.[20]).

The main characteristics of the process of hydrogenation of KPL are quite similar to those of the hydrogenation of pyruvates. These characteristics are similar positive influences of preliminary catalyst treatments at elevated temperatures in flowing hydrogen gas, the use of solvents of low polarity, and the existence of relatively high Pt surface hydrogen concentrations. All these factors strongly work for higher enantioselectivity.

5.4.2. Effect of variables on the hydrogenation of ketopantolactone

Effect of hydrogen pressure. The optical yield of KPL from hydrogenation on Cnd modified Pt-alumina increased with increasing hydrogen pressure from 1 bar (50% *ee*) to 30 bar (75% *ee*). Above 30-40 bar the *ee* values reached a plateau. This picture is similar to that observed for the hydrogenation of pyruvates (Schuerch [20], Sun [109] and Garland [67]). It was found that for high *ee*'s it is necessary to create high hydrogen concentrations on the Pt surface. Therefore *ee* values from the hydrogenation of KPL will be considerable lower when the catalyst is cooled in flowing nitrogen after heat treatment in flowing hydrogen. It was found [209,258], that *ee*'s depend considerably on the concentration of molecular hydrogen in the liquid phase rather than on the hydrogen pressure in the gas phase if the reaction proceeds under kinetic conditions rather than under diffusion control. Values of *ee* from the hydrogenation of EtPy on Pt-alumina increased in the initial period of reaction at constant pressure of 6 bar and 30^{0}C by increasing the gas-liquid mass-transfer of hydrogen. Good *ee* values could be obtained provided the catalyst is properly preconditioned and traces of oxygen and water are removed (see Johnston et al. [144] and Margitfalvi et al. [162]).

Effect of reaction temperature. Increasing the temperature of the hydrogenation of ketopantolactone in the range of 7-27^{0}C leads to a maximum *ee* of 78%, while increasing the temperature of the hydrogenation of pyruvates results in decreasing enantioselectivity.

Effect of substrate concentration and degree of conversion. In the concentration range of 0.05-2 mmol/L of KPL the influence of concentration on the rate of reaction and on *ee* is very small. On the other hand, in the hydrogenation of pyruvates there is another picture. There the rates and *ee*'s increase with extent of conversion up to 20% (Wang [211]) at least in the initial period of the reaction (Singh et al. [207,208]). No such changes occurred in the hydrogenation of KPL. That is, there is no such dependence if the reaction is carried out under "standard condition" (hydrogen pressure 70 bar). But at atmospheric pressure in toluene solution an increasing *ee* of 20-35% was obtained during the degree of conversion of 5% to 90% (Singh [207]).

Effect of modifier concentration. A broad maximum of *ee* (75-80%) has been observed as a function of the concentration of HCnd (in standard conditions of 150 mg of catalyst, 3.9 mmol KPL, 7^{0}C, 70 bar, toluene solvent).

The maximum corresponds to a low molar ratio of DHCnd to Pt of 0.065. For the hydrogenation of pyruvates this ratio is equal to 0.5 (Garland et al. [67]). Thus under standard conditions the best results obtained the product (*R*)-(-)-PL with an *ee* of 75% (if the catalyst was preliminarily treated with hydrogen but not with nitrogen!). Kunzle et al. [114] showed the enantioselective hydrogenation of KPL over Pt-alumina-Cnd in a fixed bed reactor by continuously feeding the chiral modifier, Cnd, in *ppm* concentrations to the reactant stream. The reaction proceeded at RT and 40 bar hydrogen with a rate of 94 mmol/g catalyst·h and an *ee* of 83.4%. In the same conditions EtPy was hydrogenated with 23 mmol/g cat·h and resulted in an *ee* of 89.9%.

Effect of solvents. The nature of solvent plays an important role in the asymmetric hydrogenation of KPL. As observed during the hydrogenation of pyruvates, *ee* values decreased with increasing solvent polarity (Blaser et al. [68]). The same effect was observed in the hydrogenation of KPL. A high *ee* of 78% was obtained in apolar media such as toluene. In polar solvents like ethanol *ee* values dropped almost to zero. Strangely, although acetic acid was the best solvent for enantioselective hydrogenation of pyruvates, for KPL it gave an *ee* of only 35% (Schuerch et al. [20]).

Rate acceleration. A very peculiar feature of the hydrogenation reaction of pyruvates over Pt-alumina catalysts consists in the sharp increase in the rate of reaction and in the *ee* of the product immediately after introducing a small amount of alkaloid into the reacting solution. It is the well known effect of "ligand acceleration" (Garland et al. [66]) for enantioselective hydrogenation of pyruvates. The same picture was observed for the hydrogenation of KPL. The rate increased by 4-25 times in comparison with the reaction in the absence of alkaloid modifier. Adding merely 2-20 mg of Cnd to the reaction mixture increased the rate of reaction and reduced the reaction time from 25 min to 3.5 min and produced PL with an *ee* above 70% under standard reaction conditions (0.15 g of 5% Pt-alumina catalyst, 19.5 mmol KPL, 25 ml toluene, 70 bar hydrogen, and 12^0C).

Molecular modeling of the reaction system showed that in the transient complex the basic quinuclidine N-atom of Cnd is bound to the C=O group of KPL via a hydrogen bond interaction that stabilizes the half-hydrogenated state. The interaction favors the formation of (*R*)-PL compared to (*S*)-PL. It was noted (Schuerch et al. [20]) that diastereomeric transition complexes leading to (*R*)-pantolactones are more stable by 8 kJ/mol than the corresponding complex leading to (*S*)-pantolactones. This calculation did not take into consideration the metal surface and concerned only the Modifier-Substrate interactions rather than with the Modifier-Catalyst or Substrate-Catalyst interactions. In this connection, it was suggested that the transition complex is adsorbed on the Pt surface in such a position that the quinoline rings of Cnd and the carbonyl groups of KP are adsorbed on a flat Pt(III) surface facet. This arrangement shows the similarities of the diastereomeric transition complexes when comparing the cyclic lactone, KPL, and the *alpha*-keto esters in an apolar medium, where the alkaloid is not protonated (Schuerch [20]). But the hydrogenation of pyruvates proceeds with higher *ee* values, when Cnd is protonated (Baiker [21]). For this case, a model was proposed by Schwalm et al. [96-98]. When the protonated Cnd interacts with the carbonyl substrate, the interaction can again be described as hydrogen bonding between the quinuclidine *N*-atom and the *alpha*-carbonyl O-atom.

Molecular modeling of the diastereomeric transition complexes Cnd-PL-Pt also indicated that the complex affording (*R*)-PL is energetically favored with Cnd, whereas the complex leading to (*S*)-PL formation is favored by Cn. The formation of both structures are quite similar to those proposed for the hydrogenation of MePy suggested by Schwalm et al. [98].

5.5. Effect of Solvents and Additives

In different ways solvents influence enantioselective hydrogenation of keto esters on chiral modified Pt-supported catalysts. The effect of solvent on enantioselective hydrogenation involves the solubility of hydrogen, the solubility of the reactant, and the interaction between modifier, reactant, and catalyst surface including the carrier (Singh et al. [207,208]). There are no general regularities of the effects of solvents on the actions of Pt-alumina-Cnd catalyzed enantioselective hydrogenations of pyruvic acid esters. Unfortunately, it is difficult to develop a theory of the effects of solvents but some empirical regularities can be pointed out. In a number of studies, toluene and ethanol were used as the most suitable solvents. It was found (Wehrli et al. [78]) that with increasing solvent polarity both *ee* and reaction rate were found to decrease (see Table 5.10.).

Blaser et al. [68] found that *ee*'s almost linearly decreased with increasing solvent polarity (Table 5.10.). In apolar solvents with dielectric constants in the range of 2 and 6, *ee* values obtained in the hydrogenation of MePy were maximal (Baiker [17-20], Wehrli [78]).

Almost the same dependence of optical yields was observed with the use of an empirical solvent parameter given by Reichardt [300] (see Baiker [17-19]). It is of interest that alcohols with dielectric constants of 15-30, also gave good results, although they partly reacted with the keto ester to give the hemiketal (Minder [94,95]). Using ethanol as a solvent, a good result that led to higher *ee* values in the hydrogenation of EtPy can be explained by the partial formation of AcOH during the prior oxidative treatment of the Pt-alumina catalyst (Minder et al.[94,95], Johnston et al. [144]).

The best *ee* was obtained in solutions of AcOH (dielectric constant 6.15) and with other carboxylic acids with *ee*'s up to 91%. NMR studies showed that in the presence of AcOH the quinuclidine *N*-atom of Cnd is protonated, which favors the interaction between modifier and reactant (Baiker [17-19,94,95]).

Solvent	Dielectric constant	Activity r_o (mol kg^{-1} s^{-1})	Conversion (%)	Optical yield *ee* (%)
cyclohexane	2.02	2.94	100	79.9
toluene	2.38	2.68	100	82.9
trichloroethylene	3.4	0	0	-
chlorobenzene	5.71	2.32	100	83.9
tetrahydrofuran	7.6	2.14	100	81.1
dichloromethane	9.08	2.20	100	81.4
pentan-1-ol	13.9	2.38	95	78.0
propan-2-ol	20.1	3.05	100	76.7
ethanol	24.3	2.28	100	75.0
methanol	33.6	1.08	92.7	75.1
DMSO	46.7	0	0	-
water-ethanol (1:1)	50	1.45	90.7	61.9
water	80.4	0.78	100	54.1

Table 5.10. Effect of different solvents on optical yield (*ee* %, normalized values) and reaction rate, r_0, in the enantioselective hydrogenation of ethyl pyruvate on 5% Pt-alumina catalyst modified with cinchonidine (according to Wehrli et al. [78]).

The effect of acids and bases as solvents (or as additives) in the hydrogenation of EtPy on Pt-alumina-Cnd was studied by Blaser et al. [62,68]. The addition of AcOH to common applied solvents (toluene, ethanol) or using AcOH as the solvent exhibited unusually strong positive effects on *ee*'s, up to 95%, especially with catalysts modified with MeO-DHCnd (Blaser et al. [68]).

Table 5.11. shows that all carboxylic acids as additives are suitable for enantioselective hydrogenation but *ee* and initial reaction rates decrease with increasing chain lengths of the acid molecules. Bases decrease the *ee* values somewhat.

Bartok et al. [180,302,309] found that the alumina support can react with the acetic acid producing aluminum containing oxonium cations. These cations may play a role in the enantioselection in the asymmetric hydrogenation of ethyl pyruvate catalyzed by the platinum-alumina-dihydrocinchonidine catalytic system.

Analogous to the pyruvates, the same decrease in *ee* with increasing solvent polarity was found in the enantioselective hydrogenation of ketopantolactone into pantolactone over Pt-alumina-Cnd catalyst (Schuerch et al. [20]). The most suitable solvent proved to be toluene (*ee* 78%) but in acetic acid solution the strong increase of *ee,* which was observed in the case of pyruvates, was not seen and the optical yield of pantolactone reached only up to 35% (see results in Part 5.4.).

Additive	Amount (g/L slurry)	Modifier	*ee* (%)	Rate (mol g^{-1} min^{-1})
none		MeODHCnd	77	0.08
none		MeODHCnd	73	0.08
none		DHCnd	84	0.1
CH_3COOH	80	MeODHCnd	90	0.13
CH_3COOH	80	MeODHCnd	87	0.09
HCOOH	40	MeODHCnd	91	0.01
CF_3COOH	40	MeODHCnd	87	0.03
C_6H_5COOH	40	MeODHCnd	78	0.08
$(COOH)_2$	40	MeODHCnd	71	0.01
phenol	40	DHCnd	67	0.14
phenol	40	MeODHCnd	72	0.1
quinoline	0.4	DHCnd	88	0.08
quinoline	4	DHCnd	79	0.04
Et_3N	2.8	DHCnd	79	0.05
Pyridine	0.4	DHCnd	83	0.1

Table 5.11. Effect of various acidic and basic additives to solvent toluene on the hydrogenation of ethyl pyruvate over Pt-alumina catalyst modified with 10,11-dihydrocinchonidine (DHCnd) or 10,11-dihydro-O-methylcinchonidine (MeODHCnd) at RT and 100 bar hydrogen pressure (according to Blaser et al. [68]).

Solvent	Chemical yield (%)	$[\alpha]_{589}$ (in EtOH) of the "Scalemic products" *	Optical yield of (*R*) (*ee* %)
AcOH	87	82.4	61.5
EtOH	89	83.2	62.1
iPrOH	90	92.8	69.3
PhMe	87	97.0	72.4
THF	89	97.8	73.0
AcOEt	90	104.7	78.2
iBuOH	88	101.5	75.8
iPrOiPr	89	106.3	79.4
Dioxane	89	107.1	80.0
$EtCO_2Et$	92	107.9	80.0
EtOEt	89	108.4	80.8
PhH	90	109.0	81.4
AcOiPr	91	109.5	81.0
$EtCO_2Me$	92	109.7	81.9

* "Scalemic products " are optically active mixtures, containing unequal amounts of enantiomers (see in *J. Am. Chem. Soc.*, **53**, 1922 (1988)).

Table 5.12. Asymmetric hydrogenation of methyl benzoylformate into (*R*)-(-)-methyl mandelate in different solvents over 5% Pt-C catalyst modified with cinchonidine (according to Niwa,Orito et al. [58]).

Orito et al. [59,60] also showed that the nature of the solvent proved to be important in the hydrogenation of MeBf and MePy on Pt-C and Pt-alumina catalysts modified with Cnd (Tables 5.12. and 5.13.).

The hydrogenation of MePy into MeLa proceeds somewhat less influenced by solvents. In the series of solvents, THF, MeOH, AcOEt, iPrOH, and iBuOH, the *ee*'s of MeLa increased from 64% to 73.1%. But hydrogenation without solvent proved to be somewhat more effective, giving an *ee* of 74.0%, and the addition to the reaction mixture of 0.1 g of Cnd did not affect the *ee*, but in the presence of solvents, the addition of Cnd or Qn did increase *ee*'s (Table 5.13.).

Solvent	5 % Pt-C *ee* % [32]	5 % Pt-Alumina *ee* % [33]	Supplement addition of 0.1 g alkaloid
None	74.0	-	None
MeOH	64.6	32.6	None
MeOH	71.2	86.6	Cnd
EtOH	78.7	83.2	Cnd
iPrOH	72.2	80.6	Cnd
EtOH	76.6	-	Qn
MeOH	-	82.4	Qn

Table 5.13. Asymmetric hydrogenation of methyl pyruvate (MePy) into (*S*)-(+)-methyl lactate (MeLa) on 5% Pt-C and 5% Pt-alumina modified with supplement addition of alkaloids in different solvents (according to Orito et al.[59,60]).

Addition of small amounts of triethylamine (Et_3N) or 20% of Cnd to the solvents gave large improvements in the asymmetric effect (Table 5.14.).

It was found that the initial reaction rate increases with the pK_a of bases added to the reaction mixture in the series of quinoline (pK_a= 5) < quinuclidine (pK_a = 11) < pyrrolidine (pK_a = 16.3) < cinchonidine (pK_a = 8.3).

The effect of Cnd is much higher than by the above mentioned amines despite its lower pK_a, because its specific adsorption on the Pt surface results in simultaneously increasing the reaction rate and the enantioselective effect. Such a result was explained by the "Template effect", which is a mutual adsorption of reactant and modifier molecules on the Pt catalyst surface (Sutherland et al. [141]).

Chernysheva et al.[253] were the first to have obtained evidence of inhibition from the product of the reaction of hydrogenation of ethyl acetoacetate on Raney nickel catalyst modified with (2*R*,3*R*)-tartaric acid into ethyl (*R*)-(-)-3-hydroxybutyrate. Similar results were obtained also in the enantioselective hydrogenation of ethyl acetoacetate on Raney Cobalt modified with (2*R*,3*R*)-tartaric acid [220].

Solvent	Conversion (%)	*ee* (%)	Catalyst
none	68	74.0	5% Pt-C-Cnd
MeOH	72	64.6	5% Pt-C-Cnd
MeOH*	-	69.6	5% Pt-C-Cnd
EtOH	72	70.1	5% Pt-C-Cnd
EtOH*	-	74.0	5% Pt-C-Cnd
iPrOH	66	73.0	5% Pt-C-Cnd
THF	68	53.7	5% Pt-C-Cnd
AcOEt	65	68.8	5% Pt-C-Cnd
Benzene	85	80.8	5% Pt-alumina-Cnd
Benzene*	84	85.2	5% Pt-alumina-Cnd
MeOH	79	32.6	5% Pt-alumina-Cnd
MeOH*	86	44.9	5% Pt-alumina-Cnd
MeOH**	84	86.6	5% Pt-alumina-Cnd

* Addition of 0.2 mL Et_3N ** Addition of 0.1g Cnd

Table 5.14. Effect of solvents and additives on the enantioselective hydrogenation of MePy into (*S*)-(+)-MeLa on 5% Pt-C-Cnd and 5% Pt-alumina-Cnd catalysts (0.5 g catalysts, 70 bar, RT) (according to Orito [59,60]).

Nitta et al.[134] also confirmed these effects using Ni-silica catalysts modified with (2*R*,3*R*)-tartaric acid in the hydrogenation of methyl acetoacetate (Figure 5.9.) They showed the effects on optical selectivity of the concentration of methyl (*R*)-(-)-3-hydroxybutyrate, which had been previously added to the reaction mixture.

These observations confirm that the methyl (*R*)-(-)-3-hydroxybutyrate, the product of the reaction, inhibits a fraction of the active centers of the selective reaction on the surface of catalyst and thus diminishes enantioselectivity. An especially strong inhibiting effect was observed after the addition of strongly adsorbing compounds like thiophene to the reaction mixture [106a].

The use of supercritical fluids (SCF's) as solvents is of interest for enantioselective hydrogenation. Baiker [103-105,285] was the first to employ SC-solvents in the enantioselective hydrogenation of ethyl pyruvate over Pt-alumina catalyst modified with Cnd in a solution of ethane under supercritical conditions. The reaction was carried out in an autoclave at 60 bar and 60^{0}C. It was found that the hydrogenation rate was higher than in toluene solution but the *ee* was approximately the same as in the common solvents. The temperature dependence of *ee* in the formation of (*R*)-(-)-EtLa in SCF was also the same as in common solvents [103].

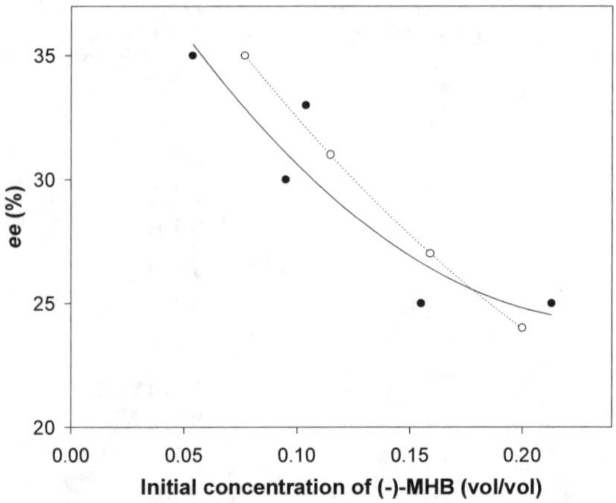

Figure 5.9. Effect of the amounts of previously added methyl (*R*)-(-)-3-hydroxybutyrate to methyl acetoacetate before hydrogenation on Ni-silica catalyst modified with (2*R*,3*R*)-tartaric acid (•) (according to Nitta et al. [134]) and the hydrogenation of ethyl acetoacetate on Raney nickel catalyst modified with (2*R*,3*R*)-tartaric acid (o) (according to Chernysheva et al [253]).

Under supercritical conditions the diffusivity of solutes in SCF is higher than in commonly applied liquids, the viscosity is lower, and mass-transfer is enhanced (Baiker [101,103,303-305]). The comparison of ethane in the supercritical state (SC) and ethanol as solvents in the enantioselective hydrogenation of EtPy over Pt-alumina-Cnd showed that under identical conditions the hydrogenation rate in SC-ethane is increased by a factor of 3.5 with the same *ee*-values. A further advantage of SC-ethane is that the high *ee* value remains constant with increasing catalyst to substrate ratio, which can be important for practical uses of the catalyst in continuous fixed-bed reactors for hydrogenation (see below Figure 5.10.). According to Baiker,Wandeler et al. [101-103, 303-306] the enantioselective hydrogenation of EtPy was studied in a fixed bed reactor over Pt-alumina-Cnd catalyst using supercritical ethane as solvent. Application of SC-ethane revealed good enantioselectivity and very fast conversion of EtPy into EtLa with a TOF 15 sec^{-1} at RT. The phase behavior in the range of 15-50^{0}C and at pressure up to 140 bar showed the multicomponent phase high-pressure reaction system. The rate of hydrogenation in the system composed of ethane-EtPy-hydrogen increased significantly [102,103] when the pressure was increased above 60 bar at 40^{0}C and the density of reaction mixture was 0.3 g/cm^3.

It was found that for enantioselective hydrogenation over modified platinum catalysts the most suitable SC-solvents are ethane and propane as indicated in Table 5.15. [101]. Application of SC-CO_2 in hydrogenation on chiral modified Pt-catalysts proved to be less suitable because CO_2 is partly hydrogenated into CO, which poisons the catalyst [101]. But polymer-supported colloidal Pd nanoparticles as catalysts in supercritical CO_2 ($scCO_2$) revealed TOF's as high as 4×10^{6} h^{-1} at 15 bar hydrogen, and 50^0C (Niessen et al. [306]).

Solvent	P_{crit} (bar)	$^oC_{crit}$
Xenone	58.0	16.59
CO_2	72.9	31.0
Ethane	48.2	32.3
Propane	96.8	42.0
Butane	36	153
$CClF_3$	28.8	39.0
CF_4	44.6	58.0
CHF_3	62	44.9
$CHCl_3$	63.8	143

Table 5.15. Supercritical fluids (SCF's) most suitable for asymmetric catalytic hydrogenation.

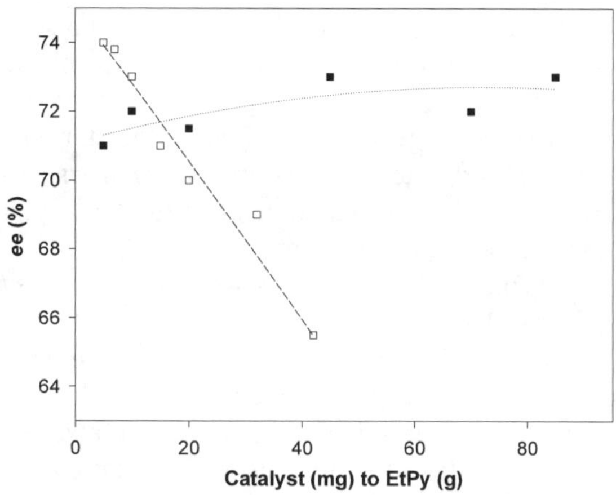

Figure 5.10. Comparative study of enantioselective hydrogenation of EtPy into (*R*)-(-)-EtLa on Pt-alumina-Cnd catalyst as a function of the catalyst to substrate weight ratio in ethanol solution (□, at 60 bar, 22^0C as subcritical solvent) and in the supercritical solvent ethane (■, at 60 bar, 40^0C) (according to Baiker [101-103]).

Table 5.15. shows the other SC-solvents that can be applied in asymmetric hydrogenation. They must be rather inert and can be used in the suitable ranges of pressure and temperatures for practical application.

Figure 5.10. shows an example of the application of SC-ethane as solvent in the hydrogenation of EtPy over Pt-alumina-Cnd catalyst. For all practical purposes the *ee* values of the resulting EtLa did not change with increasing amounts of catalyst if the reaction was carried out in SC-ethane, whereas these values fall sharply in ethanol.

Enantioselective hydrogenation of EtPy on Pt-alumina-Cnd catalyst at 60 bar, 40°C, in supercritical solvent ethane (subcritical condition for ethane are 48.2 bar and 32.2^{0}C) can be compared with hydrogenation in ethanol at 60 bar, 22^{0}C as subcritical solvent (SCF are 61 bar and 241^{0}C) (according to Baiker [102,103]).

5.6. Reaction variables

The kinetics of enantioselective hydrogenation of pyruvates on Pt-alumina (or Pt-silica) catalysts modified with Cnd (or DHCnd) was studied by the groups of Baiker [85,91,102,103], Blaser [66,67,74,75,78], Wells [141,143,146], and Blackmond [207,208,211,212].

A kinetic model for the system of Pt-alumina-Cnd-EtPy was elaborated by Garland [66]. It was based on a two cycle mechanism that proposed increasing the reaction rate on modified centers leading to (*R*)-(-)-ethyl lactate as a result of "ligand acceleration" [227], which is characterized by a high rate (and enantioselectivity) on selective centers, and a slow rate on non-selective centers, both reactions being equilibrated. A mechanistic scheme proposed, was based mainly on a concept suggested earlier by Klabunovskii [25,220,307] for enantioselective hydrogenation of *alpha*-keto esters on tartaric acid-modified Ni catalysts that suggested that two kinds of centers exist on the surface of the catalyst, namely, selective centers leading to chiral products and non-selective centers leading to racemic products.

In kinetic studies zero order was found with respect to the *alpha*-keto ester and first order with respect to hydrogen [78,141]. Apparent activation energies for hydrogenation on Pt-alumina and Pt-silica were found to be 31 kJ/mol and 38 kJ/mol, respectively (Baiker [21]).

Catalytic systems including cinchona alkaloids (or amino alcohols with similar molecular structures) as modifiers proved to be the only successful compounds in asymmetric hydrogenations that provided strong adsorption on the surface of Pt catalyst owing to their aromatic part and their ability to form a hydrogen bond between the tertiary *N*-atom of the quinuclidine moiety of the alkaloid and the oxygen atom of the *alpha*-keto group of the reactant-pyruvate molecule. In the enantioselective hydrogenation of EtPy on Pt-alumina-DHCnd, Garland et al. [66] found the effect of the lowest concentrations of modifier, DHCnd, added during the reaction to a solution of toluene or ethanol. The modifier concentrations varied between 0 and 0.23 mmol/L corresponding to zero to one DHCnd molecule per one exposed platinum atom on the surface of the catalyst.

Rather important factors influencing enantioselectivity are temperature and hydrogen pressure. Reaction temperatures in the range of 20-50^{0}C are suitable; at higher temperatures optical yields drop drastically.

Hydrogen pressures above 5-10 bar gave good results, but according to recent studies of Blackmond's group [209,257] good stirring of the reaction mixture and the resulting concentration of hydrogen in the reaction mixture determined by the mass transfer coefficient of the system was the important factor rather than the hydrogen pressure in the gas phase. Blackmond's group [207-209,211,212,257] found that the kinetic parameters affected enantioselectivity and the main factor was the high rate of mixing of the reacting liquid phase. Increasing the mass transfer coefficients by stirring the reaction mixture from 400 rpm to 1000 rpm [209,257] assisted in transferring the gas into the liquid, brought the reaction into kinetic control, and resulted in increased enantioselectivity.

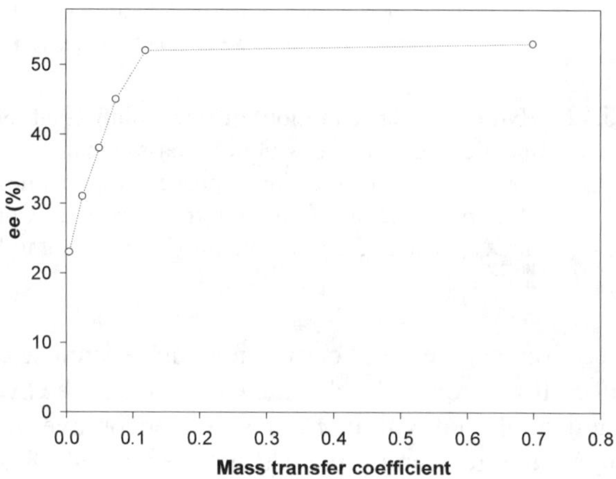

Figure 5.11. Enantioselectivity at 50% conversion as a function of the gas-liquid mass-transfer coefficient (K) in the hydrogenation of ethyl pyruvate on Pt-alumina modified with Cnd at constant pressure in gas phase of 5.73 bar and 30^{0}C (according to Blackmond, Sun et al. [209,257]).

Figures 5.11. and 5.12. demonstrate that enantioselectivity and hydrogen concentration of the solution, vary as a function of the mass transfer coefficient at constant hydrogen pressure. When the rate of mass transfer becomes high enough, owing to the high speed of the stirring in the reactor, the hydrogen concentration approaches $[H_2]^{sat}$ in solution. Further increasing in coefficient causes no additional increase in enantioselectivity. Hydrogen pressure determined the value of $[H_2]^{sat}$ at a given temperature and, therefore, may set the upper limit on enantioselectivity [209,257]. Thus the hydrogen pressure is a valid kinetic parameter only when efficient mixing allows the solution to approach its solubility limit of hydrogen.

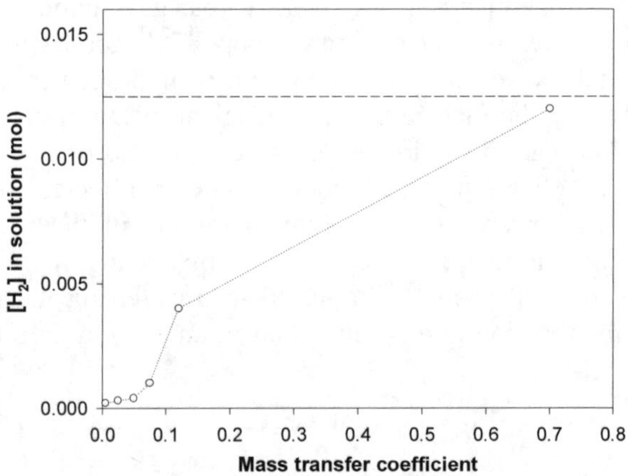

Figure 5.12. Hydrogen concentration of the solution at 50% conversion as a function of the gas-liquid mass-transfer coefficient the $[H_2]^{sat}$ concentration in the liquid phase (broken line) in hydrogenation of ethyl pyruvate on Pt-alumina modified with Cnd at constant pressure in gas phase of 5.73 bar and 30°C (according to Blackmond, Sun et al. [209,257])

The effect of degrees of conversion on enantioselectivity is also of importance. It was found [207,208] that *ee* increased in the hydrogenation of EtPy on Pt-alumina-DHCnd with increasing conversion over a wide range of reaction temperatures reaching a plateau between *ee* 54-58% at high degrees of conversion. The increase of the stirring rate in an autoclave from 400 to 1000 rpm, increases significantly the optical yield of resulting ethyl lactate from 35% *ee* (*R*) to 60%. Wang et al [211] found the significant increase of *ee* and reaction rate is in accordance with the term "ligand acceleration" that was used by Sharpless [81] to describe the significant rate increase observed in the homogeneous asymmetric dihydroxylation of olefines catalyzed by osmium complexes containing cinchona alkaloid ligands.

The initial transient period (ITP) of the enantioselective hydrogenation of activated ketones (ethyl pyruvate, pyruvaldehyde dimethyl acetal and methyl benzoylformate) on Pt-alumina modified by cinchonidine or dihydrocinchonidine under mild conditions in toluene was studied by Balazsik et al [308]. The results suggest that the ITP is affected not only by impurities but also by the competitive adsorptions of reactants, modifier and solvent. Consequently, ITP may be considered an intrinsic feature of this type of enantioselective hydrogenations.

In the enantioselective hydrogenation of ethyl benzoylformate to (*R*)-ethyl mandelate over dihydrocinchonidine modified Pt/Al_2O_3 in acetic acid 98% *ee* was achieved. The difference between the rates of racemic and enantioselective hydrogenation was less than in the case of ethyl pyruvate, which indicates that ligand acceleration is not an absolutely necessary prerequisite for achieving high enantioselectivity in the enantioselective heterogeneous hydrogenation of *alpha*-keto esters [309,310].

The repeated use of Pt-alumina modified by dihydrocinchonidine was studied by Balazsik and Bartok [311] for the enantioselective hydrogenation of ethyl pyruvate under mild experimental conditions in toluene and AcOH. In toluene, depending on the reaction conditions, an increase of *ee* by 10-20% was observed on the reused catalysts. The same effect, however, was not found in AcOH. The phenomenon was attributed to an intrinsic feature of the Pt-alumina-cinchona system, in which the restructuring of the Pt surface may play an important role.

Using electrospray ionization mass spectrometry technology on products from pyruvate hydrogenations Bartok's group identified aluminum containing oxonium cations and new adducts, which were formed by non-covalent interactions [301,312-316]. In the study of enantioselective hydrogenation of EtPy on Pt-alumina, Pt-black, and Pt-(black + alumina) modified with DHCnd in AcOH solution under the mild conditions of RT and 1 bar hydrogen EtLa was produced with an *ee* of 92% accompanied by the following ratios: EtPy/DHCnd > 43,000, EtPy/ $Pt_{surface}$ > 1000, and an extremely low DHCnd/$Pt_{surface}$ ratio (0.0072). It was found experimentally that the quinoline ring of DHCnd gradually hydrogenated during the reaction decreasing the enantioselectivity (Bartok et al. [180]). This reduction of the modifiers during the hydrogenation reaction was published in other articles, too [317,318].

LeBlond et al. [257] applied unprecedented mild conditions for the enantioselective hydrogenation of keto esters, 5.8 bar instead of the commonly used 100 bar, over 1% Pt-alumina catalyst with 25% Pt dispersion, modified with DHCnd. In their standard experiment 1.4 g of 1% Pt-alumina, 400 ml AcOH, and 0.1 mol of EtPy were evacuated and stirred before applying hydrogen.

They achieved 94% *ee*'s for (*R*)-(-)-MeLa and (*R*)-(-)-EtLa under mild conditions (17°C, 5.8 bar hydrogen), and an *ee* of 92% was obtained at 1.3 bar owing to the elaborations of control for low concentrations of the modifiers (DHCnd, DHQ) at the surface of the catalyst and for implementation of optimal conditions: TOF 4 s^{-1} (at 5.8 bar), turnover numbers EtPy/DHCnd > 28 000, EtPy/$Pt_{surface}$ > 5500.

But for hydrogenation of ethyl 2-oxo-4-phenylbutyrate, which is used in the preparation of pharmaceuticals, the optimal ratio of $Pt_{surface}$/DHCnd was 1 at an *ee* of 91%. It is of interest that the regularities of the heterogeneous enantioselective hydrogenation should be different in solutions and in gas-phase reactions.

Hutchings and Wells [153] reported the first gas-phase enantioselective - hydrogenation in a flow reactor, the hydrogenation of a pyruvate esters over Pt supported on *gamma*-alumina, *alpha*-alumina or silica, premodified with the solution of Cnd in dichloromethane. Over 2.5% Pt on *gamma*-alumina or on silica the hydrogenation of MePy at 40°C resulted in (*R*)-(-)-MeLa with *ee's* of 46-51%, while the corresponding values in toluene were 50-70%. Over 1% Pt on *alpha*-alumina (this support has no intraparticle pore structure) the *ee* was only 25%. Thus, these results demonstrate that enantioselective hydrogenations have different behaviors in gas-phase and in liquid-phase.

Because the enantioselective reaction produces a chiral product, which itself can serve as modifier and influence selectivity, it was of interest to investigate the effect of the degree of conversion on the optical yield. Such experiments were studied in the asymmetric hydrogenation of *alpha*-keto esters by Chernysheva et al.[253] and Nitta [134]. It was found [253] that the significant value of the relative adsorption coefficient of (-)-MHB resulted in it inhibiting the enantioselective hydrogenation of MAA (see Figure 5.9.). On the other hand, a decrease in *ee* with degree of conversion could indicate in-stability of modified centers of catalyst [169,220,307].

In the case of the Pt-alumina-Cnd-EtPy system, increasing conversion did not change the *ee* even at high conversions. This is in contrast to the Ni catalysts modified with tartaric acid and suggests a different mechanism of enantioselectivity. Also, in both cases the possibility of nonlinear effects must be pointed; that is, increasing enantio-purity of a product of low optical purity upon its exposure to a chiral catalyst (Blackmond [210], Avalos [4], Kagan [5,6]).

5.7. On the mechanism of hydrogenation

The mechanism of enantioselective hydrogenation on nickel catalysts modified with optically active compounds was suggested for the first time by Balandin's group [7-9]. They proposed that the reaction proceeded through an intermediate complex involving catalyst metal atom, (Ni), modifier molecule, (M), and substrate molecule, (S): {Ni-M-S}. Similar models were proposed later by Yasumori et al [224-226] and Izumi and Tai [213-218].

The aims of studies to elucidate the mechanism have been to account for the observed enantioselectivity and the reaction rate enhancement. The first attempts to account for the detailed mechanism of action of Pt-alumina-alkaloid-pyruvate system were made by Wells' group [141-152].

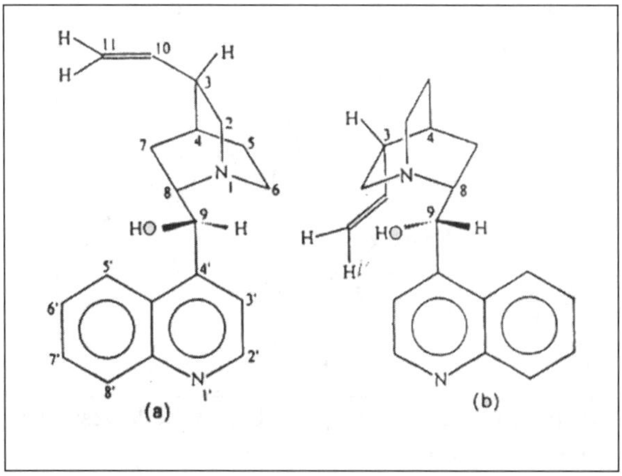

Scheme 5.20. Cinchonidine (a) and cinchonine (b) molecules.

Sutherland et al. [141] developed the 'Template model' based on a geometric explanation of enantioselectivity assuming adsorption of the cinchona alkaloid modifier and methyl pyruvate reactant on the surface of a Pt(III) facet such that the alkaloid modifier is adsorbed by its quinoline ring in such a way that there is a place for adsorption of the pyruvate molecule to form an intermediate active complex which produces, after hydrogenation of the C=O bond in pyruvate, either the (*R*)-(-)-methyl lactate, in the case of using Cnd as modifier, or the (*S*)-(+)-methyl lactate using Cn as modifier (see the modifier structures in Scheme 5.20.).

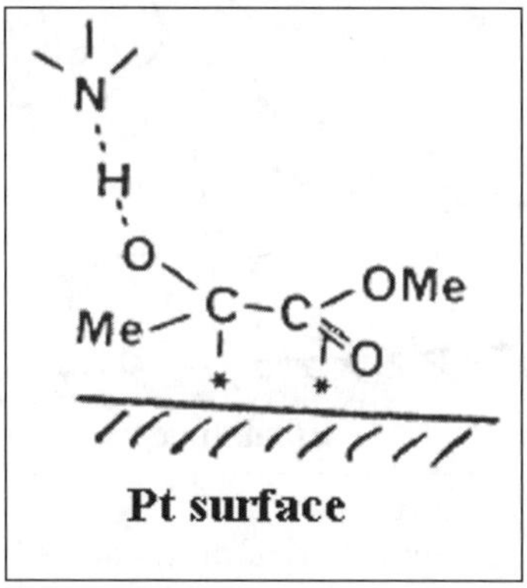

Scheme 5.21. Hydrogen bond between the *N*-base of the alkaloid and the half-hydrogenated state of the pyruvate molecule [141].

This first 'Template model' was later reexamined using LEED and improved to the '1:1 model' (Simons, Wells et al.[149]) as a simpler and better argument which shows that an H-bond is established between the N1-nitrogen of the quinuclidine bicycle and the half-hydrogenated form of pyruvate molecule (Scheme 5.21.).

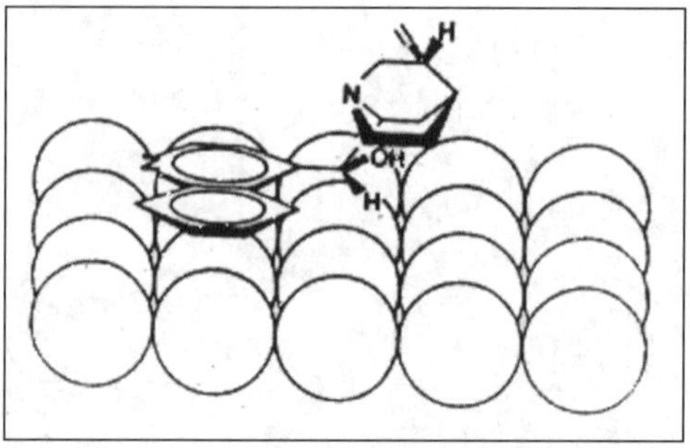

Scheme 5.22. Cinchonidine molecule adsorbed in the lowest energy conformation on the Pt surface.

Cnd adsorbs on Pt via the aromatic quinoline ring, assuming flat adsorption parallel to the Pt surface over two adjacent Pt atoms. H-D exchange between adsorbed Cnd and a deuteriated solvent showed that all of the H atoms of the quinoline moiety simultaneously exchanged, suggesting that the modifier molecules are chemisorbed on the Pt surface with the *pí*-electron system of the quinoline ring. Scheme 5.22. shows the region of the surface that the modifier molecule obscures. The quinoline ring is spatially well suited to lie above two adjacent surface Pt atoms and the quinuclidine bicycle lies just above the surface, forming an 'overhang'.

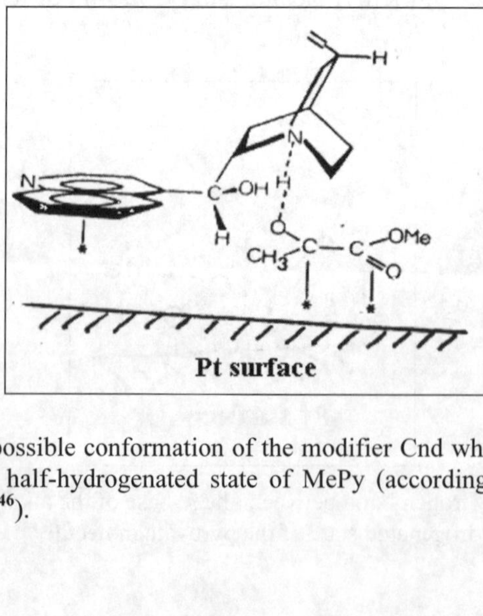

Scheme 5.23. A possible conformation of the modifier Cnd when it is H-bonded to the half-hydrogenated state of MePy (according to Bond, Wells et al.[146]).

Buergi and Baiker [115] used an *ab initio* method to calculate the interactions in the complexes between Cnd-modifier and MePy-reactant complexes that are active in the enantioselective hydrogenation over Pt-alumina catalysts. It was found that the *trans*-MePy complex, resulting in (*R*)-(-)-EtLa upon hydrogenation, is more stable by 7.5 kJ/mol than the corresponding complex leading to the (*S*)-(+)-enantiomer, which corresponds to an *ee* of 92%, in good agreement with the experimental results. Scheme 5.23. shows the schematic representation of the intermediate complex [Cnd-MePy] at the surface of a Pt catalyst.

Each adsorbed Cnd molecule utilizes an L-shaped region of the surface (Scheme 5.24.). Enantioselectivity could be attributed to the establishment of a closely packed 'adlayer' of Cnd molecules.

Scheme 5.24. shows the proposed adsorbed state of the cinchonidine molecule which provides the templated surface for enantioselective hydrogenation. With the half-hydrogenated state of methyl pyruvate stabilized by hydrogen bonding, the geometrical requirements are achieved by rotation about the C8-C9 bond in the cinchonidine molecule.

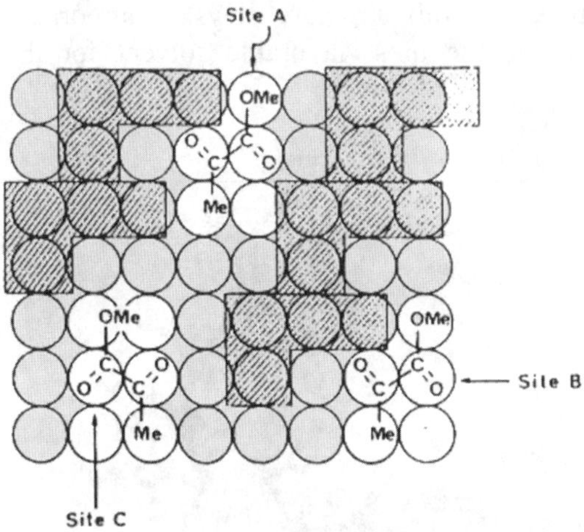

Scheme 5.24. Array of cinchonidine molecules adsorbed on Pt (100) together with two methyl pyruvate molecules.
Site A: asymmetric hydrogenation site according to the 'Template model' resulting (*R*)-(-)-MeLa.
Site B: the '1:1 interaction model' resulting in (*R*)-(-)-MeLa
Site C: racemic hydrogenation site, because there is no interaction with adsorbed cinchonidine molecules (according to Webb, Wells [105]).

Site A in Scheme 5.24. shows the regular adsorption of five Cnd molecules with their 'L-shaped forms' on a Pt area consisting of 7x7 Pt atoms accompanied by adsorption of the methyl pyruvate molecule to be hydrogenated and leading to (*R*)-(-)-methyl lactate.

Site B shows another orientation of MePy leading to (*S*)-(+)-MeLa, the 'adlayer' of Cnd molecules exert steric hindrance for adsorption of the methyl and methoxy groups and therefore make this model less probable. This was confirmed by experiments with Pt-catalyst modified with Cnd, which resulted in predominantly (*R*)-(-)-MeLa. In this reaction Cnd obscures rather more then one surface Pt atom and the adsorbed Cnd is 'L-shaped' as viewed from the liquid phase and shown on Scheme 5.24. Site C is a *racemic* hydrogenation site because there is no interaction in this case with the adsorbed Cnd molecules.

Baiker et al. [17-19,21] studied the interaction of the *N*-atom of the quinuclidine moiety using quantum chemistry techniques at both *ab initio* and semi-empirical levels as well as molecular mechanics (Schwalm, Baiker et al. [96-98]). The quinuclidine center can act as a nucleophilic center or, after protonation, as an electrophilic center. As a model of both cases the interaction of methyl pyruvate with NH_3 and NH_4^+ were studied theoretically and it was found that the {Pyruvate-NH_4^+} complex is much more stable by 10.5 kJ/mol due to favorable electrostatic interactions. This indicated that in protic solvents, e.g. in AcOH, this structure plays an important role and, indeed, AcOH proved to be the most favorable solvent for this enantioselective reaction.

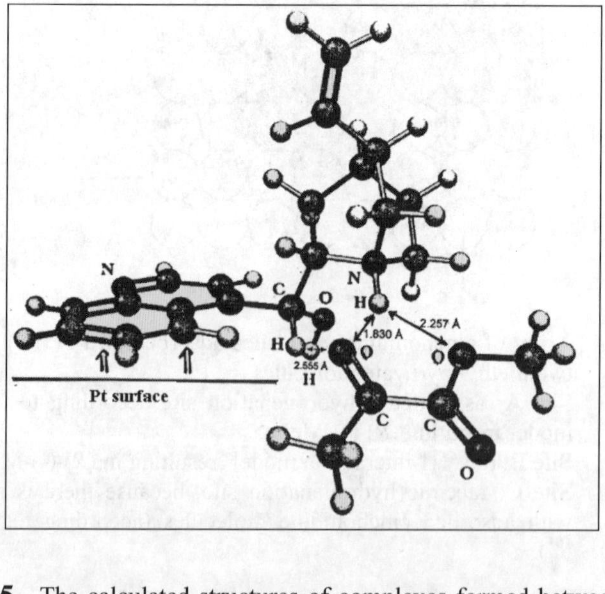

Scheme 5.25. The calculated structures of complexes formed between protonated cinchonidine and pyruvate adsorbed on a platinum surface (according to Schwalm et al. [98]).

The interactions shown in Scheme 5.25. give a reason for the almost 20-fold rate increase in enantioselective hydrogenation of pyruvate on Pt catalysts that are modified with Cnd at RT and 10 bar hydrogen, because in this catalytic system:

i) the modifier promotes the rate as an alicyclic *N*-base that stabilizes the half-hydrogenated state by H-bonds

ii) the modifier when adsorbed on the Pt surface enhances the surface coverage of hydrogen

iii) the modifier is adsorbed on the Pt surface by its quinoline aromatic part and enhances the activity of the Pt catalyst.

The next part of the specific behavior of the cinchona alkaloids as modifiers resides in their stable conformational structure. The molecules of cinchona alkaloids, e.g. cinchonidine or cinchonine (see Scheme 5.20.), consists of two relatively rigid parts: an aromatic quinoline ring and an aliphatic bicyclic quinuclidine ring, both connected through the hydroxyl-bearing chiral carbon atom, C9.

Conformational changes of the molecule, e.g. Cnd, are possible owing to rotation about the C4'-C9 and C9-C8 bonds. Calculations of the molecular mechanical energy as functions of the two dihedral angles, C3'-C4'-C9-C8 and C4'-C9-C8-N1, showed that of the six possible forms (in unprotonated and in *N*-protonated forms) there are two minimum energy conformations, a 'closed' conformation in which the N1 atom in the quinuclidine bicycle points towards the quinoline ring (the so called 'closed form') and an 'open' conformation in which the N1 atom points away from the quinoline ring (the so called 'open form') (see Scheme 5.25.).

Baiker [17-19,117,118] proposed a mechanism for the reduction of the keto-group, which explains enantioselectivity as an interaction in the hydrogen bonded complex of the alkaloid and the keto ester. According to Margitfalvi [155-162] the keto-compounds and the Cnd molecules form a complex in which one side of the keto-group is shielded by the aromatic ring of Cnd molecule, leaving the other side open for hydrogenation. But this model requires the Cnd molecule in 'closed form', in which the *N*-atom of the quinuclidine moiety points towards the quinoline aromatic rings. This model is just opposite to the 'open' form conformation in which this N-atom of the quinuclidine moiety points away from the quinoline rings. But it was found experimentally that only the 'open' forms of the alkaloids can induce enantioselectivity in the hydrogenation reaction, and the 'closed forms' can not play any important role in this process [17-19]. Baiker [118] found an $N^{+}\cdots H\cdots O$ hydrogen bond between the keto ester and Cnd using vibrational spectra. McBreen's group [319] used vibrational spectra techniques for the analysis of the interaction between a chiral modifier, 1-(1-naphthyl)ethylamine (a modifier molecule with more simple structure than an alkaloid) and the reactant, MePy. In the absence of the modifier and at low MePy concentration, the MePy molecule tends to adopt 'enediolate' geometry in which the carbonyl groups are in a *cis* conformation and the ester molecule bonds to the surface through both carbonyl oxygen atoms .

Another model proposes nucleophilic catalysis involving the *N*-atom of the Cnd and the C=O group of the pyruvate [205] . This became unpopular in favor of the hydrogen bonding model. One exception is the works of Bartok's group. They found that the DHCD/Pt_{surf} ratio necessary for achieving maximum enantioselectivity is highly solvent-dependent and the large difference may be a reflection of the difference in the reaction mechanism, and they assumed a switch from the nucleophilic mechanism in toluene to the hydrogen bonding mechanism in acetic acid [178,179].

The enantioselective hydrogenation of fluoroketones on cinchonidine modified Pt-alumina in toluene with and without trifluoroacetic acid was also studied, and it was proposed that the compounds responsible for the chiral induction were the intermediate complexes, the structures of which depended on whether hydrogenation was performed with or without TFA [320]. An unexpected change in the sense of enanatioselectivity was observed in the hydrogenation of ethyl pyruvate (EtPy) over *beta*-isocinchonine modified Pt-alumina catalysts, where (*R*)-(-)-ethyl lactate was the major product in toluene, while (*S*)-(+)-EtLt was formed in AcOH [321].

The hydrogenation of trifluoromethylcyclohexyl ketone in toluene or ethanol on Pt-alumina modified by cinchona alkaloids was studied, and depending on the solvents inversion of the enantioselectivity was observed. It was assumed that the structures of intermediates responsible for chiral induction depend mostly on the acidic or non-acidic nature of the hydrogenation medium [99,100,116,322].

In contrast with other reaction mechanism proposals that are based on weak hydrogen bonding, Vayner, Houk et al.[323] proposed that the key interaction is a strong bond between the amine, Cnd, and the carbonyl group of the ester, forming a zwitterion adduct at the platinum surface in acetic acid media [118,324].

It was noted that in the "closed' form there exists significant steric hindrance for the quinuclidine-*N* lone pair to participate in interaction with pyruvate, whereas in the 'open' conformation of the Cnd N-lone pair of quinuclidine is more readily accessible to the reactant. Scheme 5.26. shows the model of interaction of Cnd in the 'open' conformation with methyl pyruvate as calculated by Schwalm et al.[98] for the intermediate complex.
Another important aspect is the role of adsorption of the intermediate complex on the Pt active center. In the intermediate complex the pyruvate molecule is bound to the modifier via stabilizing hydrogen bond interactions, N-H···O, between the protonated quinuclidine-N atom and the O-atom of the *alpha*-carbonyl group of pyruvate, or O-H···N-bond for unprotonated system, such as for reactions in toluene solution. In this case, the H-atom can come from dissociatively adsorbed dihydrogen from the Pt-surface. The same

model is valid for the hydrogenation of ketopantolactone into (*R*)-pantolactone (Scheme 5.27.).

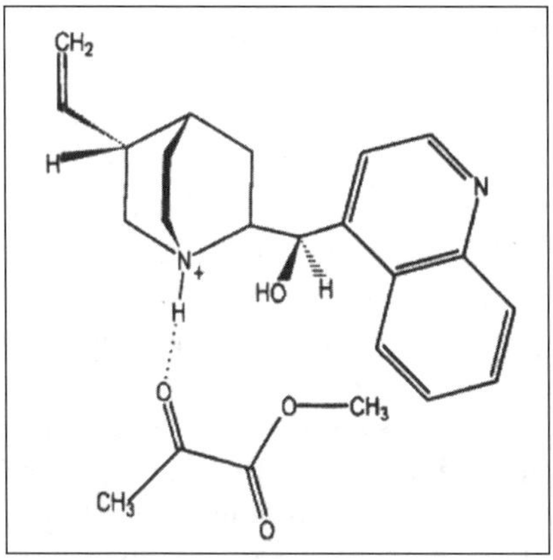

Scheme 5.26. Transition complex formed between cinchonidine and the half-hydrogenated state of methyl pyruvate resulting (*R*)-(-)-methyl lactate (according to Baiker et al. [17]).

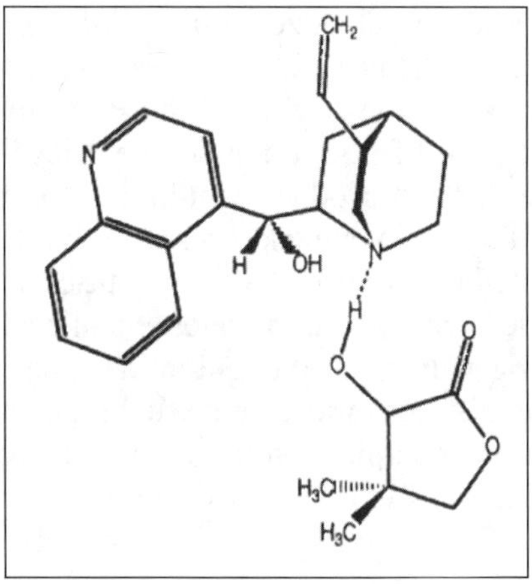

Scheme 5.27. Transition complex formed between cinchonidine and the half-hydrogenated state of ketopantolactone resulting (*R*)-pantolactone (according Schwalm et al. [98]).

Calculations for this model showed the transition state for (*R*)-pantolactone to be more stable by 8.4 kJ/mol than the corresponding complex leading to (*S*)-pantolactone. According to these calculations, it is possible to produce an *ee* of 90%, but experimentally only 79% was achieved (Schuerch, Baiker et al. [20]).

Buergi and Baiker [117] studied the influence of different solvents for the conformations of Cnd in the enantioselective hydrogenation of ketopantolactone. They claimed three conformations of Cnd at RT: 'closed (1)', 'closed (2)' and 'open (3)'. The latter structure is the most stable in polar solvents and increases with solvent polarity, as suggested by experiments on the enantioselective hydrogenation to pantolactone with a maximal *ee* of 78% in toluene solution, which is the solvent with the lowest dielectric constant. The increase in dielectric constant in the series of cyclohexane, hexane, diethyl ether and THF, decreases *ee* up to 50% and in EtOH and water even to 15%, in accordance with a decrease in the population of the 'open (3)' conformation of Cnd.

Margitfalvi et al. [160] suggested that the catalytic systems: Pt– and Pd–cinchona alkaloid–keto esters, are the first examples of a new class of heterogeneous catalytic reactions of supramolecular catalysis, where the prochiral group is part of a conjugated double bond system.

The role of the Pt surface in the formation of intermediates {Pt-surface-modifier-Cnd-reactant-pyruvate} considered in the above cited publications is not enough. According to Blaser et al.[69], an enantioselective center is formed by adsorption of the cinchona alkaloid on well defined platinum ensembles. The effect of crystallite size on the *ee* of the product was discussed earlier in Part 5.2 of this Chapter. The modification process involved only a part of the active centers on the surface of the Pt catalyst, while the absolute configuration of the C8 of the quinuclidine bicycle determined which configuration of the lactate product would be formed preferably. The decisive interaction of the pyruvate takes place with N1, but probably C9 also plays an important role in enantioselectivity because substitution of the OH at C9 for OAc or H decreases or annihilates enantioselectivity, as also was observed after hydrogenation of the quinoline ring in Cnd. As it was proposed by Blaser et al.[69], enantioselective hydrogenation proceeds through the modifier-reactant '1:1 complex' on the Pt surface, as was proposed in the case of the hydrogenation of ethyl acetoacetate on Ni-catalyst modified with (2*R*,3*R*)-tartaric acid (Izumi and Tai [213,219], Yokozeki [325], and Klabunovskii [9,220], see, also Chapter 4).

Thus, the main question of the mechanism of reaction consists of the mode of conformation of the alkaloid modifier on the surface of the catalyst. As mentioned above, there are two mechanistic concepts, which can be divided into path I and path II. According to path I, alkaloid modifier (e.g. Cnd) interacts with EtPy adsorbed on the surface of Pt by means of adsorption of the quinoline ring in the 'open' conformation according to the Scheme 5.28. as "***a***" (Augustine et al.[205,206]), "***b***" (Baiker, Blaser et al.[77]), "***c***" (Blaser et al.[74,75]) or "***f***" (Simons et al.[149]). According to path II, intermediate complex, Cnd-EtPy, is formed in solution in 'closed', "***d***" and "***e***" conformations (Margitfalvi et al.[159,160]).

The first group of conformations (***a, b, c*** and ***f***) adsorb on the Pt surface by multicenter *pí*-bonds of the quinoline ring and the *sigma, pí*-system of the pyruvate. Another group of conformations (***d*** and ***e***) suggest formation of the intermediate in solution *before* adsorption on the Pt surface.

To choose between these possibilities Bartok et al.[321,326-329] studied the mechanism of hydrogenation using derivatives of cinchonine (instead of cinchonidine used as the common effective modifier in all works earlier) and synthesized *alpha*-isocinchonine as a novel modifier. Scheme 5.29. shows the stable conformations of cinchonine in the 'open' conformation (***a***) and the 'closed' conformation (***b***) and the conformation of *alpha*-isocinchonine, α-iCn (***c***).

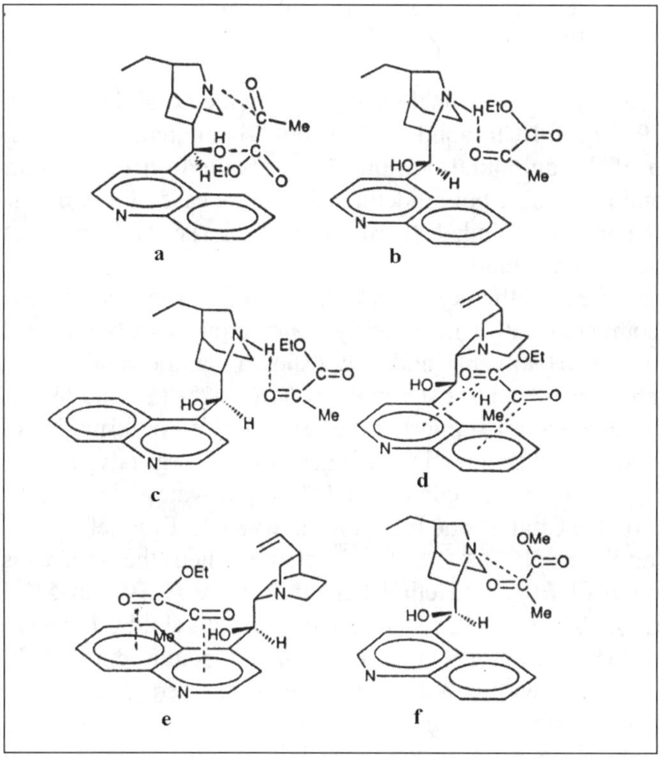

Scheme 5.28. The structures of the [cinchonidine-ethyl pyruvate] intermediate complexes (***a***, ***b***, ***c*** and ***f*** are in 'open'-conformations, while ***d*** and ***e*** are in 'closed' conformations (according to Bartok et al.[326]).

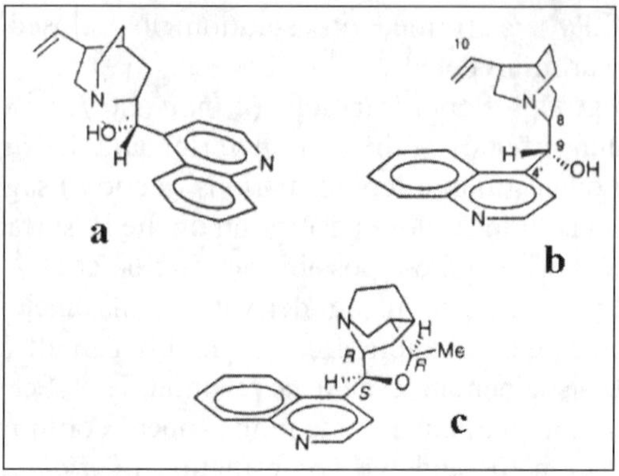

Scheme 5.29. The structure of cinchonine (Cn) in 'open' conformation (***a***), Cn in 'closed' conformation (***b***) and *alpha*-isocinchonine (***c***) (according to Bartok et al.[326]).

As can be seen, conformational changes in Cn are possible by rotation around the C4'-C9 and C8-C9 bonds. It should be noted, that in the case of adsorption of the quinoline portion of the alkaloid on Pt, rotation along the C'4-C9 bond is possible but hindered. But in the case of α-iCn, rotation along the C8-C9 bond is excluded, therefore, α-iCn exists only in the 'anti-open' steric hindered conformation.

The effect of the action of α-iCn as a modifier with fixed conformation was compared with Cn in the hydrogenation of EtPy on 5% Pt-alumina (RT, 50 bar, AcOH solvent), and it was shown that the catalyst modified with Cnd, Cn, and α-iCn lead to EtLa with 90% (*R*), 67% (*S*), and 69% (*S*), respectively. These *ee* values supported the existence of conformations (b) and (c) in the reaction and discounted the suggestion of Margitfalvi et al.[156-161] on the formation of an intermediate {alkaloid-pyruvate} in solution *before* adsorption on the catalyst and his criticism of the '1:1 model'.

Bartok et al.[169,321,326,327,329,330] also studied the mechanism of the hydrogenation of EtPy and ethyl benzoylformate (EtBf) on 5% Pt-alumina using (8*R*,9*S*,10*R*)-*alpha*-iCn and isocinchona alkaloids, namely *alpha*-isoquinidine [(8*R*,9*S*,10*R*)-α-iQnd] and *gamma*-isoquinidine [(8*R*,9*S*,10*S*)-γ-iQnd], as modifiers with new 10*R* or 10*S* chiral centers, possessing more rigid conformations (Scheme 5.30.).

Using NMR spectra combined with X-ray crystallography and molecular calculations it was shown by Bartok et al.[169] that Cnd, Cn, Qn, and Qnd must exist in their 'anti-open' conformations as modifiers. Rotations around the C4'-C9 and C8-C9 bonds are hindered. Therefore, the alkaloids exist also in the 'anti-open' conformations but they do not have the C8-C9 free rotation and form a rigid structure during adsorption on the surface of Pt-alumina catalyst.

H-D exchange studies of the isoalkaloids confirmed that adsorbed alkaloids undergo H-D exchange only inside the quinoline ring and on the C9 atom. In the quinuclidine skeleton H-D exchange appeared only in the case of Cn and iCn but not in the case of *gamma*-iQnd; therefore, good values of *ee* from the hydrogenation of EtPy were observed with the modifiers Cn (88%, *S*), Qn (90%, *R*), Qnd (89%, *S*), and with the isoalkaloids *alpha*-iCn (88%, *S*) and *alpha*-iQnd (85%, *S*) except *gamma*-iQnd (22%, *S* for EtPy and 9%, *S* for hydrogenation of EtBf). The methyl group at the C10 in *gamma*-iQnd exerts inhibition of rotation around the C9-C10 bond and the OMe group in the quinoline ring which gives an unfavorable steric hindrance in adsorption resulting in diminishing *ee* of the resulting EtLa.

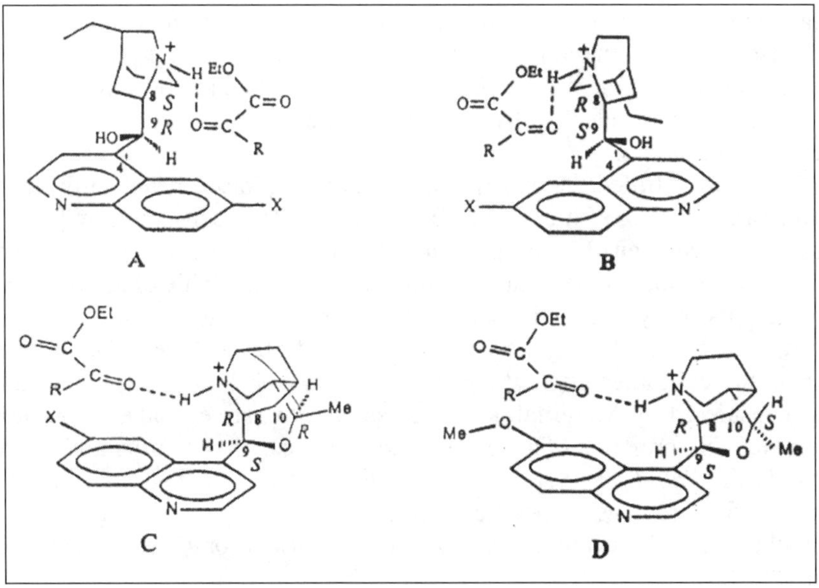

Scheme 5.30. Structures of the '1:1 cinchona alkaloid pyruvate' intermediates:
A: Cinchonidine, Cnd, (X=H) and quinine, Q, (X=OMe) in 'anti-open'
conformations.
B: Cinchonine, Cn, (X=H) and quinidine, Qnd , (X=OMe) in 'anti-open' conformation
C: *alpha*-iCn (X=H) and *alpha*-iQnd (X=OMe) in 'anti-open' complex
D: *gamma*- iQnd in 'anti-open' complex conformation
(Bartok et al.[169]).

Some important conclusions can be drawn from the experimental confirmation of the proposed 1:1 [modifier-reactant] chiral complex based on the 'anti-open' conformation of the alkaloid-modifier, which fully excludes the concept of the closed conformation in the reacting complex as well as the formation of the [modifier-pyruvate] complex in solution before reaction. Also the new C10 center in the isoalkaloids does not allow the formation of the intermediate form, so the isoalkaloids can be considered novel effective modifiers. Chiral induction is not affected by the new chiral C10 in the isoalkaloids but the Me group at C10 exerts a significant effect on enantioselectivity diminishing the steric adsorption on catalyst surface.

Thus the '1:1 model' proposed earlier by Wells [149] and Baiker [17-19,21] has gained experimental proof using the isoalkaloids as modifiers. Margitfalvi et al.[156,159,162] criticized the 'Template model' because of some experimental uncertainties and put forward a new mechanism based on the "Shielding effect" suggesting that enantioselectivity proceeds through the formation of a weak complex {Cnd-MePy}, not on the surface of the heterogeneous catalyst, but in solution in the liquid phase, such that the modifier, cinchonidine, shields one side of the prochiral site of the reactant and allows the other side to undergo enantioselective hydrogenation after its adsorption on Pt-catalyst (see, Margitfalvi et al. [155-157]).

They fulfilled the experiments and calculations for the confirmations of this aspect of the 'shielding effect' in the system consisting of [Cnd (in closed conformation)-Methyl pyruvate] leading to the (*R*)-(-) methyl lactate. Also they considered these data as confirmation of the 'Shielding effect' from experiments of hydrogenation of the complex in the 'closed form' of methyl pyruvate with chiral Troger's base at 10^0C and 50 bar hydrogen. The molecular model of interaction of *alpha*-keto esters over Pt-alumina-Cnd catalyst was considered by Margitfalvi et al.[331] as an alternative model. The model suggested the formation of a weak complex between the modifier and the substrate in the *liquid phase rather than on the surface of the catalyst*, in which the modifier provides a specific shielding effect, owing to the fact that the alpha-keto ester can interact with the metal surface only on its unshielded side.

If the reactivity of the substrate in the shielded [substrate-modifier] complexes is higher than that of the free keto ester, enantioselectivity can occur. According to Margitfalvi [157-162] the quinuclidine *N*, which aims towards the keto carbonyl group in the keto ester, provides the increased reactivity of the keto carbonyl group. This shielding model, can explain enantioselectivity and rate acceleration effects for almost all substrates used in many examples, for example, hydrogenation of MePy, MBf, ketopantolactone, and trifluoroacetophenone over Pt-alumina-Cnd catalysts.

On the other hand, Tungler et al. [332] considered the selective characteristics of heterogeneous chiral modified catalysts in the enantioselective and diastereselective hydrogenation of C=C, C=O, and C=N double bonds. Because of the absence of interaction between these groups and the tertiary *N*-atom of the Troger's base (Scheme 5.31.), rate acceleration in hydrogenation could not be observed and resulted in an *ee* of only 38.3%.

From this fact the conclusion was drawn that the enantioselective step proceeded in the liquid phase with the use of the Troger's base as the chiral matrix in the 'closed form' of the alkaloid molecule interacting with the reactant molecule followed by hydrogenation of this complex on the Pt surface.

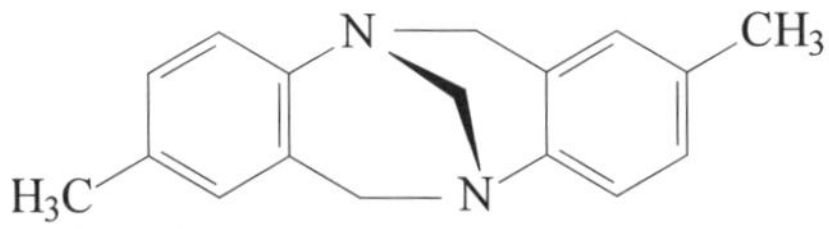

Scheme 5.31. Troger's base.

But according to Margitfalvi [156] the formation of the 'Troger's base-methyl pyruvate' intermediate, with its following hydrogenation on Pt just can be considered as an indication against his suggestion of a 'liquid phase complex formation'. Rather, it is indicative of formation in solution of a di-astereomeric compound like the reaction of isophorone with proline studied by Tungler et al.[260] and leading to a modest enantioselective effect. Thus only participation of a heterogeneous catalyst with subsequent formation of stereospecific adsorption on the metal surface can conditions be created for high asymmetric effects.

But as Johnston and Wells [144] noted, this model, which is based on the 'shielding effect', could not be considered true because that complex {Cnd-Pyruvate} can be achieved not only in the liquid phase but also in the gas-phase where the interaction of reactant and modifier before adsorption can not possibly occur. Moreover it is possible to conduct reaction in the liquid phase on a catalyst surface and obtain high enantioselective by such minute quantities of Cnd even though its concentration in solution is extremely low. On the other hand, results received by Bartok [169,321,326] indicated that the reaction can proceed through formation of a complex of pyruvate with Cnd only in the 'open' form adsorbed on the Pt surface owing to bonding to the surface of the quinoline aromatic ring.

The aspects of the mechanism of action of the catalytic system Ptalumina - Cnd-MePy discussed by Margitfalvi et al. [159,162] seemed to be reasonable but they can be applied equally well to processes proceeding on the surface of the catalyst rather than only in the solution above the catalyst.

Additional crucial evidence against a mechanism based on such associations formed in solution is supported by the fact that under these conditions on Pt-alumina-Cnd catalysts C=C prochiral bonds, which are unable to form associations like {Cnd-MePy}, can be enantioselectively hydrogenated with good *ee*'s.

Indeed, Lipkin [29] and Nakamura [30] noted for the first time that the Pt-alkaloid system can be used for the hydrogenation of C=C bonds in methyl-

and naphthylcinnamic acids. Attempts to hydrogenate asymmetrically the C=C bonds in olefinic acids were described also by Schwab et al. [1] and by Terent'ev et al. [2] using tiglic acid and *alpha*-phenylcinnamic acid over Pt and Ni-catalysts supported on optically active quartz crystals but resulting in very small *ee's*, about 0.1%. Klabunovskii and Patrikeev [12,221,223] considered the mechanism of this reaction and suggested that small chiral effects were due to weak asymmetric adsorption of reactant on the surface of optically active quartz crystals (see, Chapter 2).

Therefore, for a long time the modified Pt and Ni metal catalysts were considered effective only for the enantioselective hydrogenation of the carbonyl C=O bond in 2- and 3-oxocarboxylic esters with the particular mechanisms of reactions much different from the others. In 1958 Isoda et al.[333] tried to hydrogenate enantioselectively the azlactone derivative of 2-acetamidocinnamic acid to (*S*)-phenylalanine and described striking results of high efficacy of the action of Pd catalysts modified with (*S*)-tyrosine. The catalyst was prepared by hydrogen treatment of Pd-(*S*)-tyrosine complex at high hydrogen pressure and high temperature and could hydrogenate the azlactone into the amino acid with an *ee* of 65%. Unfortunately, these high results were not confirmed later in the literature.

Then it was shown that catalysts modified with Cnd were effective in the hydrogenation of tiglic acid ((*E*)-2-methylbut-2-enoic) acid on Pd-silica with an *ee* of 22% (Bartok et al.[172]). The detailed studies of Nitta et al. [120,124,125] have shown that enantioselective hydrogenations of (*E*)-2-phenylcinnamic acid into (*S*)-(+)-2,3-diphenylpropionic acid can proceed very effectively with an *ee* above 70% on a catalyst of 5% Pt-titania in the presence of Cnd in polar solutions, especially in the dispersion range below 0.2. In the solvents with low polarities, the optimal Pd dispersions are much lower than those observed in polar and protic solvents. The catalysts with the lowest dispersions, however, exhibit poor selectivities, irrespective of the solvent employed. In the hydrogenation of an (*E*)-2-methylpent-2-enoate and (*E*)-2-methylbut-2-enoic acid, the dispersion dependencies are much milder and independent of the solvent [130].

Baiker et al. [86,107] also hydrogenated 2-methylpent-2-enoic acid on Pt-alumina-Cnd at 20^{0}C and 50 bar with an *ee* of 52%.

As it have been shown previously (see, Chapter 4) Ni catalysts modified with chiral acids revealed very little effectiveness in enantioselective hydrogenation of C=O bonds in keto esters and C=C bonds in unfunctionalized substrates (see Izumi, Tai [213-215]).

The MePy molecule mostly exists in an anti-carbonyl conformation and cannot interact with modifier-tartaric acid on Ni catalysts through two hydrogen bonds. Therefore on RNi modified with (2*R*,3*R*)-tartaric acid, hydrogenation of MePy leads also to a poor *ee*: 2% (*S*) (rate 0.07

mmol/g·min), whereas MAA hydrogenates with an *ee* of 86% at a rate of 0.48 mmol/g·min (Tai et al.[218], Okamoto et al.[219]).

But using special techniques and reaction variables it was possible to improve the effectiveness of modified Ni catalysts, e.g. using ultrasonic treatment (Tai et al.[216] and Torok et al [170]), and to receive very high optical yields from the hydrogenation of *alpha*-keto carboxylic acids with *ee*'s of 94%. Other examples of the ultrasonic treatments exist [334-341].

Smith et al. [203] succeeded in obtaining impressive data from the hydrogenation of prochiral C=C bonds in carboxylic acids on modified Ni catalyst. They suggested that the failure of C=C containing compounds to display significant enantioselective effect in former works can be explained by the absence of an appropriate binding site to form a complex with the catalysts and with chiral groups of modifiers like intermediate {Catalyst-Modifier-Substrate} in accordance with a scheme suggested by Klabunovskii et al.[3,9,220] for the hydrogenation of ethyl acetoacetate on a number of metal catalysts (Ni, Cu, Ru, Pd, and LaNi) modified with (2*R*,3*R*)-tartaric acid.

Later, using RNi modified with (2*R*,3*R*)-tartaric acid and NaBr Bartok et al.[172] succeeded in the enantioselective hydrogenation of Na-salts of (*E*)- and (*Z*)-2-phenylcinnamic acid, resulting in the (*E*)-substrate producing the 2,3-diphenylpropanoic acid with an *ee* of 17.0% whereas the (*Z*)-and (*E*)-isomers of the free acid resulted in an *ee* of only 0.47% and 0.2%, respectively. The explanation of such differences of behaviors of (*E*)- and (*Z*)-acids consists in their difference of stereochemical and electronic properties. In the (*E*)-acid *the beta*-phenyl group and the -CO_2H group are in conjugative interaction with the C=C bond and are in a coplanar arrangement

The use of Pd catalysts with Cnd as modifier proved to be more effective. Following the first works [29,30], Perez et al.[119] found that hydrogenation of (*E*)- and (*Z*)-isomers of 2- and 3-methylcinnamic acids and 2-phenylcinnamic acid over 5% Pd-C in the presence of cinchona alkaloids in polar solutions was rather effective. It was found that hydrogenation of (*E*)-2-methylcinnamic acid is less effective (*ee* 3.3% *R*) whereas (*E*)-2-phenylcinnamic acid gave an *ee* of 13.3% *S* in the presence of an equivalent amount of substrate and quinine as modifier. But in the case of a small addition of 3% Cnd, the hydrogenation of (*E*)-2-phenylcinnamic acid resulted to the best *ee* of 30.5% *S*.

Nitta et al.[120-128] studied the effect of variables of hydrogenation of substituted cinnamic acids and found striking results. The best *ee* approached 80% [120,126]. It was found that titania was the best support for 5% Pd in the hydrogenation of (*E*)-2-phenylcinnamic acid in the presence of a small amounts of Cnd.

The use of solvents with larger dielectric constants result in hydrogenation over 5% Pd-alumina of higher *ee*'s, up to 53% [121]. All these features are quite different from properties of the Pt-alumina-cinchonidine system in the hydrogenation of the pyruvates and indicate different mechanisms of these two reactions.

It was found that reactions with *unmodified* catalyst proved to be much faster than those with Cnd modified catalyst, which also is quite different from the results reported for Pt-alumina-Cnd catalysts in the hydrogenation of *alpha*-keto esters. Reuse of catalysts resulted in almost complete loss of *ee* and indicates elution and absence of Cnd from the surface of the catalyst.

Pd-species containing the chiral modifier Cnd are precipitated during preparation of the catalyst only on the exterior surface of the support. Therefore nonporous and ultrafine particles seem to be more suitable as the support materials for this process. This was the case for titania as a support. The same results were found in the hydrogenation of *alpha*-keto esters on Pt-alumina-Cnd catalysts [69]. The best results were obtained using alumina support with relatively small surface area, high pore volume, and rather large pores. Thus the best *ee* obtained in the hydrogenation of the C=C bond in 2-phenylcinnamic acids on 5% Pd-TiO_2-Cnd catalyst was 72% in polar solution (DMF + 10% water) at 10^0C and 1 bar at 100% conversion. In the beginning of the reaction at 20% conversion the *ee* reached 80%.

The effect of modifier structure was studied in the hydrogenation of EtPy in MeOH and in (DMF + water) [126]. Heinz et al.[105] showed that a Pt-alumina catalyst modified with (*S*)-(-)-1-(1-naphthyl)ethylamine gave an *ee* of 82% in the hydrogenation of EtPy, whereas modification with Cnd under the same conditions gave only *ee*'s of 73-75%; but in the hydrogenation of the C=C bond in cinnamic acids *ee* values reached only 4-12% [123].

The best modifier for the hydrogenation of substituted cinnamic acids proved to be DHCnd (*ee*, 61% *S*, in DMF + water mixture). Most cinchona alkaloids with the same absolute configuration similar to Cnd also gave the (*S*)-products. It was interesting that other cinchona alkaloids, like quinine (*ee* 10% *S*), norcinchol (*ee* 50% *S*), and *N*-benzilcinchonidinium chloride (*ee* 57% *S*), proved to be less effective. Cinchonine also proved to be much less effective as a modifier and resulted in a product with 28% *R*. These data suggest that the absolute configuration at C8 and C9 in the alkaloid molecule determine the dihydroacid configuration as in the case of hydrogenation of *alpha*-keto esters on Pt-alumina-Cnd catalysts. As was found by Beamer et al.[50,51], the hydrogenation of 2-methylcinnamic acid on Pd supported on silica formed in the presence of the alkaloids Cnd, Cn, Qn, and Qnd, resulted only in the (+)-dihydroacid, indicating the crucial role of the C3 and C4 centers of the cinchona alkaloid molecules in chiral recognition, whereas the C8 and C9 centers in the above mentioned diasteromeric alkaloids affect only the numerical values of the *ee*. Cnd and Qn proved to be more active than Cn and Qnd. Thus the mechanism of asymmetric hydrogenation of the C=C bond is significantly different from hydrogenation of the C=O bond in oxo-carboxylic acids. The data obtained suggested that the mechanism of interaction of phenylcinnamic acid with modifier Cnd shows differences between both mechanisms of hydrogenation of C=O and C=C bonds.

In the system {Cnd-2-phenylcinnamic acid} the alkaloid interacts with the acid *via* two hydrogen bonds [126]. Interaction of the C9-OH group is the crucial influence in the chiral induction. Substitution of OH for OMe will weaken the interaction between modifier and the substrate. A second hydrogen bond between the N-atom in the quiniclidine group and the carboxylic group was revealed by experiments of modification of Pd-titania catalyst with a mixture of (Cnd + 9-MeO-DHCnd). Therefore esterification of the carboxylic group resulted in complete loss of enantioselectivity.

Smith et al [32,33] reported that the hydrogenation of 2-methylcinnamic acid and 2-methylpent-2-enoic acid on 1% Pd-silica catalysts modified with 1-(*S*)-*endo*-bornyloxytrimethylsilane in MeOH at 25^{0}C results in chiral products with *ee*'s of 22.5% and 11.6%. These results noted that modified Pt and Pd catalysts can hydrogenate enantioselectively systems with 1,3-conjugated bonds, but the transfer of these aspects for hydrogenation of 1,4-double bonds as in methyl acetoacetate (MAA) has definite difficulties because it needs to assume the enol form (En) to react (Scheme 5.32.).

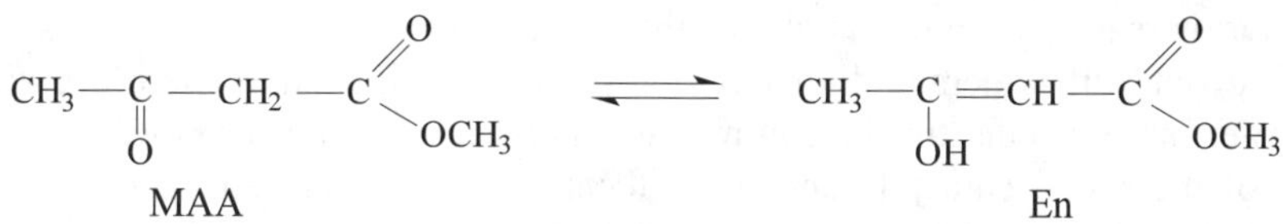

Scheme 5.32.

Although this enol form (En) is confirmed by MNDO calculations (Yokozeki et al.[325]), the reaction of the keto-form (MAA) can not be excluded and, moreover, the formation of the enol (En) as reacting species in the hydrogenation of methyl acetoacetate is not confirmed experimentally.

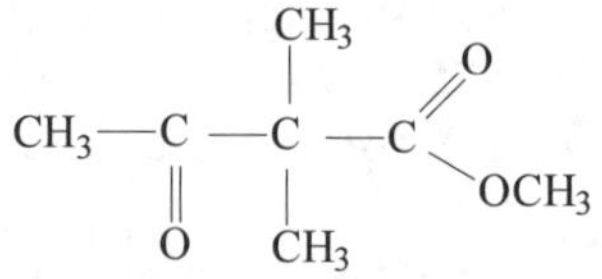

Scheme 5.33.

First, the correlation of *ee* of the product with the concentration of enol (En) in the original 3-oxocarboxylic ester (MAA) is absent; in different solvents the enol content is much different; second, it was shown that methyl 2,2-dimethyl-3-oxobutyrate (DMB) (Scheme 5.33.), which cannot form an enol, also can be hydrogenated asymmetrically with an *ee* of 30% on catalyst Ni-Tart (see Smith [203]), while the hydrogenation of methyl acetoacetate under the same conditions results in methyl 2-hydroxybutyrate with an *ee* of 38%.

CONCLUSION

Enantioselective heterogeneous hydrogenations in the liquid phase of *alpha*-keto esters on Pt-alumina catalysts modified with cinchona alkaloids as chiral ligands known as the "Orito reaction", have become common. Over 130 publications considering this system have appeared since 1989, and this reaction is continued to be studied intensively by many groups, such as Baiker and Blaser (Switzerland), Wells, Hutchings and Blackmond (England), Nitta (Japan), Sheldon (Netherlands), Tungler, Bartok and Margitfalvi (Hungary), Reschetilowski (Germany), Smith and Augustine (USA) and Zhou (China). A lot of experimental and theoretical work has dealt with different catalytic systems. Based on studies of the mechanism of reaction, new types of modifiers were elaborated. Optical yields of the resulting hydroxy ester products reached above 95%. It was stated that this reaction is structure sensitive. Enantioselectivities increase with increase of the crystalline size of the supported or bulk platinum catalysts.

Owing to the last experimental and theoretical works in the study of the Orito reaction some insight in the mechanism of the reaction has been gained but full understanding has not yet been achieved. Therefore much more effort is needed in the future, especially in studying the role of the nature of the metal catalysts and the effects of triple intermediate complexes, [Metal-Modifier-Reactant], in enantioselectivity as shown in the case of the enantioselective hydrogenation of *alpha*-keto esters on tartaric acid modified Ni, Cu, and Co catalysts [7].

References

1 Schwab, G.M., Rudolph, L. (1932) Katalytische Spaltung von Racematen durch Rechts und Links Quarz, *Naturwissenschaften,* **20**, 363-364.

2 Terent'ev, A.P., Klabunovskii, E.I., Patrikeev, V.V. (1950) Asymmetric synthesis by means of catalysts supported on right and left quartz, *Dokl. Akad. Nauk SSSR*, **74**, 947-950, *Chem. Abstr.* **45**, 3798c (1951).

3 Klabunovskii, E.I. (1968) *Stereospecific catalysis (in Russian),* Nauka, Moscow, 368 pp.

4 Avalos, M., Babiano, R., Cintas, P., Jimenes, J.L., Palacios, C. (1997) Non-linear stereochemical effects in asymmetric reactions, *Tetrahedron Asym.* **8**, 2997-3017.

5 Girard, C., Kagan, H.B. (1998) Nonlinear effects in asymmetric synthesis and stereoselective reactions: ten years of investigation, *Angew. Chem. Int. Ed. Engl.* **37**, 2922-2959.

[6] Kagan, H.B., Girard, C., Gillaneux, D., Rainford, D., Samuel, O., Zhang, S.Y., Zhao, S.H. (1996) Nonlinear effects in asymmetric catalysis: some recent aspects, *Acta Chem. Scand.* **50**, 345-352.

[7] Fridman, Ya.D., Klabunovskii, E.I. (1980) Dependence of enantioselectivity of metallic catalysts modified with complex-forming agents on stability constants of complex compounds, *Kinetika i Kataliz* (*Kinetics and Catalysis*) **21**, 1199-1207, *Chem. Abstr.* **94**, 21057e (1981).

[8] Petrov, Yu.I., Klabunovskii, E.I., Balandin, A.A. (1967) Asymmetric hydrogenation over skeletal nickel, modified with optically active compounds, *Kinetika i kataliz (Kinetics and Catalysis)* **8**, 814–820, *Chem. Abstr.* **68** , 58786f (1968).

[9] Klabunovskii, E.I., Petrov, Yu.I. (1967) On mechanism of asymmetric hydrogenation, *Dokl. Acad. Nauk SSSR.* **173**, 1125–1128, *Chem. Abstr.* **67**, 81620k (1967).

[10] Fish, M.J., Ollis, D.J. (1978) Heterogeneous catalysis by optically selective surfaces, *Catal. Rev.-Sci. Eng.* **18**, 259-295.

[11] Fish, M.J., Ollis, D.J. (1977) Characterization of enantioselective hydrogenation catalysts: transient electrochemical oxidation of D-(+)-tartaric acid on Nickel, *J. Catal.* **50**, 353-363.

[12] Klabunovskii, E.I., Patrikeev, V.V. (1951) On mechanism of asymmetrizing action of metal catalysts deposited on right and left quartz, *Dokl. Akad. Nauk SSSR*, **78**, 485-487, *Chem. Abstr.* **45**, 7860a (1951).

[13] Nakagawa, S., Sugimura, T., Tai, A. (1997) Almost perfect enantiodifferentiating hydrogenation of methyl 3-cyclopropyl-3-oxopropionate over tartaric acid modified Raney Nickel catalyst, *Chem. Lett.* 859-860.

[14] Jannes, G., Dubois, V. (eds.) (1995) *Chiral Reactions in Heterogeneous Catalysis*, Plenum Press, NewYork.

[15] Noyori, R. (1994) *Asymmetric catalysis in organic synthesis*, Wiley, New York.

[16] Brunner H., Zettlemeyer, A. (1993) *Handbook of enantioselective catalysis with transition metal complexes,* VCH, Weinheim.

[17] Baiker, A. (1977) Progress in asymmetric heterogeneous catalysis: design of novel chirally modified Pt metal catalysts, *J. Mol. Catal. A. Chem.* **115**, 473-493.

[18] Baiker, A. (2000) Transition state analogues-a guide for the rational design of enantioselective heterogeneous hydrogenation catalysts, *J. Mol. Catal. A. Chem.* **163**, 203-220.

[19] Baiker, A (2000) Design of New Chiral Modifiers for Heterogeneous Enantioselective Hydrogenation: A Combined Experimental and Theoretical Approach, in *Chiral catalyst immobilization and recycling*, De Vos, D.E., Vankelecom, I.F.J., Jacobs, P.A. (eds.) Wiley-VCH, Weinheim, pp. 155-171.

[20] Schuerch, M., Schwalm, O., Mallat, T., Weber, J., Baiker, A. (1997) Enantioselective hydrogenation of ketopantolactone, *J. Catal.* **169**, 275-286.

[21] Baiker, A. (1993) Enantioselective hydrogenation on solid catalysts, in *DGMK-Conference on Selective Hydrogenation and Dehydrogenation*, Kassel, *DGMK Tagungsberichte,* **9305**, 119-129.

[22] Blaser, H.U. (1992) The chiral pool as a source of enantioselective catalysts and auxilliaries, *Chem. Rev.* **92**, 935-952.

[23] Webb, G., Wells, P.B. (1992) Asymmetric hydrogenation, *Catal. Today* **12**, 319–337.

[24] Wells, P.B., Wells, R.P.K. (2000) Enantioselective hydrogenation catalyzed by platinum group metals modified by natural alkaloids, in *Chiral catalyst immobilization and recycling,* De Vos, D.E., Vankelecom, I.F.J., Jacobs, P.A. (eds.) Wiley-VCH, Weinheim, pp. 123-154.

[25] Klabunovskii, E.I. (1991) On some regularities in the action of enantioselective hydrogenation catalysts (Dedicated to the memory of professor Piero Pino), *J. Organomet. Chem.* **417**, 181-191.

[26] Blaser, H.U. (1991) Enantioselective synthesis using chiral heterogeneous catalysts, *Tetrahedron Asymm.* **2**, 843 -866.

[27] Blaser, H.U., Mueller, M. (1991) Enantioselective catalysis by chiral solids: approaches and results, *Stud. Surf. Sci. Catal.* **59**, 73-92.

[28] Blaser, H.U., Jalett, H.P., Mueller, M., Studer, M. (1997) Enantioselective hydrogenation of alpha-ketoesters using cinchona modified platinum catalysts and related systems: A review, *Catal. Today*, **37**, 441-463.

[29] Lipkin, D., Stewart, T.D. (1939) The preparation and asymmetric reduction of *beta*-methyl cinnamic acid, *J. Amer. Chem. Soc.* **61**, 3295-3296.

[30] Nakamura, Y. (1940) Formation of optically active amines by the reduction of oximes in the presence of optically active organic acids. Asymmetric synthesis, III., *J. Chem. Soc. Jpn. (Pure Chem. Sect.)* **61**, 1051-1054.

[31] Tjandra, S., Smith, G.V., Ostgard, D., Musoiu, M., Wiltowski, T., Notheisz, F., Bartok, M., Stoch, J. (1992) Silicon modification of Pd and Pt catalysts, *Chemical Industries Series* (*Catalysis of Organic Reactions*), (Dekker), **47,** 137-141.

[32] Smith, G.V., Tjandra, S., Musoiu, M., Wiltowski, T., Notheisz, F., Bartok, M., Hannus, I., Ostgard, D., Malhotra,V. (1996) Modified activities and selectivities of silated-oxidized-reduced Pd and Pt catalysts, *J. Catal*. **161**, 441-450.

[33] Smith, G.V., Song, R. (1997) New enantioselective heterogeneous catalysts, *Catal. Lett.* **45**, 73-78.

[34] Bayston, D.J., Polywka, M.E.C. (2000) Catalytic hydrogenation, hydroformilation, in *Chiral catalyst immobilization and recycling,* De Vos, D.E., Vankelecom, I.F.J., Jacobs, P.A. (eds.) Wiley-VCH, Weinheim, pp. 211-234.

35 Pugin, B., Blaser, H.U. (1999) Catalyst immobilization: solid supports, in *Comprehensive Asymmetric Catalysis,* Jacobsen, E.N., Pfaltz, A., Yamamoto, H. (eds.), Springer, pp. 1367-1386.

36 Raja, R., Thomas, J.M., Jones, M.D., Johnson, B.F.G., Vaughan, D.E.W. (2003) Constraining asymmetric organometallic catalysts within mesoporous supports boosts their enantioselectivity, *J. Amer. Chem. Soc.* **125**, 14982–14983.

37 Sento, T., Shimazu, S., Ichikuni, N., Uematsu, T. (1999) Asymmetric hydrogenation of itaconates by hectorite-intercalated Rh-DIOP complex, *J. Mol. Catal. A. Chem.* **137**, 263-267.

38 Balandin, A.A. (1969) Modern state of the multiplet theory of heterogeneous catalysis, *Advances in Catalysis*, Acad. Press, New York, London, V.19, pp. 1-210.

39 Sento, T., Shimazu, S., Ichikuni, N., Uematsu, T. (1998) New clay-supported chiral Rh-complexes: interlayer modification with structural tuning guests and asymmetric hydrogenation, *Chem. Lett.* 1191-1192.

40 Malmstrom, T., Andersson, C. (1999) Enantioselective hydrogenation in water catalysed by rhodium phosphine complexes bound to polyacrylic acid, *J. Mol. Catal. A. Chem.* **139**, 259-270.

41 Herault,D., Saluzzo,C., Duval,R., Lamaire,M. (2002) Enantiopure beds: a tool for asymmetric heterogeneous catalysis, *J. Mol. Catal. A. Chem.* **182-183**, 249-256.

42 Mastrorilli,P., Rizzutti,A., Romanazzi,G., Suranna,G.P., Gobetto,R., Nobile,C.F. (2002) Synthesis and catalytic activity of new supported Rhodium(1) complexes for the enantioselective hydrogenation of methyl (*Z*)-acetamidocinnamate, *J. Mol. Catal. A. Chem.* **180**, 177-185.

43 Pavlov,V.A., Spitsyna,N.I., Klabunovskii,E.I. (1983) Enantioselective hydrogenation in media of cholesteryltridecanoate as chiral liquid crystal matrix, *Izv. AN SSSR Ser. Khim.* 1653-1656, *Chem. Abstr.* **99**, 194048j (1983).

44 Dwars,T., Haberland,J., Grassert,I., Oehme,G., Krag,U. (2001) Asymmetric hydrogenation in a membrane reactor: recycling of the chiral catalyst by using a retainable micellar system, *J. Mol. Catal. A. Chem.* **168**, 81-86.

45 Yin, M.Y., Yuan,G.L., Huang, M.Y., Jiang, Y.Y. (1999) Catalytic behavior of a wool-Pd complex in asymmetric hydrogenation of diacetone alcohol and 3-methyl-2-butanone, *J. Mol. Catal. A. Chem.* **147**, 89-92.

46 Yuan, G.L., Yin, M.Y., Jiang, T.T., Huang, M.Y., Jiang, Y.Y. (2000) Catalytic behaviors of silica-supported chitin-Platinum complex for asymmetric hydrogenation of alpha-phenylethanol, *J. Mol. Catal. A. Chem.* **159**, 45-50.

47 Studer, M., Blaser, H.U. (2001 Note to Editor. The enantioselective hydrogenation of 1-phenylethanol and of ketones using Pt and Pd supported on natural polymers: work of the group of Ying-Yang Jiang, Institute of

Chemistry, Chinese Academy of Sciences, China, *J. Mol. Catal. A. Chem.* **172**, 277-278.

48 Studer, M., Blaser, H.U. (2002) Note to Editor. The enantioselective hydrogenation of 1-phenylethanol and of ketones using Pt and Pd supported on natural polymers: work of the group of Ying-Yang Jiang, Institute of Chemistry, Chinese Academy of Sciences, China (vol 177, pg 307, 2002) *J. Mol. Catal. A. Chem.* **180**, 299-299.

49 Yin, M.Y., Yuan, G.L., Wu, Y.Q., Huang, M.Y., Jiang, Y.Y. (1999) Asymmetric hydrogenation of ketones catalysed by a silica-supported chitosan-Palladium complex, *J. Mol. Catal. A. Chem.* **147**, 93-98.

50 Padgett, R.E., Beamer, R.L. (1964) Stereospecific hydrogenation using palladium-on-silica gel catalysts, *J. Pharm. Sci.* **53**, 689-690.

51 Beamer, R.L., Fickling, C.S., Ewing, J.H. (1967) Stereospecific hydrogenation using Pd-on-poly-L-leucine, *J. Pharm. Sci.* **56**, 1029-1032.

52 Orito, Y., Niwa, S., Imai, S. (1976) Asymmetric hydrogenation of methyl acetoacetate using Nickel-Platinum-Kieselguhr catalysts modified with tartaric acid, *J. Synth. Org. Chem. Jpn.* (*Yuki Gosei Kagaku Kiokaishi*) **34**, 236-239.

53 Orito, Y., Niwa, S., Imai, S. (1976) Asymmetric hydrogenation of methyl acetoacetate using Nickel-Palladium-Kieselguhr catalysts modified with tartaric acid, *J. Synth. Org. Chem. Jpn.* (*Yuki Gosei Kagaku Kiokaishi*) **34**, 672-674.

54 Orito, Y., Niwa, S., Iwai, S. (1977) Agency of Industrial Science and Technology, Asymmetric catalyst, *Jpn. Kokai*, 77.05.692, *Chem. Abstr.* 86, P 146521t.

55 Orito, Y., Niwa, S., Imai, S. (1977) Effect of additives on the asymmetric hydrogenation of methyl acetoacetate using Nickel-Palladium-Kieselguhr catalysts modified with tartaric acid, *J. Synth. Org. Chem. Jpn.* (*Yuki Gosei Kagaku Kiokaishi*) **35**, 672-674.

56 Orito, Y., Niwa, S., Imai, S. (1977) Highly selective asymmetric hydrogenation of methyl acetoacetate with Nickel-Palladium-Kieselguhr catalyst modified with tartaric acid, *J. Synth. Org. Chem. Jpn.* (*Yuki Gosei Kagaku Kiokaishi*) **35**, 753-754.

57 Orito, Y., Imai, S., Niwa, S., Nguyen, G.H. (1979) Asymmetric hydrogenation of methyl benzoylformate using platinum-carbon catalysts modified with cinchonidine, *J. Synth. Org. Chem. Jpn.* (*Yuki Gosei Kagaku Kiokaishi*) **37**, 173-174.

58 Niwa, S., Imai, S., Orito, Y. (1982) Asymmetric hydrogenation of ethyl benzoylformate using the platinum-carbon catalyst modified with cinchonidine-influence of preparation of the catalyst on the asymmetric selectivity, *J. Chem. Soc. Jpn.* (*Nippon Kagaku Kaishi*) N1, 137-138.

59 Orito, Y., Imai, S., Niwa, S. (1979) Asymmetric hydrogenation of methyl pyruvate using Pt-C catalyst modified with cinchonidine, *J. Chem. Soc.*

Jpn. (*Nippon Kagaku Kaishi*) N8, 1118-1120, see also *Collected Papers of the 43*[rd] *Catalyst Forum, Japan,* 1978, p. 30.

[60] Niwa, S., Imai, S., Orito, Y. (1980) Asymmetric hydrogenation of alpha-keto esters using platinum-alumina catalyst modified with cinchona alkaloid, *J. Chem. Soc. Jpn.* (*Nippon Kagaku Kaishi*) N4, 670-672.

[61] Orito, Y., Imai, S., Niwa, S., Nguyen, G.H. (1980) Asymmetric hydrogenation of methyl benzoylformate using platinum-carbon catalysts modified with cinchonidine, *J. Nat. Chem. Lab. Ind.* (*Kagaku Gijutsu Kenkyusho Hokoku*) **75**, 77-78.

[62] Blaser, H.U., Jalett, H.P., Monti, D.M., Reber, J.F., Wehrli, J.T. (1988) Modified heterogeneous Pt catalysts for the enantioselective hydrogenation of *alpha*-ketoesters, *Stud. Surf. Sci. Catal.* **41**, 153-163.

[63] Wehrli, J.T., Baiker A., Monti, D.M., Blaser, H.U. (1989) Particle size effect on enantioselective hydrogenation of ethyl pyruvate over alumina-supported Pt catalyst, *J. Mol. Catal. A. Chem.* **49**, 195-203.

[64] Blaser, H.U., Jalett, H.P., Lottenbach, W.,Studer, M. (2000) Heterogeneous enantioselective hydrogenation of ethyl pyruvate catalysed by cinchona-modified Pt catalysts: effect of modifier structure, *J. Amer. Chem. Soc.* **122,** 12675-12682.

[65] Blaser, H.U., Jalett, H.P., Monti, D.M., Wehrli, J.T. (1989) Enantioselective hydrogenation of *alpha*-keto esters: temperature-programmed reduction study of liquid-phase Pt-Alumina hydrogenation catalysts, *Appl. Catal.* **52**, 19-32.

[66] Garland, M., Blaser, H.U. (1990) A heterogeneous "Ligand-accelerated" reaction: enantioselective hydrogenation of ethyl pyruvate catalysed by cinchona-modified Pt-Alumina catalysts, *J. Amer. Chem. Soc.*, **112**, 7048-7050.

[67] Garland, M., Jalett, H.P., Blaser, H.U. (1991) Mass transfer considerations for the enantioselective hydrogenation of alpha-ketoesters catalysed by cinchona modified Pt-Alumina, *Stud. Surf. Sci. Catal.* **59**, 177-184.

[68] Blaser, H.U., Jalett, H.P., Wiehl, J. (1991) Enantioselective hydrogenation of alpha-ketoesters with cinchona-modified Pt catalysts: effect of acidic and basic solvents and additives, *J. Mol. Catal. A. Chem.* **68**, 215-222.

[69] Blaser, H.U., Jalett, H.P., Monti, D.M., Baiker, A., Wehrli, J.T. (1991) Enantioselective hydrogenation of ethyl pyruvate: effect of catalyst and modifier structure, *Stud. Surf. Sci. Catal.* **67**, 147-155.

[70] Blaser, H.U., Boyer, S.K., Pittelkow, U. (1991) The heterogeneous enantioselective hydrodehalogenation of alpha, alpha-dichlorobenzazepinone-2: a unique case of substrate specificity, *Tetrahedron Asymm.* **2**, 721-732.

[71] Blaser, H.U., Jalett, H.P. (1993) Enantioselective hydrogenation of alpha-keto acids using Platinum catalysts modified with cinchona alkaloids, *Stud. Surf. Sci. Catal.* **78**, 139-146.

[72] Blaser, H.U., Imhof, D., Studer, M. (1997) Kinetic modeling of the ligand accelerated catalysis in the enantioselective hydrogenation of ethyl pyru-

vate: Influence of solvents, catalysts and additives, *Stud. Surf. Sci. Catal.* **108**, 175-182.

73 Blaser, H.U., Jalett, H.P., Spindler, F. (1996) Enantioselective hydrogenation of alpha-ketoesters: comparison of homogeneous and heterogeneous catalysts, *J. Mol. Catal. A. Chem.* **107**, 85-94.

74 Blaser, H.U., Jalett, H.P., Garland, M., Studer, M., Thies, H., Wirth-Tijani A. (1998) Kinetic studies of the enantioselective hydrogenation of ethyl pyruvate catalysed by a cinchona modified Pt-Alumina catalysts, *J. Catal.* **173**, 282-294.

75 Blaser, H.U., Garland, M., Jallet, H.P. (1993) Enantioselective Hydrogenation of Ethyl Pyruvate: Kinetic Modeling of the Modification of Pt Catalysts by Cinchona Alkaloids, *J. Catal.,* **144,** 569-578.

76 Studer, M., Burhardt, S., Blaser, H.U. (1999) Enantioselective hydrogenation of alpha-keto acetals with cinchona modified Pt catalyst, *Chem. Commun.* 1727-1728.

77 Baiker, A., Blaser, H.U. (1997) in *Handbook of heterogeneous catalysis*, Ertl, G., Knozinger, H, Weitkamp, J. (eds.), Wiley-VCH, Weinheim, Vol. 5., p. 2422.

78 Wehrli, J.T., Baiker, A., Monti, D.M., Blaser, H.U., Jalett, H.P. (1989) Enantioselective hydroganation of alpha-ketoesters: influence of reaction medium and conversion, *J. Mol. Catal. A. Chem.* **57**, 245-257.

79 Margitfalvi, J.L., Marti, P., Baiker, A., Botz, L., Sticher, O. (1990) Role of the modifier in the enantioselective hydrogenation of ethyl pyruvate over Pt-Alumina catalyst, *Catal. Lett.* **6**, 281-288.

80 Wehrli, J.T., Baiker, A., Monti, D.M., Blaser, H.U. (1990) Enantioselective hydrogenation of alpha-ketoesters: preparation and catalytic behavior of different alumina-supported Pt catalysts modified with cinchonidine, *J. Mol. Catal. A. Chem.* **61**, 207-226.

81 Margitfalvi, J.L., Jalett, H.P., Talas, E., Baiker, A, Blaser, H.U. (1991) Enantioselective hydrogenation of ethyl pyruvate on the promoted Pt-Alumina catalysts, *Catal. Lett.* **10**, 325-334.

82 Schuerch, M., Schwalm, O., Mallat, T., Baiker, A. (1996) Enantioselective hydrogenation of ketopantolactone on Pt/alumina, 4th *Int. Symp. Heterogen. Catal. Fine Chem.,* Basel, Book of Abstracts, pp. 132-133.

83 Schuerch, M., Kunzle, N., Mallat, T., Baiker, A. (1996) Enantioselective hydrogenation of ketopantolactone: effect of stereospecific product crystallization during reaction, *J. Catal.* **176**, 569-571.

84 Borszeky, K., Buergi, T., Zhaohui, Z., Mallat, T., Baiker, A. (1999) Enantioselective hydrogenation of alpha,beta-unsaturated carboxylic acids over cinchonidine modified Palladium: nature of modifier-reactant interaction, *J. Catal.* **187**, 160 - 166.

85 Mallat, T., Bodnar, Z., Minder, B., Borszeky, K., Baiker, A. (1997) Nature of initial transient period during enantioselective hydrogenation on Pt and Pd, *J. Catal.* **168**, 183-193.

[86] Borszeky, K, Mallat, T, Baiker, A (1996) Enantioselective hydrogenation of 2-methyl-2-pentenoic acid over cinchonidine-modified Pd/alumina, *Catal. Lett.* **41**, 199-202

[87] Kuenzle, N., Szabo, A., Schuerch, M., Wang, G., Mallat, T., Baiker, A. (1998) Enantioselective hydrogenation of a cyclic imidoketone over chirally modified Pt-Al_2O_3, *Chem. Commun.* 1377-1378.

[88] Schuerch, M., Heinz, T., Aeschiman, R., Mallat, T., Pfaltz, A., Baiker, A. (1998) Design of new modifiers for the enantioselective hydrogenation of ethyl pyruvate, *J. Catal.* **173**, 187-195

[89] Sipos, E., Tungler, A., Bitter, I. (2002) New chiral modifier for enantioselective heterogeneous catalytic hydrogenation, *Chemical Industries Series* (Dekker), (*Catalysis of Organic Reactions*, Morrell D.G., ed.) **89**, 653-658.

[90] Wang, G., Heinz, T., Pfaltz, A., Minder, B., Mallat, T., Baiker, A. (1994) New chiral modifier for the enantioselective hydrogenation of ethyl pyruvate over Pt-Alumina catalysts, *J. Chem. Soc., Chem. Commun.* 2047-2048.

[91] Margitfalvi, J.L., Minder, B., Talas, E., Botz, L., Baiker, A (1993) Enantioselective hydrogenation of *alpha*-ketoesters over Pt-Alumina catalyst: kinetic aspects of the rate acceleration effect induced by addition of cinchonidine, *Stud. Surf. Sci. Catal.* **75**, 2471-2474.

[92] Minder, R., Mallat, T., Baiker, A., Wang, G., Heinz, T., Pfaltz, A. (1995) A novel aminoalcohol modifier for the enantioselective hydrogenation of ethyl pyruvate on Pt-Alumina, *J. Catal.* **154**, 371-378.

[93] Talas, E., Botz, L., Margitfalvi, J.L., Sticher, O., Baiker, A. (1992) Planar chromatographic separation of cinchona alkaloids formed during enantioselective hydrogenation of ethyl pyruvate, *Planar Chromatography* **5**, 28-34.

[94] Minder, R., Mallat, T., Srabal, P., Baiker, A. (1994) Enantioselective hydrogenation of ethyl pyruvate-influence of oxidative treatment of cinchonidine-modified platinum catalyst and hemiketal formation in alcoholic solvents, *Catal. Lett.* **29**, 115-124

[95] Minder, R., Schuerch, M., Mallat, T., Baiker, A. (1995) Chiral nitrogen-compounds as new modifiers for the enantioselective hydrogenation of ethyl pyruvate, *Catal. Lett.* **31**, 143-151.

[96] Schwalm, O., Weber, J., Margitfalvi, J.L., Baiker, A. (1993) Ab initio and semiempirical investigation of the complexation of methyl pyruvate by ammonia and the ammonium cation, *J. Mol. Structure* **297**, 285-293.

[97] Schwalm, O., Minder, B., Weber, J., Baiker, A. (1994) Enantioselective hydrogenation of *alpha*-ketoesters over Pt-Alumina modified with cinchonidine: theoretical investigation of the substrate-modifier interaction, *Catal. Lett.* **23**, 271-279.

[98] Schwalm, O., Weber, J., Minder, B., Baiker, A. (1994) Theoretical investigation of the hydrogenation of *alpha*-ketoesters over Pt-Alumina modified with cinchonidine, *Intern. J. Quantum Chem.* **52**, 191-197.

[99] Mallat, T., Bodmer, M., Baiker, A. (1997) Enantioselective hydrogenation of trifluoroacetophenone over cinchonidine modified platinum, *Catal. Lett.* **44**, 95-99.

[100] Bodmer, M., Mallat, T., Baiker, A. (1998) Enantioselective hydrogenation of trifluoroacetophenone over Pt-Alumina, *Chemical Industries Series (Catalysis of Organic Reactions)* (Dekker) **75**, 75-87.

[101] Baiker, A. (1999) Supercritical fluids in heterogeneous catalysis, *Chem. Rev.* **99**, 453-473.

[102] Grunwaldt, J.D., Wandeler, R., Baiker, A. (2003) Supercritical fluids in catalysis: opportunities of *in situ* spectroscopic studies and monitoring phase behavior, *Catal. Rev. Sci. Eng.* **45**, N1, 1-96

[103] Minder, R., Mallat, T., Pickel, K.H., Steiner, K., Baiker, A. (1995) Enantioselective hydrogenation of ethyl pyruvate in supercritical fluids, *Catal. Lett.* **34**, 1-9.

[104] Minder, R., Schuerch, M., Mallat, T., Baiker, A., Heinz, T., Pfaltz, A. (1996) Enantioselective hydrogenation of ethyl pyruvate over Pt-Alumina modified by (*R*)-1-(1-naphthyl)ethylamine derivatives, *J. Catal.* **160**, 261-268.

[105] Heinz, T., Wang, G., Pfaltz, A., Minder, R., Schuerch, M., Mallat, T., Baiker, A. (1995) 1-(1-Naphthyl)ethylamine and derivatives thereof as chiral modifiers in the enantioselective hydrogenation of ethyl pyruvate over Pt-Alumina, *J. Chem. Soc. Chem. Comm.* 1421-1422.

[106] Mallat, T., Baiker, A. (1998) Reply to comment on "Nature of initial transient period during enantioselective hydrogenation on Pt and Pd" by Donna G. Blackmond, *J. Catal.* **176**, 271-279.

[107] Borszeky, K., Mallat, T., Baiker, A. (1996) Enantioselective hydrogenation of an *alpha,beta*-unsaturated carboxylic acid on Pd/alumina, 4th *Int. Symp. on Heterogeneous Catal. Fine Chem.*, Basel, Book of Abstracts, pp. 130-131.

[108] Borszeky, K., Mallat, T., Aeschiman, R., Schweizer, W.B., Baiker, A. (1996) Enantioselective hydrogenation of pyruvic acid oxime to alanine on Pd-alumina, *J. Catal.* **161**, 451-458.

[109] Wang, G.Z., Mallat, T., Baiker, A (1997) Enantioselective hydrogenation of *alpha*-ketoamides over Pt-alumina modified by cinchona alkaloids, *Tetrahedron Asymm.* **8**, 2133-2140.

[110] Borszeky, K., Mallat, T., Baiker, A. (1997) Enantioselective hydrogenation of *alpha, beta*-unsaturated acids. Substrate-modifier interaction over cinchonidine modified Pd/Al_2O_3, *Tetrahedron Asymm.* **8**, 3745-3753.

[111] Simons, K.E., Wang, G., Heinz, T., Giger, T., Mallat, T., Pfaltz, A., Baiker, A. (1995) A new class of chiral modifiers for the enantioselective

hydrogenation of *alpha*-ketoesters with Pt-Alumina, *Tetrahedron Asymm.* **6**, 505-518.

[112] Maris, M., Huck, W.R., Mallat, T., Baiker, A. (2003) Palladium-catalized asymmetric hydrogenation of furan carboxylic acids, *J. Catal.* **219,** 52-58.

[113] Huck, W.R., Buergi, T., Mallat, T., Baiker, A. (2003) Palladium catalysed enantioselective hydrogenation of 2-pyrones: evidence for competing reaction mechanism, *J. Catal.* **219**, 41–51.

[114] Kuenzle, N., Hess, R., Mallat, T., Baiker, A. (1999) Continuous enantioselective hydrogenation of activated ketones, *J. Catal.* **186**, 239-241.

[115] Buergi, T., Baiker, A. (2000) Model for enantioselective hydrogenation of *alpha*-ketoesters over chirally modified Platinum revisited: influence of *alpha*-ketoester conformation. *J. Catal.* **194**, 445-451.

[116] Von Arx, M., Mallat, T., Baiker, A. (2001) Platinum catalysed enantioselective hydrogenation of aryl substituted trifluoroacetophenones, *Tetrahedron Asymm.* **12**, 3089-3094.

[117] Buergi, T., Baiker, A. (1998) Conformational behavior of cinchonidine in different solvents: a combined NMR and *ab initio* investigations, *J. Amer. Chem. Soc.* **120**, 12920-12926.

[118] Bonalumi, N., Burgi, T., Baiker, A. (2003) Interaction between ketopantolactone and chirally modified Pt investigated by attenuated total reflection IR concentration modulation spectroscopy, *J. Amer. Chem. Soc.* **125**, 13342-13343.

[119] Perez, J.R.G., Malthete, J., Jacques, J. (1985) Hydrogenation asymetrique d'acides cinnamiques prochiraux en presence de Pd sur charbon et de bases chirales, *Compt. rend.* **300**, II, N5, 169-172.

[120] Nitta, Y., Ueda, Y., Imanaka, T. (1994) Enantioselective hydrogenation of (*E*)-phenylcinnamic acid over cinchonidine-modified Palladium catalysts, *Chem. Lett.* 1095-1098.

[121] Nitta, Y., Kobiro, K. (1995) Solvent effect on enantioselective hydrogenation of (*E*)-*alpha*-phenylcinnamic acid with cinchonidine-modified Palladium catalysts, *Chem. Lett.* 165-165.

[122] Nitta, Y., Kobiro, K., Okamoto, Y. (1997) Enantioselective hydrogenation of (*E*)-*alpha*-phenylcinnamic acid on cinchonidine modified Palladium catalyst: influence of support, *Stud. Surf. Sci. Catal.* **108**, 191-198.

[123] Nitta, Y., Kobiro, K., Okamoto, Y. (1996) Enantioselective hydrogenation of phenylcinnamic acid with cinchonidine-modified Pd-catalysts, *Proc. 70th Ann. Meet. Chem. Soc. Jpn.,* Part I, p. 573.

[124] Nitta, Y., Kobiro, K. (1996) Conversion dependence of enantioselective hydrogenation of (*E*)-*alpha*-phenylcinnamic acid on cinchonidine-modified Pd-TiO_2-catalyst, *Chem. Lett.* 897-898.

[125] Nitta, Y., Okamoto, Y. (1998) Effect of support texture on enantioselective hydrogenation of (*E*)-*alpha*-phenylcinnamic acid with cinchonidine-modified Palladium catalysts, *Chem. Lett.* 1115-1116.

[126] Nitta, Y., Shibata, A. (1998) Enantioselective hydrogenation of (*E*)-*alpha*-phenylcinnamic acid on Pd/TiO_2 catalyst modified by cinchona alkaloids: effect of modifier structure, *Chem. Lett.* 161-162.

[127] Nitta, Y. (1999) Importance of product desorption in enantioselective hydrogenation of (*E*)-*alpha*-phenylcinnamic acid with a cinchonidine-modified Pd/TiO_2 catalyst. Effect of additives, *Chem. Lett.* 635-636.

[128] Nitta, Y., Kubota, T., Okamoto, Y. (2000) Preparation of cinchonidine-modified Palladium catalysts for the enantioselective hydrogenation of (*E*)-*alpha*-Phenylcinnamic acid, *Bull. Chem. Soc. Jpn.*, 73, 2635-2641.

[129] Tungler, A., Nitta, Y., Fodor, K., Farkas, G., Mathe, T. (1999) Comparison of chiral modifiers in the Pd catalysed hydrogenation of phenylcinnamic acid and isophorone, *J. Mol. Catal. A. Chem.* **149**, 135-140.

[130] Nitta, Y., Kubota, T., Okamoto, Y. (2004) Solvent effect on the structure sensitivity in enantioselective hydrogenation of *alpha,beta*-unsaturated acids with modified palladium catalysts, *J. Mol. Catal. A. Chem.* **212**, 155-159.

[131] Nitta, Y., Sekine, P., Imanaka, T., Teranishi, S. (1981) The effect of crystalline size of nickel on the enantioselectivity of modified nickel catalysts, *Bull. Chem. Soc. Jpn.* **54**, 980-984.

[132] Nitta, Y., Sekine, F., Imanaka, T., Teranishi, S. (1982) Effect of preparation variables on enantioselectivity of supported Nickel catalysts modified with tartaric acid, *J. Catal.* **74**, 382-392.

[133] Nitta, Y., Yamanishi, O., Sekine, F., Imanaka, T., Teranishi, S. (1975) Effect of support on enantioselectivity of modified catalyst, *J. Catal.* **79**, 475-480.

[134] Nitta, Y., Sekine, F., Sasaki, J., Imanaka, T., Teranish,i S. (1983), Conversion dependence of enantioselective hydrogenation of methyl acetoacetate with modified Ni-silica catalyst, *J. Catal.* **79**, 211-214.

[135] Nitta, Y., Utsumi, T., Imanaka, T., Teranishi, S. (1984) Effect of preparation variable on morphological and catalytic properties of precipitated Ni-Silica catalysts, *Chem. Lett.* 1339-1342.

[136] Nitta, Y., Imanaka, T., Teranishi, S. (1985) Preparation chemistry of precipitated Ni-silica catalysts for enantioselective hydrogenation, *J. Catal.* **96**, 429-438.

[137] Nitta, Y., Kawabe, M., Kojita, H., Imanaka, T. (1986) Crystalline size effect in enantioselective hydrogenation on modified Ni-silica catalysts, *Chem. Express*, **1**, 631-634.

[138] Nitta, Y., Utsumi, T., Imanaka, T., Teranishi, S. (1986) Effect of aditives on morphological and catalytic properties of Ni-silica catalysts for enantioselective hydrogenation, *J. Catal.* **101**, 376-388.

[139] Nitta, Y., Imanaka, T. (1988) Effect of pore diffusion in liquid-phase enantioselective hydrogenation with Ni-silica catalysts, *Bull. Chem. Soc. Jpn.* **61**, 295-297.

[140] Nitta, Y., Kawabe, M., Sun, L., Ohmachi, Y., Imanaka, T. (1989) Preparation of uniformly dispersed Nickel-silica catalysts from synthetic Nickel-crysotile, *Appl. Catal.* **53**, 15-28.

[141] Sutherland, I.M., Ibbotson, A., Moyes, R.B., Wells, P.B. (1990) Enantioselective hydrogenation I. Surface conditions during methyl pyruvate hydrogenation catalysed by cinchonidine-modified Pt-silica (EUROPT-1), *J. Catal.* **125**, 77-88.

[142] Meheux, P.A., Ibbotson, A., Wells, P.B. (1991) Enantioselective hydrogenation II. Variation of activity and optical yield with experimental variables in methyl pyruvate catalysed by cinchona-modified Platinum-silica (EUROPT-1), *J. Catal.* **128**, 387-396.

[143] Carley, A.F., Rajumon, M.K., Roberts, M.W., Wells, P.B. (1995) XPS and LEED Studies of 10,11-dihydrocinchonidine absorption at Pt (III). *J. Chem. Soc., Farad. Trans.*, **91**, 2167-2172.

[144] Johnston, P., Wells, P.B. (1995) Cinchona-modified Pt as a catalyst for enantioselective hydrogenation: criticism of the "Template model" - a reply, *J. Catal.* **156**, 180-183.

[145] Slipszenko, J.A., Griffits, S.P., Johnson, P., Simons, K.E., Vermeer, W.A.H., Wells, P.B. (1998) Enantioselective hydrogenation V. Hydrogenation of butane-2,3-dione and 3-hydroxybutan-2-one catalyzed by cinchona-modified Pt, *J. Catal.* **179**, 267-276.

[146] Bond, G., Meheux, P.A., Ibbotson, A., Wells, P.B. (1991) Origin of enhanced rate in the Platinum-catalysed enantioselective hydrogenation of methyl pyruvate, *Catal. Today,* **10**, 371-378.

[147] Griffiths, S.P., Johnston, P., Vermeer, W.A.H., Wells, P.B. (1994) Novel alkaloid modifiers for enantioselective heterogeneous catalysis, *J. Chem. Soc. Chem. Comm.* 2431-2432.

[148] Griffiths, S.P., Wells, P.B., Griffin, K.G., Johnston, P. (1998) Heterogeneous enantioselective hydrogenation catalysed by Pt modified by some morphine alkaloids, *Chemical Industries Series* (*Catalysis of Organic Reactions*, Herkes F.E., ed.), (Dekker) **75**, 89-100.

[149] Simons, K.E., Meheux, P.A., Griffiths, S.P., Sutherland, I.M., Johnston, P., Wells, P.B., Carley, A.F., Rajumon, M.K., Roberts, M.W., Ibbotson, A. (1994) A model for the enantioselective hydrogenation of pyruvate catalysed by alkaloid-modified Platinum, *Recl. Trav. Chim. Pays-Bas,* **113**, 465-474.

[150] Vermeer, W.A.H., Fulford, A., Johnston, P., Wells, P.B. (1999) New enantioselective reactions catalysed by cinchonidine-modified Platinum, *Chem. Commun.* 1153-1154.

[151] Allan, E., Jackson, S.D., Korn, S., Webb, G., Young, N. (1996) The enantioselective hydrogenation of coumaranone over chiral centers generated on a Pt-silica catalyst, *Chemical Industries Series* (*Catalysis of Organic Reactions,* Malz, R.E., Jr., ed.) (Dekker) **68,** 191-198.

[152] Simons, K.E., Meheux, P.A., Ibbotson, A., Wells, P.B. (1993) Effect of mixtures of modifiers on optical yield in enantioselective hydrogenation. A test of the template model, *Stud. Surf. Sci. Catal.* **75,** 2317-2320.

[153] Von Arx, M., Dummer, N., Willock, D.J., Taylor, S.H., Wells, R.P.K., Wells, P.B., Hutchings, G.J. (2003) Observation of high enantioselectivity for the gas phase hydrogenation of methyl pyruvate using supported Pt catalysts pre-modified with cinchonidine, *Chem. Commun.* 1926–1927.

[154] Feast, S., Bethell, D., Page, P.C.B., King, F., Rochester, C.H., Siddiqui, M.R.H., Willock, D.J., Hutchings, G.J. (1995) Gas-phase catalytic asymmetric reaction using chirally modified microporous catalyst, *J. Chem. Soc., Chem. Comm.* 2409–2411.

[155] Margitfalvi, J.L. (1995) On the role of modifier-substrate interaction in different heterogeneous catalytic asymmetric hydrogenation reactions, *Chemical Industries Series* (*Catalysis of Organic Reactions,* Scaros M.G., Prunier M.L., eds.) (Dekker) **62**, 189-200.

[156] Margitfalvi, J.L., Hegedus, M. (1996) Enantioselective hydrogenation of *alpha*-ketoesters over cinchona Pt-Alumina catalysts. New interpretation of the rate acceleration and the induction of enantio-differentiation, *J. Mol. Catal. A. Chem.* **107**, 281-289.

[157] Margitfalvi, J.L., Hegedus, M., Tfirst, E. (1996) Enantioselective hydrogenation of *alpha*-ketoesters over cinchona-Pt-Alumina. Kinetic evidence for the substrate-modifier interaction in the liquid phase, *Tetrahedron Asymm.* **7**, 571-580.

[158] Margitfalvi, J.L., Talas, E., Yakhyaeva, L., Tfirst, E., Bertoti, I., Toth, L. (2002) Enantioselective hydrogenation of ethyl pyruvate over Pt colloids, *Chemical Industries Series* (*Catalysis of Organic Reactions,* Morrell D.G., ed.) (Dekker) **89,** 393-404.

[159] Margitfalvi, J.L., Hegedus, M., Tfirst, E. (1996) Enantio-differentiation over heterogeneous catalysts. The shielding effect model, S*tud. Surf. Sci. Catal.* **101**, 241-250.

[160] Margitfalvi, J.L., Talas, E., Tfirst, E., Kumar, C., Gergely, A. (2000) The role of cinchona alkaloids in enantioiselective hydrogenation reactions: are they modifiers or hosts involved in supramolecular heterogeneous catalysis? *Appl. Catal. A. Gen.* **191**, 177–191.

[161] Margitfalvi, J.L., Tfirst, E., Hegedus, M., Talas, E. (1998) Enantioselective hydrogenation of *alpha*-keto esters over cinchona-Pt-Alumina catalysts. Kinetic and molecular modeling, *Chemical Industries Series* (*Catalysis of Organic Reactions*, Herkes, F.E., ed.) (Dekker) **75,** 531-542.

[162] Margitfalvi, J.L., Hegedus, M. (1995) Cinchona-modified Pt as a catalyst for enantioselective hydrogenation: criticism of the "Template model", *J. Catal*. **156**, 175-179.

[163] Klabunovskii, E.I., Sheldon, R.A. (1997) Catalytic approaches to the asymmetric synthesis of *beta*-hydroxyacid esters, *Cattech* 1, 153-160.

[164] Tarnai, T., Tungler, A., Mathe, T., Petro, J., Sheldon, R.A., Toth, G. (1995) A new chiral auxiliary in enantioselective hydrogenation: (-)-dihydrovinpocetine. Part I. Hydrogenation of isophorone, *J. Mol. Catal. A. Chem.* **102**, 41-47.

[165] Tungler, A., Kajtar, M., Mathe, T., Toth, G., Fogassy, E., Petro, J. (1989) Enantioselective hydrogenation of *alpha,beta*-unsaturated ketones, *Catal. Today* **5**, 159-171.

[166] Tungler, A., Fuercht, A., Karancsi, Z.P., Toth, G., Mathe, T., Hegedus, L., Sandi, A. (1999) Diastereoselective and enantoselective heterogeneous catalytic hydrogenation of aminocinnamic acid derivatives, *J. Mol. Catal. A. Chem.* **139**, 239-244.

[167] Klabunovskii, E.I., Sheldon, R.A. (2004) On configuration correlations of 2-hydroxycarboxylic acid and their esters. Note., *J. Mol. Catal. A. Chem.* **212**, 53-54.

[168] Torok, B., Felfoldi, K., Szakonyi, G., Balazsik, K., Bartok, M. (1998) Enantiodifferentiation in asymmetric sonochemical hydrogenation, *Catal. Lett.* **52**, 81-84.

[169] Bartok, M., Felfoldi, K., Szollosi, G., Bartok, T. (1999) Rigid cinchona conformers in enantioselective catalytic reactions: new cinchona-modified Pt catalysts in the Orito reaction, *Catal. Lett.* **61**, 1-5.

[170] Torok, B., Felfoldi, K., Szakonyi, G., Molnar, A., Bartok, M. (1998) Ultrasonics in chiral metal catalysis. Effect of presonication on the asymmetric hydrogenation of ethyl pyruvate over Pt catalysts, *Chemical Industries Series* (*Catalysis of Organic Reactions,* Herkes, F.E., ed.) (Dekker) **75**, 129-140

[171] Bartok, M., Wittmann, G.W., Gondos, G., Smith, G.V. (1987) Homogeneous and heterogeneous catalytic asymmetric reactions I. Asymmetric hydrogenation of the prochiral C=C bond on a modified Raney nickel catalyst, *J. Org. Chem.* **52**, 1139-1141.

[172] Bartok, M., Wittman, G., Bartok, G.B., Gondos, G. (1990) Homogeneous and heterogeneous catalytic asymmetric reactions IV. Hydrogenation of the Ni complexes of cinnamic acid salts over MRNi catalysts, *J. Organomet. Chem.* **386**, 385-395.

[173] Wittman, G., Gondos, G., Bartok, M., Smith, G.V. (1985) Asymmetric hydrogenation of (*E*) and (*Z*)-*alpha*-phenylcinnamic acid on Raney Nickel modified with (*RR*)-tartaric acid and NaBr, *2nd Czech. Conf. Heterogeneous Catal., 1985,* Bechine-near-Tabor, Summaries, p. 197.

[174] Bartok, M., Szollosi, G., Balazsik, K., Bartok, T. (2002) Heterogeneous asymmetric reactions. Part 25. On the pretreatment and prehydrogenation of Pt-alumina catalyst in the hydrogenation of ethyl pyruvate, *J. Mol. Catal. A. Chem.* **177**, 299–305.

[175] Szollosi, G., Somlai, C., Szabo, P.T., Bartok, M. (2001) Heterogeneous asymmetric catalysis. Part 21. Amino acid derived modifiers in the

enantioselective hydrogenation of ethyl pyruvate over supported Platinum catalyst, *J. Mol. Catal. A. Chem.* **170**, 165-173.

[176] Felfoldi, K., Balazsik, K., Bartok, M. (2003) Heterogeneous asymmetric catalysis. Part 32. High enantioselectivities in the hydrogenation of activated ketones on cinchona alkaloid modified Platinum-alumina catalysts, *J. Mol. Catal. A. Chem.* **202**, 163-170.

[177] Felfoldi, K., Szori, K., Bartok, M. (2003) Heterogeneous asymmetric reactions. Part 35. Enantioselective hydrogenation of 2-oxoglutaric acid over cinchona modified Pt/Al_2O_3 catalysts, *Appl. Catal. A. Gen.* **251**, 457–460.

[178] Bartok, M., Balazsik, K., Notheisz, F. (2002) Heterogeneous asymmetric reactions. Part 33. Novel interpretation of the enantioselective hydrogenation of *alpha*-ketoesters on Pt/alumina-cinchona alkaloid catalyst system, *React. Kinet. Catal. Lett.* **77,** 363-370.

[179] Bartok, M., Balazsik, K., Bartok, T., Kele, Z. (2003) Heterogeneous asymmetric reactions. 34. New data in the enantioselective hydrogenation of ethyl pyruvate catalysed by cinchona alkaloid modified Pt/Al_2O_3 in toluene, *Catal. Lett.* **87,** 235-240.

[180] Bartok, M., Balazsik, K., Szollosi, G., Bartok, T. (2002), Electrospray ionization-mass spectrometry in the enantioselective hydrogenation of ethyl pyruvate catalysed by dihydrocinchonidine modified Pt/Al_2O_3 in acetic acid, *J. Catal.* **205,** 168-176.

[181] Balazsik, K., Torok, B., Szakonyi, G., Bartok, M. (1999) Homogeneous and heterogeneous asymmetric reactions. Part X. Enantioselective Hydrogenations over K-10 Montmorillonite-Supported Noble Metal Catalysts with Anchored Modifier, *Appl. Catal. A. Gen.* **182**, 53-63.

[182] Torok, B., Balazsik, K., Felfoldi, K., Bartok, M. (2000) Effect of coacid acidity on the cinchona-modified Pt-catalyzed enantioselective hydrogenations, *Stud. Surf. Sci. Catal.* **130,** 3381-3386.

[183] Kun, I., Torok, B., Felfoldi, K., Bartok, M. (2000) Heterogeneous asymmetric reactions. Part 17. Asymmetric hydrogenation of 2-methyl-2-pentenoic acid over cinchona modified Pd/Al_2O_3 catalysts, *Appl. Catal. A. Gen.* **203,** 71-79.

[184] Gondos, G., Wittmann, G., Bartok, M., Orr, J.C (1993) Chiral hydrogenation of estrone-3-methyl ether on modified Raney nickel catalysts, *Steroids,* **58**, 533-535.

[185] Szori, K., Torok, B., Felfoldi, K., Bartok, M. (2001) Enantioselective Hydrogenation of alpha-Ketoesters over a Pt/Al_2O_3 Catalyst: Effect of Steric Constraints on the Enantioselection, *Chemical Industries Series* (*Catalysis of Organic Reactions,* Ford, E.M., ed.) (Dekker) **82,** 489-495.

[186] Szori, K., Sutyinszki, M., Felfoldi, K., Bartok, M., (2002) Heterogeneous asymmetric reactions. Part 28. Efficient and practical method for the preparation of (*R*)- and (*S*)-*alpha*-hydroxy esters by the enantioselective

heterogeneous catalytic hydrogenation of *alpha*-ketoesters, *Appl. Catal. A. Gen.* **237**, 275-280.

[187] Torok, B., Szollosi, G., Rozsa-Tarjani, M., Bartok, M. (1998) Preparation, Characterization and Application of K-10 Montmorillonite Modified with Chiral Ammonium Halides, *Mol. Cryst. Liq. Cryst.,* **311**, 289-294.

[188] Balazsik, K., Torok, B., Szakonyi, G., Bartok, M. (1999) Homogeneous and heterogeneous asymmetric reactions. Part X. Enantioselective Hydrogenations over K-10 Montmorillonite-Supported Noble Metal Catalysts with Anchored Modifier, *Appl. Catal. A. Gen.* **182**, 53-63.

[189] Torok, B., Bartok, M., Dekany, I. (1999) The Structure of Chiral Phenylethyl-Ammonnium Montmorillonites in Ethanol-Toluene Mixtures, *Colloid & Polymer Science* **277**, 340-346.

[190] Balazsik, K., Torok, B., Kiricsi, I., Dekany, I., Bartok, M. (1999) Characterization of Cinchonidine Doped Montmorillonite Supported Noble Metal Catalysts by Thermoanalytical Methods, *J. Therm. Anal.* **56**, 337-343.

[191] Torok, B., Balazsik, K., Dekany, I., Bartok, M. (2000) Preparation and Characterization of New Chirally Modified Laponites, *Mol. Cryst. Liq. Cryst.* **341**, 339-344.

[192] Szollosi, G., Kun, I., Mastalir, A., Bartok, M., Dekany, I. (2001) Preparation, characterization and application of platinum catalysts immobilized on clays, *Solid State Ionics* **141**, 273-278.

[193] Szollosi, G., Mastalir, A., Bartok, M. (2001) Effect of ion exchange by an organic cation on platinum immobilization on clays, *React. Kinet. Catal. Lett.* **74**, 241-249.

[194] Torok, B., Balazsik, K., Kun, I., Szollosi, G., Szakonyi, G., Bartok, M. (1999) Homogeneous and Heterogeneous Asymmetric Reactions. Part 13. Clay-Supported Noble Metal Catalysts in Enantioselective Hydrogenations, *Stud. Surf. Sci. Catal.* **125**, 515-522.

[195] Szollosi, G., Kun, I., Bartok, M. (2001) Heterogeneous Asymmetric Reactions. Part 24. Heterogeneous Catalytic Enantioselective Hydrogenation of the C=N Group Over Cinchona Alkaloids Modified Palladium Catalyst, *Chirality* **13**, 619-624.

[196] Morgenschweis, K., Polkiehn, E., Reschetilowski, W. (1996) Controlling the enantioselective hydrogenation of ethyl pyruvate using zeolites as catalyst support, *Stud. Surf. Sci. Catal.* **108**, 167-174.

[197] Reschetilowski, W., Boehmer, U., Morgenscweis, K. (1995) Asymmetrische Fluessigphasenhydrierung von Ethylpyruvat am chiral modifizierten Pt-haltigen ZSM-35-Zeolith, *Chem. Ing. Techn.* **67**, 205-207.

[198] Boehmer, U., Morgenschweis, K., Reschetilowski, W. (1995) Liquid phase asymmetric hydrogenation on Pt-containing zeolite Y catalysts, *Catal. Today* **24**, 195-199.

[199] Reschetilowski, W., Boehmer, U., Morgenschweis, K. (1997) Langzeitverhalten chiral modifizierter Pt-Zeolith-Katalysatoren bei der enantioselecti-

ven Hydrierung von Ethylpyruvat, *Wissenschtliche Zeitschr. Techn. Univ. Dresden* **46**, N5, 39-46.

[200] Reschetilowski, W., Boehmer, U., Wiehl, J. (1994) Pt-containing zeolites for enantioselective catalytic hydrogenation of ethyl-pyruvate-effects of zeolite structures, solvent and modifier concentration, *Stud. Surf. Sci. Catal.* **84**, 2021-2028.

[201] Reschetilowski, W. (2000) Enantioselective hydrogenation of ethyl pyruvate. Long-term perfomance of chirally modified Pt-zeolite catalysts, *Catal. Today* **20**, 1-7.

[202] Smith, G.V., Cheng, J., Song, R. (1996) Enantioselective hydrogenation of prochiral C=C bonds over noble metal catalysts supported by *beta*-cyclodextrin polymer, *Chemical Industries Series* (*Catalysis of Organic Reactions,* Malz, R.E., Jr., ed.), (Dekker), **68,** 479-483.

[203] Smith, G.V., Musoiu, M. (1979) Asymmetric hydrogenation over modified Raney Nickel, *J. Catal.* **60**, 184-192.

[204] Smith, G.V., Notheisz, F. (1999) *Enantioselective hydrogenations*, in *Heterogeneous Catalysis in Organic Chemistry*, Acad. Press, pp. 97-117.

[205] Augustine, R.E., Tanielyan, S.K., Doyle, L.K. (1993) Enantioselective heterogeneous catalysts I. A working model for the catalyst:modifier:substrate interactions in chiral pyruvate hydrogenations, *Tetrahedron Asymm.* **4**, 1803-1827.

[206] Augustine, R.E., Tanielyan, S.K. (1996) Enantioselective heterogeneous catalysis II. Examination of the formation of the individual (*R*) and (*S*) lactates in the cinchonidine modified platinum hydrogenation of pyruvate, *J. Mol. Catal. A Chem.* **112**, 93-104.

[207] Singh, U.K., Landau, R.N., Sun, Y., LeBlond, C., Blackmond, D.G., Tanielyan, S.K., Augustine, R.L. (1995) Enantioselective catalysis: influence of conversion and bulk diffusion limitations on selectivity in the hydrogenation of ethyl pyruvate, *J. Catal.* **154**, 91–97.

[208] Blackmond, D.G. (1998) Comments on "Nature of Initial Transient Period During Enantioselective Hydrogenation on Pt and Pd" by Mallat, T., Bodnar, Z., Minder, B., Borszeky, K., and Baiker, A. *J. Catal.* **176**, 267-270.

[209] Sun, Y.K., Landau, R.N., Wang, J., LeBlond, C., Blackmond, D.G. (1996) A re-examination of pressure effects on enantioselectivity in asymmetric catalytic hydrogenation, *J. Amer. Chem. Soc.* **118**, 1348–1353.

[210] Blackmond, D.G. (1997) Mathematical models of nonlinear effects in asymmetric catalysis: new insights based on the role of reaction rate, *J. Amer. Chem. Soc.*, **119**, 12934-12939.

[211] Wang, J., Sun, Y., LeBlond, C., Landau, D.G., Blackmond, D.G. (1996) Asymmetric hydrogenation of ethyl pyruvate: relationship between conversion and enantioselectivity, *J. Catal.* **161**, 752-758.

[212] Sun, Y., Wang, J., LeBlond, C., Landau, D.G., Blackmond, D.G. (1996) Asymmetric hydrogenation of ethyl pyruvate: diffusion effects on enantioselectivity, *J. Catal.* **161**, 759-765.

[213] Izumi, Y., Tai, A. (1977) *Stereo-differentiating Reactions. The Nature of Asymmetric Reactions*, Kodansha Ltd., Tokyo, and Academic Press, New York, p. 334

[214] Tai, A., Harada, T. (1986) Asymmetrically modified Nickel catalysts, in *Tailored Metal Catalysts,* Iwasawa, Y. (ed.), D. Reidel Publ., Amsterdam, pp. 265-324.

[215] Izumi, Y. (1983) Modified Raney Nickel (MRNi) catalyst: heterogeneous enantio-differentiating (asymmetric) catalysts, *Adv. Catal. Rel. Subjects* **32**, 215-271.

[216] Tai, A., Kikukawa, T., Sugimura, T., Inoue,Y., Osawa, T, Fujii, S. (1991) Highly efficient enantio-differentiating hydrogenation over an ultrasonicated Raney Nickel catalyst modified with tartaric acid, *J. Chem. Soc., Chem. Comm.* 795-796.

[217] Ozaki, H., Tai, A., Kabatake, S., Watanabe, H., Izumi, Y. (1978) Enanatio-face-differentiating (asymmetric) hydrogenation of *beta*-ketoester with modified Raney Nickel catalyst (MRNi). XXXI. A comparative study of reaction rates and optical yields, *Bull. Chem. Soc. Jpn.* **51**, 3559-3563.

[218] Tai, A., Harada, T., Hiraki, Y., Murakami, S. (1983) Stereochemical investigation on asymmetrically modified Raney Nickel catalyst. Mode of interaction between modifying reagent and substrate in the enantioface-differentiating process, *Bull. Chem. Soc. Jpn.* **56**, 1414-1419.

[219] Okamoto, S., Harada, T., Tai, A. (1979), Enantioface-differentiating (asymmetric) hydrogenation of keto ester with modified Nickel catalyst. XXXII. Structural requirements of modifying reagent and substrate, *Bull. Chem., Soc., Jpn.* **52**, 2670-2673.

[220] Klabunovskii, E.I., Vedenyapin, A.A. (1980) *Asymmetric catalysis. Hydrogenation on metals (*russian*)*, Nauka, Moscow, p. 221

[221] Klabunowskii, E.I. (1963) *Asymmetric Syntheses* (german) VEB Deutscher Verlag der Wissenschaften, Berlin, p. 213

[222] Vedenyapin A.A., Klabunovskii E.I., Talanov Yu.M., Areshidze G.Kh. (1978) Effect of the particle size of Ruthenium in Ru-silica catalyst for its catalytic and asymmetrizing properties, *Izv. Acad. Sci. SSSR. Ser. Khim.* 2628–2630, *Chem. Abstr.* **86**, 89128u (1977).

[223] Terent'ev, A.P., Klabunovskii, E.I., Patrikeev, V.V. (1950) Asymmetric synthesis by means of catalysts supported on right and left quartz, *Dokl. Akad. Nauk SSSR*, **74**, 947-950, *Chem. Abstr.* **45**, 3798c (1951).

[224] Yasumori, I., Inoue, Y., Okabe, K. (1975) Mechanism of asymmetric hydrogenation on nickel surface coordinated with an optically active molecule, in *Proc. Int. Symp. Rel. Heterogen. Homogen. Catal.,* Brussels, 1974, Elsevier, pp. 41-50.

[225] Yasumori, I. (1978) Mechanism of enantioface-differentiating hydrogenation on modified nickel surfaces, *Pure Appl. Chem.* **50**, 971-978.

[226] Inoue, Y., Okabe, K., Yasumori, I. (1981) X-ray photoelectron spectra of adsorbed methyl acetoacetate and coordinated tartaric acid, aspartic acid and alanine on the Nickel surface, *Bull. Chem. Soc. Jpn.* **54,** 613-614.

[227] Jacobsen, E.N., Marko, I., France, M.B., Svendsen, J.S., Sharpless, K.B. (1989) Kinetic role of the alkaloid ligands in asymmetric catalytic dihydroxylation, *J. Amer. Chem. Soc.* **111**, 737-739.

[228] Eliel, E. L. (1962) *Stereochemistry of carbon compounds*, McGraw Hill, New York, London.

[229] Zuo, X.B., Liu, H., Guo, D., Yang, X. (1999) Enantioselective hydrogena tion of pyruvates over polymer-stabilized and supported platinum nanoclusters, *Tetrahedron*, **55**, 7787-7804.

[230] Tungler, A., Mathe, T., Fodor, K., Sheldon, R.A., Gallezot, P. (1996) A new chiral auxiliary in enantioselective hydrogenations: (-)-dihydrovinpocetine. Hydrogenation of ethyl pyruvate II, *J. Mol. Catal. A. Chem.* **108**, 145-151.

[231] Fraga, M.A., Mendes, M.J., Jordao, E. (2002) Examination of the surface chemistry of activated carbon on enantioselective hydrogenation of methyl pyruvate over Pt/C catalysts, *J. Mol. Catal. A. Chem.* **179**, 243-251.

[232] Farkas, G., Hegedus, L., Tungler, A., Mathe, T., Figueiredo, J.L., Freitas, M. (2000) Effect of carbon support properties on enantioselective hydrogenation of isophorone over palladium catalysts modified with (-)-dihydroapovincaminic acid ethyl ester, *J. Mol. Catal. A. Chem.* **153**, 215-219.

[233] Farkas, G., Sipos, E., Tungler, A., Sarkany, A., Figueiredo, J.I. (2001) Enantioselective hydrogenation of isophorone over Pd catalysts in the presence of (-)-dihydroapovincaminic acid ethyl ester, *J. Mol. Catal. A. Chem.* **170**, 101-107.

[234] Sakuraba, H., Inomato, N., Tanaka, Y. (1989) Asymmetric induction of ketones with crystalline cyclodextrin complexes of amine-boranes, *J. Org. Chem.* **54**, 3482-3484.

[235] Li, C., (2004) Chiral synthesis on catalysts immobilized in microporous and mesoporous materials, *Catal. Rev.* 46, 419-492.

[236] Corma, A., del Pino, M., Iglesias, M., Sanchez, F. (1992) Synthesis and characterization of new chiral Rh(I) complexes with *N,N'*-ligands and *N,P*-ligands–a study of anchoring on the modified zeolites and catalytic properties of heterogenized complexes, *J. Organomet. Chem.* **431,** 233-246.

[237] Carmona, A., Corma, A., Iglesias, M., San Jose, A., Sanchez, F. (1995) Optically-active complexes of transition-metals (Rh(I), Ru(II), Co(II) and Ni(II)) with 2-aminocarbonylpyrrolidine ligands–selective catalysts for hydrogenation of prochiral olefins, *J. Organomet. Chem.* **492,** 11-21.

[238] Corma, A., Fuerte, A., Iglesias, M., Sanchez, F. (1996) Preparation of new chiral dioxomolybdenum complexes heterogenized on modified USY-zeolites. Efficient catalysts for selective epoxidation of allilic alcohols, *J. Mol. Catal. A. Chem.* **107**, 225-234

[239] Corma, A., de Dios, M.I., Iglesias, M., Sanchez, F. (1997) Regiospecific hydrosilylation of styrene by rhodium complexes heterogenised on modified USY-zeolites, *Stud. Surf. Sci. Catal.* **108,** 501-507.

[240] Sabater, M.J., Corma, A., Domenech, A., Fornes, V., Garcia, H. (1997) Chiral Salen Manganese complex encapsulated within Zeolite Y: a heterogeneous enantioselective catalyst for the epoxidation of alkenes, *Chem. Commun.* 1285–1286.

[241] Ogunwumi, S.B., Bein, T. (1997) Intrazeolite assembly of a chiral manganese salen epoxidation catalysts, *Chem. Commun.* 901-902.

[242] Gbery, G., Zsigmond, A., Balkus, K.J. (2001) Enantioselective epoxidations catalyzed by zeolite MCM-22 encapsulated Jacobsen's catalyst, *Catal. Lett.* **74,** 77-80.

[243] Sundarababu, G., Leibovich, M., Corbin, D.R., Scheffer, J.R., Ramamurphy, V. (1996) Zeolite as a host for chiral induction, *Chem. Commun.* 2159-2160.

[244] Hutchings, G.J. (1999) New approaches to rate enhancement in heterogeneous catalysis, *Chem. Commun.* 301-306.

[245] Parvulescu, V.I., Parvulescu, V., Coman, S., Radu, C., Macovei, D., Angulescu, E., Russu, R. (1995) Modified Ru exchanged zeolites for enantioselective hydrogenation, *Stud. Surf. Sci. Catal.* **91,** 561-570.

[246] Coman, S., Bendic, C., Hillebrand, M., Angulescu, E., Parvulescu, V.I., Petride, A (1998) Diastereoselective hydrogenation of cyclic beta-ketoesters over modified Ru-Zeolite catalysts, *Chemical Industries Series* (*Catalysis of Organic Reactions,* Herkes F.E., ed.) **75,** 169-181.

[247] Cattech highlights (1997) Zeolites and chirality, *Cattech,* **1**, 149-151.

[248] Mastalir, A., Szollosi, G., Kiraly, Z., Razga, Z. (2002) Preparation and characterization of platinum nanoparticles immobilized in dihydrocinchonidine-modified montmorillonite and hectorite, *Appl. Clay Sci.* **22**, 9-16.

[249] Gross, L.H., Rys, P. (1974) Heterogeneous catalytic asymmetric hydrogenation, *J. Org. Chem.* **39**, 2429-2430.

[250] Sachtler, W.M.H. (1982) What makes a catalyst selective? *The Second Rideal Lecture, Farad. Disc. Roy. Soc. Chem.* **72**, 7-31.

[251] Fu, L., Kung, H.H., Sachtler, W.M.H. (1987) Particle size effect on enantioselective hydrogenation of methyl acetoacetate over silica-supported nickel catalyst, *J. Mol. Catal. A. Chem.* **42**, 29-36.

[252] Molvinger, K., Lopez, M., Court, J. (1999) Asymmetric reduction and hydrogenation over heterogeneous catalysts prepared by reacting nickel-boride with norephedrine, *J. Mol. Catal. A. Chem.* **150**, 267-273.

[253] Chernysheva, V.V., Murina, I.P., Klabunovskii, E.I. (1983) Study of inhibition of enantioselective hydrogenation of ethyl acetoacetate, *Izv. AN SSSR. Ser. Khim.* 757-760, *Chem. Abstr.* 99,21647a (1983)

[254] Neupokoev, V.I., Klabunovskii, E.I., Petrov, Yu.I. (1973) Asymmetric hydrogenation of ethyl acetoacetate on dissymmetric cobalt, *Kinetika i Kataliz (Kinetics and Catalysis)* **14**, 447-451, *Chem. Abstr.* **79**, 18008z (1973).

[255] Solladie-Cavallo, A., Hoernel, F., Schmitt, M., Garin, F. (2003) Heterogeneous asymmetric hydrogenation and deuteration of 2-methyl-2-pentenoic acid over Pd supported catalysts: proton/deuterium exchange, *J. Mol. Catal. A. Chem.* **195**, 181-188.

[256] Zuo, X.B., Liu, H., Liu, M. (1998) Asymmetric hydrogenation of alpha-ketoesters over finely dispersed polymer–stabilized platinum clusters *Tetrahedron Lett.* **39**, 1941-1944.

[257] LeBlond, C., Wang, J., Liu, J., Andrews, A.T., Sun, Y.K. (1999) Highly enantioselective heterogeneously catalyzed hydrogenation of *alpha*-ketoesters under mild conditions, **121**, 4920-4921.

[258] Tungler, A., Mathe, T., Tarnai, T., Fodor, K., Toth, G., Kajtar, J., Kolossvary, I., Herenyi, B., Sheldon, R.A. (1995) (-)-Dihydro-apovincaminic acid ethyl ester, preparation and use as a chiral modifier in enantioselective heterogeneous catalytic hydrogenations, *Tetrahedron Asymm.* **6**, 2395-2402.

[259] Fodor, K., Tungler, A., Mathe, T., Szabo, S., Sheldon, R.A. (1998) Effect of poisoning and metal adsorption on the enantioselective heterogeneous catalytic hydrogenation of isophorone and ethyl pyruvate, *Chemical Industries Series* (*Catalysis of Organic Reactions,* Herkes F.E., ed.),((Dekker) **75,** 115-127.

[260] Tungler, A., Tarnai, T., Hegedus, L., Fodor, K. (1998) Palladium-mediated heterogeneous catalytic hydrogenations. Selectivity of liquid-phase reactions for the fine chemicals industry, *Platinum Metals Rev.* **42**, 108-115.

[261] Boennemann, H., Braun, G.A. (1996) Enantioselective hydrogenation on Platinum colloids, *Angew. Chem. Int. Ed. Engl.* **35**, 1992–1995.

[262] Koehler, J.U., Bradley, J.S. (1997) Enantioselective hydrogenation of ethyl pyruvate with colloidal platinum catalysts: the effect of acidity on rate, *Catal. Lett.* **45**, 203-208.

[263] Koehler, J.U., Bradley, J.S. (1998) A kinetic probe of the effect of a stabilizing polymer on a colloidal catalyst: accelerated enantioselective hydrogenation of ethyl pyruvate catalysed by poly(vinylpyrrolidone)-stabilized platinum colloids, *Langmuir* **14**, 2730-2735.

[264] Collier, P.J., Iggo, J.A., Whymann, R. (1999) Preparation and characterization of solvent-stabilised nanoparticulate platinum and palladium and their catalytic behaviour towards the enantioselective hydrogenation of ethyl pyruvate, *J. Mol. Catal. A. Chem.* **146**, 149-157.

[265] Morawsky, K., Preusse, U., Vorlop, K.D. (2002) Increase of the long-term stability of chiral modified platinum catalysts used in the enantioselective hydrogenation of ethyl pyruvate, *Chemical Industries Series* (*Catalysis of Organic Reactions,* Morrell D.G., ed.) **89,** 21-32.

[266] Koeckritz, A., Bischoff, S., Morawsky, V., Pruesse, U., Vorlop, K.D. (2002) A novel strategy for heterogenisation of homogeneous and colloidal chiral catalysts and their application in enantioselective reactions, *J. Mol. Catal. A. Chem.* **180,** 231-243.

[267] Huang, Y., Li, Y., Hu, J., Cheng, P., Chen, H., Li, R., Li, X., Yip, C.W., Chan, A.S.C. (2002) Enantioselective hydrogenation of ethyl pyruvate catalysed by polyvinylpyrrolidone-stabilized and supported Rhodium nanocluster, *J. Mol. Catal. A. Chem.* **189**, 219-224.

[268] Ma, H., Chen, H., Zhang, Q., Li, X. (2003) Enantioselective hydrogenation of ethyl pyruvate catalysed by TS-1 supported Rhodium nanoclusters, *J. Mol. Catal. A. Chem.* **196**, 131-135.

[269] Zhang, J., Yan, X.P., Liu, H.F. (2001) Enantioselective hydrogenation of trifluoroacetophenone over polymer-stabilized platinum nanoclusters, *J. Mol. Catal. A. Chem.* **175**, 125-130.

[270] Zuo, X., Liu, H., Tian, J. (2000) Asymmetric hydrogenation of alpha-diketones over polymer-stabilized and supported Platinum clusters, *J. Mol. Catal. A. Chem.* **157**, 217-224.

[271] Zuo, X., Liu, H., Yue, C. (1999) Enantioselective hydrogenation of methyl pyruvate over polymer-stabilized and supported Iridium clusters, *J. Mol. Catal. A. Chem.* **147**, 63-72.

[272] Huang, Y., Chen, J., Chen, H., Li, R., Li, Y., Min, L., Li, X. (2001) Enantioselective hydrogenation of ethyl pyruvate catalysed by PVP-stabilized Rhodium nanoclusters, *J. Mol. Catal. A. Chem.* **170**, 143-146.

[273] Skvarchenko, V.R., Shalaev, V.K., Klabunovskii, E.I. (1974) Advances in tripticene chemistry, *Uspekhi Khimii Russ. Chem. Rev.)* **43**, 1983–2013, *Chem. Abstr.* **82**, 186 231d (1975).

[274] Soai, K., Shibata, T., Morioka, H., Choji, K. (1995) Asymmetric autocatalysis and amplification of enantiomeric excess of a chiral molecule, *Nature* **378**, 767-768.

[275] Torey, C., Henin, F., Muzart, J. (1996) Enantioselective hydrogenation of alpha,beta-unsaturated ketones over Pd on charcoal in the presence of (-)-ephedrine, *Tetrahedron Asymm.* **7**, 975-976.

[276] Wells, P.B., Simons, K.E., Slipszenko, J.A., Grifffits, S.P., Ewing, D.F. (1999) Chiral environments at alkaloid-modified platinim surfaces, *J. Mol. Catal. A. Chem.* **146**, 159-166.

[277] Huck, W.R., Buergi, T., Mallat, T., Baiker, A. (2003) Asymmetric hydrogenation on platinum: nonlinear effect of coadsorbed cinchona alkaloids on enantiodifferentiation, *J. Mol. Catal. A. Chem.* **216**, 276-287.

[278] Farkas, G., Fodor, K., Tungler, A., Mathe, T., Toth, G., Sheldon, R.A. (1999) New chiral auxiliaries in enantioselective heterogeneous catalytic

hydrogenations: (-)- and (+)-dihydroapovincaminic acid. Comparison with (-)-dihydroapovincaminic acid ethyl ester. III., *J. Mol. Catal. A. Chem.* **138**, 123-127.

[279] Sipos, E., Tungler, A., Bitter, I. (2003) (*S*)-Proline based chiral modifiers, *J. Mol. Catal. A. Chem.* **198**, 167-173.

[280] Sipos, E., Farkas, G., Tungler, A., Figueiredo, J.I. (2002) Enantioselective hydrogenation of isophorone with titania supported Pd catalysts modified with (-)-dihydroapovincaminic acid ethyl ester effect of the support and the reduction method, *J. Mol. Catal. A. Chem.* **179**, 107–112.

[281] Sipos, E., Fogassy, G., Tungler, A., Samant, P.V., Figueiredo, J.I. (2004) Enantioselective hydrogenation with highly mesoporous carbon supported Pd catalysts, *J. Mol. Catal. A. Chem.* **212**, 245–250.

[282] Torok, B., Felfoldi, K., Balazsik, K., Bartok, M. (1999) New synthesis of a useful C3 chiral building block by a heterogeneous method: enantioselective hydrogenation of pyruvaldehyde dimethyl acetal over cinchona modified Pt-Al_2O_3 catalysts, *Chem. Commun.* 1725-1726.

[283] Balazsik, K., Szori, K., Felfoldi, K., Torok, B., Bartok, M. (2000) Asymmetric synthesis of alkyl 5-oxotetrahydrofuran-2-carboxylates by enantioselective hydrogenation of dialkyl 2-oxoglutarates over cinchona modified Pt-alumina catalysts, *Chem. Commun.* 555-556.

[284] Perosa, A., Tundo, P., Selva, M. (2002) Multiphase heterogeneous catalytic enantioselective hydrogenation of acetophenone over cinchona-modified Pt/C, *J. Mol. Catal. A. Chem.* **180**, 169-175.

[285] Hess, R., Mallat, T., Baiker, A. (2003) Enantioselective hydrogenation of aromatic ketones: structural effects, *J. Catal.* **218**, 453-456.

[286] von Arx, M., Mallat, T., Baiker, A. (1999) Unprecedented selectivity bahaviour in the hydrogenation of an alpha,beta-unsaturated ketone: hydrogenation of ketoisophorone over alumina supported Pt and Pd, *J. Mol. Catal. A. Chem.* **148**, 275-283.

[287] Toukonitty, E., Maki-Arvela, P., Kuusisto, J., Nieminen, V., Paeivaearinta, J., Hotokka, M., Salmi, T., Murzin, D.Yu. (2003) Solvent effect in enantioselective hydrogenation of 1-phenyl-1,2-propanedione, *J. Mol. Catal. A. Chem.* **192**, 135-151.

[288] Sonderegger, O., Buergi, T., Baiker, A. (2003) Asymmetric hydrogenation of cyclohexane-1,2-dione over cinchonidine-modified platinum, *J. Catal.* **215,** 116-121.

[289] Gobolos, S., Tfirst, E., Margitfalvi, J., Hayes, K.S. (1999) Asymmetric synthesis of (*S*)-alkylamines via reductive transamination of ketones over carbon-supported Palladium catalysts, *J. Mol. Catal. A. Chem.* **146**, 129-138.

[290] Blaser, H.U., Hoenig, H., Studer, M., Wedemeyer-Exl., C. (1999) Enantioselective synthesis of ethyl nipecotinate using cinchona modified heterogeneous catalysts, *J. Mol. Catal. A. Chem.* **139**, 253-257.

[291] Klabunovski, E.I., Godunova, L.F., Karakhanov, R.A., Beilinson, E.Yu., Rybina, I.A., Rubtsov, I.A. (1969) On hydrogenation of optically active furan carboxylic acids esters. Comm. 1. Hydrogenation of 2-methylbutyl-furanoate and bornylfuranoate on Ni and Pd catalysts, *Izv. Acad. Sci. SSSR, Otdel. Khim. Nauk*, 298–303, *Chem. Abstr.* **70**, 114356w (1969).

[292] Godunova, L.F., Klabunovskii, E.I., Karakhanov, R.A., Beilinson, E.Yu., Rybina, I.A., Rubtsov, I.A. (1969) On hydrogenation of optically active furan carboxylic acids esters. Comm. 2. Hydrogenation of menthyl furanoate on Pd catalysts, *Izv. Acad. Sci. SSSR, Otdel. Khim. Nauk,* 303–307, *Chem. Abstr.* **71,** 2772e (1969).

[293] Fogassy, G., Tungler, A., Levai, A., Toth, G. (2002) Enantioselective hydrogenation of exocyclic alpha,beta-unsaturated ketones. Part II. Hydrogenation in the presence of (*S*)-proline, *J. Mol. Catal. A. Chem.* **179**, 101-106.

[294] Fogassy, G., Tungler, A., Levai, A. (2003) Enantioselective hydrogenation of exocyclic alpha,beta-unsaturated ketones, Part III. Hydrogenation with Pd in the presence of cinchonidine, *J. Mol. Catal. A. Chem.* **192**, 189-194.

[295] Hapiat, F., Agboussou, F., Mortreux, A. (1994) Synthesis of new chiral arene Ru(II) aminophosphine-phosphinite complexes and use in asymmetric hydrogenation of an activated keto compounds, *Tetrahedron Asymm.* **5**, 515-518.

[296] Carpentier, J.F., Agboussou, F., Mortreux, A. (1995) Catalytic asymmetric hydrogenation of activated keto-compounds in some homogeneous and silica supported di(4-carboxylato)aminophosphine-phosphinite)diRhodium complexes, *Tetrahedron Asymm.* **6**, 39-42.

[297] Roucoux, A., Suisse, I., Devocelle, M., Carpentier, J.F., Agboussou, F., Mortreux, A. (1996) Rh(I)-bis(aminophosphine)complexes as catalysts for asymmetric hydrogenation of activated ketones, *Tetrahedron Asymm.* **7**, 379-382.

[298] Roucoux, A., Agboussou, F., Mortreux, A., Petit, F. (1993) New alkyl-arylamidophosphine-phosphinites as chiraldiphosphines for asymmetric hydrogenation of activated keto compounds, *Tetrahedron Asymm.* **4**, 2279-2282.

[299] Niwa, S., Imamura, Y., Otsuka, K. (1987) Asymmetric hydrogenation of alpha-oxolactones in the presence of Pt-C catalyst containing cinchona alkaloids. Agency Ind. Sci. & Techn., *Jpn. Kokai Tokkyo Koho, Jpn.* 62,158288,14.07.1987, *Chem. Abstr.* 1988, 109, 128815f.

[300] Reichardt, C. (1988) *Solvents and Solvent Effects in Organic Chemistry,* VCH, Weinheim.

[301] Bartok, M., Szabo, P.T., Bartok, T., Szollosi, G., Balazsik, K. (2001) Identification of new types of aluminium compounds by electrospray ionization mass spectrometry: oxonium cations, *Rapid Commun. Mass Spectrom.* **15**, 65-69.

[302] Bartok, M., Balazsik, K., Szollosi, G., Bartok, M. (2001) Solvent and support effects in the case of acetic acid and alumina: Oxonium cations in asymmetric hydrogenation of ethyl pyruvate over dihydrocinchonidine modified platinum, *Catal. Commun.* **2**, 269-272.

[303] Wandeler, R., Kunzle, N., Schneider, M.S., Mallat, T., Baiker, A. (2001) Continuous enantioselective hydrogenation of ethyl pyruvate in "supercritical" ethane: relation between phase behavior and catalytic performance, *J. Mol. Catal. A. Chem* **200**, 377-388.

[304] Wandeler, R., Kunzle, N., Schneider, M.S., Mallat, T., Baiker, A. (2001) Continuous Platinum-catalysed enantioselective hydrogenation in "supercritical" solvents, *Chem. Commun.* 673–674.

[305] Wandeler, R., Baiker, A. (1999) In situ monitoring of heterogeneously catalyzed reactions under supercritical conditions, *Chimia* **53**, 566-569.

[306] Niessen, H.G., Eichhorn, A., Woelk, K., Bargon, J. (2002) Homogeneous hydrogenation in supercritical fluids mediated by colloidal catalysts, *J. Mol. Catal. A. Chem.* **182-183**, 463-470.

[307] Klabunovskii, E.I. (1991) Asymmetric hydrogenation on metals, *Russ. Chem. Rev.*, **60**, 980-995.

[308] Balazsik, K., Bartok, M. (2004) New data on enantiomeric excess versus conversion during enantioselective hydrogenation of activated ketones on a platinum catalyst, *J. Catal.* **224**, 463-472.

[309] Sutyinszki, M., Szori, K., Felfoldi, K., Bartok, M. (2002) Hetero-geneous asymmetric reactions. 29. Enantioselective hydrogenation of ethyl benzoylformate over dihydrocinchonidine-modified platinum-alumina catalyst in acetic acid, *Catal. Lett.* **81,** 281-284.

[310] Sutyinszki, M., Szori, K., Felfoldi, K., Bartok, M. (2002) 98% Enantioselectivity in the asymmetric synthesis of a useful chiral building block by heterogeneous method: enantioselective hydrogenation of ethyl benzoylformate over cinchona modified Pt/Al_2O_3 catalysts, *Catal. Commun.* **3**, 125-127.

[311] Balazsik, K., Bartok, M. (2004) Heterogeneous asymmetric reactions-Part 40. New data on repeated use of Pt-alumina catalyst during enantioselective hydrogenation of ethyl pyruvate, *J. Mol. Catal. A. Chem.* **219,** 383-389.

[312] Bartok, M., Bartok, T., Szollosi, G., Felfoldi, K. (1999) Electrospray ionization mass spectrometry in the heterogeneous catalyzed organic reactions: unknown compounds in the pyruvate hydrogenation, *Catal. Lett.* **61,** 57-60.

[313] Bartok, M., Szabo, P.T., Bartok, T., Szollosi, G. (2000) Identification of ethyl pyruvate and dihydrocinchonidine adducts by electrospray ionization mass spectrometry, *Rapid Commun. Mass Spectrom.* **14,** 509-514.

[314] Bartok, M., Bartok, T., Szollosi, G., Felfoldi, K. (2002) Structure characterization of acetyl pyridinium-ethyl pyruvate adducts by electrospray ionization mass spectrometry, *J. Mass Spectrom.* **37**, 1034-1038.

[315] Bartok, T., Szollosi, G., Felfoldi, K., Bartok, M., Thiel, J. (2000) New results on the mass spectra of cinchona alkaloids., *J. Mass Spectrom.* **35,** 711-717.

[316] Bartok, T., Felfoldi, K., Szollosi, G., Bartok, M., Dega Szafran, Z., Thiel, J. (2000) Mass spectra of iso-cinchona and halogenated cinchona alkaloids, *Eur. J. Mass Spectrom.* **6,** 347-355.

[317] Szollosi, G., Forgo, P., Bartok, M. (2003) Hydrogenation of cinchona alkaloids over supported Pt catalyst, *Chirality,* **15**, S82-S89.

[318] Bartók, M., Kele, Z., Sutyinszki, M., Bucsi, I., Felfoldi, K. (2004) Investigation of chiral reactions: the structural detection of new hydrogenated isocinchona alkaloids from mixtures without isolation using electrospray ionization tandem mass spectrometry, *Rapid Commun. Mass. Spectrom.* **18,** 1352-1360.

[319] Lavoie, S., Laliberte, M.A., McBreen, P.H. (2003) Adsorption states and modifier–substrate interactions on Pt (III) relevant to the enantioselective hydrogenation of alkyl pyruvates in the Orito reaction, *J. Amer. Chem. Soc.* **125**, 15756-15757.

[320] Varga, T., Felfoldi, K., Forgo, P., Bartok, M. (2004) Heterogeneous asymmetric reactions. Part 38. Enantioselective hydrogenation of fluoroketones on Pt-alumina catalyst, *J. Mol. Catal. A. Chem.* **216**, 181-187.

[321] Bartok, M., Sutyinszki, M., Felfoldi, K., Szollosi, G. (2002) Unexpected change of the sense of the enanatioselective hydrogenation of ethyl pyruvate catalysed by a Pt-Alumina-cinchona alkaloid system, *Chem. Commun.* 1130-1131.

[322] Felfoldi, K., Varga, T., Forgo, P., Bartok, M. (2004) Enantioselective hydrogenation of trifluoromethylcyclohexyl ketone on cinchona alkaloid modified Pt-alumina catalyst, *Catal. Lett.* **97**, 65-70.

[323] Vayner, G., Houk, K.N., Sun, Y.K (2004) Origins of enantioselectivity in reduction of ketones on cinchona alkaloid modified platinum, *J. Amer. Chem. Soc.* **126**, 199-203.

[324] Jacoby, M. (2004) Chiral catalysis at surfaces, *Chem. Eng. News,* **82**, 37–41.

[325] Yokozeki, M., Shimokoshi, K., Miyazaki, E. (1985) Enantioselective hydrogenation of methyl acetoacetate over a modified Nickel surface: MNDO calculations of stability for intermolecular interactions, *J. Phys. Chem.* **89**, 2397-2400.

[326] Bartok, M., Felfoldi, K., Torok, B., Bartok, T. (1998) A new cinchona-modified Pt catalyst for the enantioselective hydrogenation of pyruvate: the structure of the 1:1 alkaloid-reactant complex, *Chem. Commun.* 2605-2606.

[327] Szollosi, G., Felfoldi, K., Bartok, T., Bartok, M. (2000) Heterogeneous asymmetric reactions, 22. Beta-Isocinchona alkaloids in enantioselective hydrogenations, *React. Kinet. Catal. Lett.* **71,** 99-108.

[328] Bartok, M., Torok, B., Balazsik, K., Bartok, T. (2001) Heterogeneous asymmetric reactions. 23. Enantioselective hydrogenation of ethyl pyruvate over cinchonine- and alpha-isocinchonine-modified platinum catalysts, *Catal. Lett.* **73**, 127-131.

[329] Bartok, M., Sutyinszki, M., Felfoldi, K. (2003) Enantioselective hydrogenation of ethyl pyruvate catalyzed by alpha- and beta-isocinchonine modified Pt/Al_2O_3 in acetic acid, *J. Catal.* **220,** 207-214.

[330] Bartok, M., Felfoldi, K., Szollosi, G., Bartok, T. (1999) Heterogeneous asymmetric reactions. 14. Epicinchona alkaloids in the enantioselective hydrogenation of ethyl pyruvate over Pt/alumina. What determines the sense of enantioselection? *React. Kinet. Catal. Lett.* **68**, 371-377 (1999).

[331] Margitfalvi, J.L., Tfirst, E. (1999) Enantioselective hydrogenation of alpha-ketoesters over cinchona-Pt-Alumina catalysts, *J. Mol. Catal. A. Chem.* **139**, 81-95.

[332] Tungler, A., Fogassy, G. (2001) Catalysis with supported Palladium metal, selectivity in the hydrogenation of C=C, C=O and C=N bonds, from chemo- to enantioselectivity, *J. Mol. Catal. A. Chem.* **173**, 231-247.

[333] Isoda, T., Ichikawa, A, Shimamoto, T. (1958) Asymmetric reduction by means of catalytic complex Tyrosine-Palladium and Raney Nickel with adsorbed amino acid, II., *Rikagaku Kenkyusho Hokoku, Inst. Phys. Chem. Res.* (Tokyo) **34**, 134-142 and 143-151, *Chem. Abstr.* 1960, 287d.

[334] Torok, B., Balazsik, K., Felfoldi, K., Bartok, M. (2001) Asymmetric reactions in sonochemistry, *Ultrason. Sonochem.* **8,** 191-200.

[335] Torok, B., Balazsik, K., Torok, M., Felfoldi, K., Bartok, M. (2002) Heterogeneous asymmetric reactions 20. Effect of ultrasonic variables on the enantiodifferentiation in cinchona-modified platinum-catalyzed sonochemical hydrogenations, *Catal. Lett.* 81, 55-62.

[336] Balazsik, K., Torok, B., Felfoldi, K., Bartok, M. (1999) Homogeneous and Heterogeneous Asymmetric Reactions. Part 11. Sonochemical Enantioselective Hydrogenation of Trifluoromethyl Ketones over Platinum Catalysts, *Ultrasonics Sonochem.* **5**, 149-155.

[337] Torok, B., Szollosi, G., Balazsik, K., Felfoldi, K., Kun, I., Bartok, M. (1999) Ultrasonics in Heterogeneous Metal Catalysis. Sonochemical Chemo- and Enantioselective Hydrogenations over Supported Platinum Catalysts, *Ultrasonics Sonochem.* **6**, 97-103.

[338] Torok, B., Balazsik, K., Szollosi, G., Felfoldi, K., Bartok, M. (1999) Ultrasonics in Asymmetric Syntheses. Sonochemical Enantioselective Hydrogenation of Prochiral C=O Groups over Platinum Catalysts, *Chirality* **11**, 470-474.

[339] Torok, B., Balazsik, K., Torok, M., Szollosi, G., Bartok, M. (2000) Asymmetric sonochemical reactions. Enantioselective hydrogenation of α-ketoesters over platinum catalysts, *Ultrasonics Sonochem.* **7**, 151-155.

[340] Torok, B., Balazsik, K., Torok, M., Felfoldi, K., Bartok, M. (2002) Heterogeneous asymmetric reactions 20. Effect of ultrasonic variables on the

enantiodifferentiation in cinchona-modified platinum-catalyzed sonochemical hydrogenations, *Catal. Lett.* **81**, 55-62.

[341] Torok, B., Felfoldi, K., Szakonyi, G., Bartok, M. (1997) Sonochemical Enantioselective Hydrogenation of Ethyl Pyruvate over Platinum Catalysts, *Ultrasonics Sonochem.,* **4,** 301-304.

[342] Smith, G.V., Musoiu, M. (1979) Asymmetric hydrogenation over modified Raney Nickel, *J. Catal.* **60**, 184-192.

Chapter 6

ELECTROCHEMICAL ENANTIOSELECTIVE REDUCTIONS

Abstract

This chapter considers the results received from enantioselective electrochemical reactions mainly from hydrogenations on metal (Pt, Ni), graphite, and Hg cathodes modified with chiral compounds (polyamino acids, alkaloids, and amino alcohols). Enantioselectivities of hydrogenation reactions were not effective, *ee* values did not exceed 50%, except in the cases of dehydrohalogenation and oxidation, from which optical yields reached above 90%.

6.1. General

Electrochemical reactions have some advantages in comparison with catalytic organic reactions. They do not need high hydrogen pressures in reaction vessels because hydrogen amounts and rates of reaction can be controlled by electrode potentials. This is a great advantage compared to the usual organic and catalytic reactions. But unfortunately, electrochemical enantioselective reactions in general have proved not to be very effective. The efficiency of enantioselectivity depends of many factors, such as the nature of the chiral inductor and its concentration, the electrode material, and the values of the potentials.

There are several methods for accomplishing electrochemical enantioselective reactions with the use of chiral inductors. They are the use of

a) chiral solvents
b) chiral supporting electrolytes
c) chiral electrode surface active material and
d) chiral modified catalytically active electrodes (see reviews [1-4]).

The first two methods, a) and b), are not very effective and need rather large amounts of chiral compounds as chiral inductors, although in some cases they give the products of reduction with *ee*'s up to 20% [1-5]. On the other hand processes c) and d) are more effective. They require only very small amounts of optically active compounds as asymmetric inductors and gave rather high optical yields.

The formation of optically active alcohol [6-8] and its pinacolinization [9] was observed in the reduction of acetophenone on a Hg cathode in the presence of (+) or (-)-ephedrine. (-)-Ephedrine resulted in the (*R*)-(+)-1-phenylethanol with an *ee* of 4.2% and (+)-ephedrine resulted in the (*S*)-(-)-1-phenylethanol with an *ee* of 4.6%.

The electrochemical reductions of prochiral carbonyl compounds like acetophenone and other carbonyl compounds in the presence of a number of alkaloids as chiral inductors have been described [10]. Thus, the reduction of phenylglyoxalic acid into mandelic acid was studied on a Hg cathode in the presence of a number of optically active alkaloids and other compounds which were strongly adsorbed on the cathode (strichnine, brucine, nicotine) and weakly adsorbed on the cathode (codeine, morphine) [11]. It was found that *ee* values increased with increasing adsorption of depolarizing molecules on the cathode, and the maximal *ee* was found in the presence of strychnine (*ee* was 22% at 0^{0}C).

It is interesting to note that quinidine promoted the formation of (*R*)-(-)-mandelic acid with low enantioselectivity (*ee* 2%), while quinine, cinchonine, and cinchonidine were ineffective [12]. If the alkaloid is not very strongly adsorbed on the cathode, it can be added in large amounts and play the role as the supporting electrolyte. If the alkaloid is strongly adsorbed, it needs only a small amounts to achieve a good effect [13].

Double asymmetric reduction was observed in the reduction of (-)-menthyl phenylglyoxilate into (*R*)-(-)-mandelate on a Hg cathode with an *ee* of 20.6% [14,15] and in the diastereoselective reduction of (*S*)-(-)-*N*-(*alpha*-methylbenzyl)benzoylformamide into the (*R,S*)-diastereomer with an *ee* of 12.5% [16].

Analogous to the reduction of phenylglyoxalic acid, strichnine proved to be moderately effective in the reduction of the oxime of this acid [17].

In acetate buffer and in the presence of strichnine the oxime of phenylglyoxalic acid was reduced into phenylglycine with either *R* or *S* configuration depending on the value of the cathode potential. The configuration of *R* in the potential interval of -0.95 V to -1.14 V was changed to *S* with an *ee* of 11.1% in the potential interval of -1.20 V to -1.35 V [17]. The polarographic reduction leads to phenylglycine with an *ee* of 17.1% [18,19].

Increasing the concentration of strichnine also reverses the stereochemical course of the electroreduction from an *ee* of 5% *R* at 0.1mM to an *ee* of 8% *S* at 3 mM [13]. The same effect was observed in the reduction of phenylglyoxalic acid to mandelic acid, in which an *ee* of 14% *R* resulted at 0.1 mM strychnine but an *ee* of 1% *S* resulted at 3.0 mM strychnine [13].

The 2-, 3-, and 4-acylpyridines were chosen as substrates on the Hg electrode because they are reducible at the potential ranges where alkaloids used as chiral inductors are electroinactive [10]. It was found that the stereochemical reduction was more effective in the case of 2-acetylpyridine (buffer acetate-EtOH, 0.5 mmol strychnine, 0^{0}C), the *ee*'s proved to be a function of the electrode potential, and the maximal *ee*'s were observed at very small concentrations of alkaloid. All characteristics indicate that the reactions proceeded on the surface of the electrode in the presence of the alkaloid as chiral inductor. The reduction proved to be very sensitive for the structure of alkaloids. In the cases of reductions of 2-acetylpyridine the following *ee* values were found with alkaloids: strychnine (47.5%), brucine (27%), cinchonine (2.5%), cinchonidine, (2%), yohimbine (5%), and sparteine (2%). The alkaloids quinine, quinidine, reserpine, and eserine proved to be inactive. In the reduction of 4-acetylpyridine only strychnine (*ee* 40%), brucine (18%), and yohimbine (4%) were stereochemically active (all alkaloids gave (+)-rotating products except yohimbine which gave (-)-rotating products). The reduction of 3-acetylpyridine resulted in only racemic products probably because it could not form an enol as a reactive intermediate.

The hydrogenolysis of carbon-halogen bonds were rather successful. The hydrogenolysis of the C-Cl bond in *N*-Methyl-3,3-dichloro-4,4-diphenylsuccinimide at a Hg electrode resulted in the (-)-monochloro-compound with an *ee* of 3% (in the presence of emetine), 2.7% (presence of yohimbine) and 26.5% (strichnine), while the hydrogenolysis of the C-Br bond in 1,1-dibromo-2,2-dimethylcyclopropane produced an *ee* of 44.3%. More negative working potentials gave higher *ee*'s [20,21].

Very innovative methods were attempted in the use of asymmetrically modified electrodes, although the first experiments were not very successful.

A graphite electrode modified chemically with L-phenylalanine gave an *ee* of 9.7% in the reduction of phenylglyoxalic acid into mandelic acid [22] and an *ee* of 2.5% in the oxidation of methyl *p*-tolyl sulfide into its sulfoxide [23]. Later this oxidation reaction was improved [24].

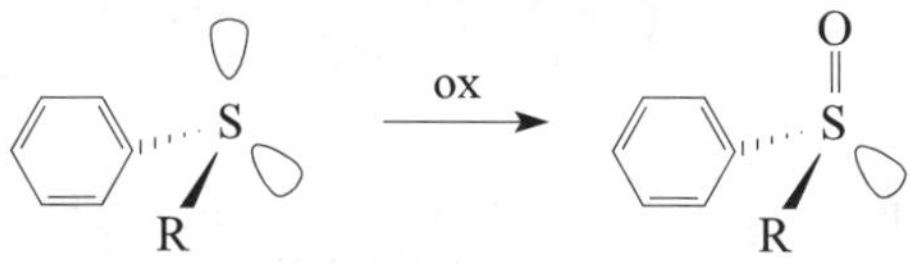

Scheme 6.1.

Oxidation of prochiral sulfides to chiral sulfoxides (Scheme 6.1.) was performed using a Pt electrode coated with a chiral polymer film consisting of the pyrrol-species [Ph-CH(Me)-NH-CO-CH_2CH_2-Pyrrol]$_n$. The reaction occurred in an acetonitrile solution containing 0.1mol nBu_4N-BF_4 at 1.3V vs. SCE (saturated calomel electrode) and gave the results shown below in Table 6.1.

A new set of electrodes were described that consisted of a graphite electrode covered with an optically active polyamino acid, such as poly-L-valine [1,25]. With this electrode a product with an *ee* of 43% from the reduction of 4-methylcoumarin was obtained [26], while a more effective result, an *ee* of 54%, in the oxidation of cyclohexyl phenyl sulfide on a Pt electrode coated with poly-L-valine was obtained [27].

R in sulfoxide	Chemical yield (%)	Optical yield (%)
Me	24	4 (S)
*n*Bu	15	67 (S)
*t*Bu	15	> 99 (S)
Hexyl	22	50 (S)

Table 6.1. Oxidation of prochiral sulfides into chiral sulfoxides.

According to Abe et al. [5] the reduction on graphite-poly-L-valine electrode was extended to prochiral compounds other then olefins. As shown below carbonyl compounds were reduced in reactions 1-3 (Scheme 6.2.) and a cyclopropane derivative was hydrogenolysed in reaction 4.

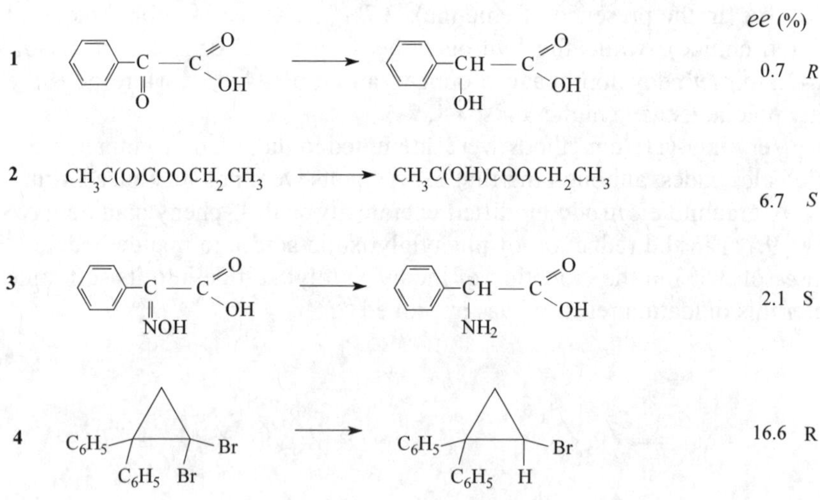

Scheme 6.2.

On a poly-L-valine-coated graphite electrode, the asymmetric reduction of the C=C bond in 2-methylcoumarin and in citraconic acid produced *ee*'s of 43% and 25%, respectively [25,19]. Earlier, methyldihydrocoumarin was obtained only with an *ee* of 18.6% using spartein as an asymmetric inductor [26].

The low enantioselectivity in these cases can be explained by low coverage of chiral species on electrode surfaces. Therefore, some attempts were undertaken to use metal hydrogenating catalysts as electrode materials. But at first [28] only *ee* values of 2-6% were obtained in the reduction of 2-oxo compounds into alcohols using Raney-Ni powder electrodes modified with

(2*R*,3*R*)-tartaric acid (like Ni catalysts modified with hydroxy or amino acids elaborated earlier [29,30], see also Chapter 4 of this book).

Twice modified Raney-Ni powder was deposited on a Ni plate electrode. Reduction of 2-hexanone or 2-octanone was performed in MeOH + LiCl electrolite at -1.0 V vs SCE (saturated calomel electrode). Modification of the electrode with (2*R*,3*R*)-(+)-tartaric acid resulted in (*S*)-(+) alcohols with *ee*'s of about 9%, and modification with (2*S*,3*S*)-(-)-tartaric acid produced (*R*)-(-)-alcohols with the same *ee*'s [28].

Electrodes made from Pt covered with Ni powder or Ni-black and modified with (2*R*,3*R*)-(+)-tartaric acid were more effective [31,32]. In the reduction of ethyl acetoacetate *ee* values of 8-12% were obtained.

A cathode containing a Raney-Ni-Tartaric acid catalyst coated with CdS particles proved to be rather effective. Irradiation of this electrode by a xenon-lamp during reduction of methyl acetoacetate in EtOH solution leads to methyl 3-hydroxybutyrate with an *ee* of 67% [33].

According to Kambe et al.[33], the mechanism of reaction consists not in the reduction of methyl acetoacetate by photochemically produced hydrogen, but in the electrochemical reduction of the C=O bond and proton by photogenerated electrons. Thus, the hydrogen production sites on the surface of Raney Ni are different from methyl 3-hydroxybutyrate production sites.

Some special methods of enantioselective electrochemical reactions should be mentioned. D-Alanine was prepared with an *ee* close to 100% using the electrochemical reduction of pyruvic acid using an electrode on which amino acid oxidase and electron mediator were immobilized [34].

A strange method of preparation of optically active compounds without any chiral inductors was described using an electrochemical cell with electrodes of special asymmetric configuration made of barium titanate [35]. Reduction of fumaric acid resulted in (*R*)-(+)-malic acid with an *ee* of 17%, or (*S*)-(-)-malic acid, with 18% *ee*, depending on the orientation of the cathode. In this case perhaps the mica plates are the asymmetric inductors present in the electrochemical cell construction materials; however, the effect on the configuration of product of the orientation of the cathode inside the electrolytic cell remains incomprehensible.

In 1986 Takahashi et al. [36] alleged that the electrolytic reduction of 2-oxocarboxylic acids into 2-hydroxycarboxylic acids at a Hg-cathode, perpendicular to a magnetic field of 0.168T (1680 Gauss) resulted in chiral products with an *ee* up to 25%. These experiments were checked by Bonner [36] using a more powerful magnetic field of 7.03T (70300 Gauss) and found the products were totally racemic.

References

1 Tallec, A. (1985) Syntheses asymetrique par voie electrochimique, *Bull. Soc. Chim. Fr.* 743-761.

2 Blaser, H.U. (1992) The chiral pool as the source of enantioselective catalysis and auxiliaries, *Chem. Rev.* **92**, 935-952.

3 Blaser, H.U (1991) Enantioselective synthesis using chiral heterogeneous catalysts, *Tetrahedron Asymm.* **2**, 843-866.

4 Blaser, H.U. and Mueller, M. (1991) Enantioselective catalysis by chiral solids: Approaches and results, *Stud. Surf. Sci. Catal.* **59,** 73-92.

5 Abe, S., Fuchigami, T. and Nonaka, T. (1983) Electrochemical asymmetric reduction of prochiral carbonyl compounds, oximes and a gem-dihalide on a poly-L-Val-coated graphite electrode, *Chem. Lett.* 1033-1036.

6 Horner, L. and Degner, D. (1968) Asymmetrische Induktion bei der Reduktion von Acetophenon an der Quecksilberkathode mit chiralen Leitsalzen, *Tetrahedron Lett.* 5889-5892.

7 Horner, L. and Brich, W. (1977) Elektroreduktion von prochiralen Aryl-ketonen an Grenzflaechen modifizierten Kohleelektroden, *Ann.* 1354-1364.

8 van Tilborg, W.J.M. and Smit, C.J. (1978) Asymmetric induction in electropinacolization, *J. Roy. Netherl. Chem. Soc.* **97,** 89-90.

9 Kopilov, J., Kariv, E. and Miller, L.L. (1977) Asymmetric cathodic reduction of acetophenones, *J. Amer. Chem. Soc.* **99**, 3450-3454.

10 Kariv, E., Terni, H.A. and Gileaodi, E. (1973) Asymmetric induction by alkaloids in electrolyte reductions, *J. Electrochem. Soc.* **120**, 639-641.

11 Jubault, M., Raoult, E. and Peltier, D. (1973) Synthese asymetrique par voie electrochimique.Reduction de l'acide phenylglyoxalique en presence d'alkaloides, *Compt. rend. C,* **277**, 583-585.

12 Jubault, M., Raoult, E. and Peltier, D. (1974) Synthese asymetrique par voie electrochimique. Etude de l'influence de differents parametres sur le rendement optique, *Electrochem. Acta* **19**, 865-874.

13 Jubault, M. (1980) Effect of alkaloid concentration in asymmetric electrosynthesis, *J. Chem. Soc., Chem. Comm.* 953-954.

14 Jubault, M., Raoult, E. and Peltier, D. (1977) Asymmetric synthesis. Reduction of (-)-menthyl phenyglyoxilate, *Electrochem. Acta,* **22,** 67-73.

15 Jubault, M., Raoult, E. and Peltier, D. (1981) Asymmetric synthesis by electrochemical routes.VI. Studies in double induction, *Electrochem. Acta*, **26**, 287 -290.

16 Boulmedais, A. and Jubault, M. (1988) Electrochemical induction of (*S*)-*N*-(1-methylbenzyl)benzoylformamide, *Bull. Soc. Chim. Fr.* 610-612.

[17] Jubault, M., and Raoult, E. (1977) Effect of cathodic potential on the electrochemical synthesis of optically active amino acids, *J. Chem. Soc., Chem. Comm.* 250-251.

[18] Jubault, M., Raoult, E. and Peltier, D. (1979) Asymmetric synthesis by electrochemical routes. IV. *Electrochem. Acta,* **24**, 1219-1227.

[19] Mairanovskii, S.G. (1989) Electrosynthesis of optically active organic compounds, *Second republic meeting on asymmetric reactions,* Telavi, Gruz. SSR., 24-26 October, Abstracts. p. 55.

[20] Tallec, A., Hazard, R., LeBouc, A., and Grimshaw, J. (1986) Asymmetric electrochemical synthesis by reduction of a *gem*-dichlorosuccinimide, *J. Chem. Res. (S)* 342 –343, (see also *Tetrahedron* 1982, **38**, 93-98).

[21] Mazur, S. and Ohkubo, K. (1973) Enantioselective reduction in a chiral double layer, *J. Amer. Chem. Soc.* **97**, 2911-2912.

[22] Watkins, B.F., Behling, J.R., Kariv, E. and Miller, L.L. (1975) A chiral electrode, *J. Amer. Chem. Soc.* **97,** 3549-3550.

[23] Firth, B.E., Miller, L.L., Mitani, M., Rogers, T., Lennox, J. and Murray, R.W. (1976) Anodic and cathodic reactions on a chemically modified edge surface of graphite, *J. Amer. Chem. Soc.* **98**, 8271-8272; (see also Firth, B.E. and Miller, L.L. (1976) Oxidation on DSA and chirally modified DSA and SnO_2 electrodes, *J. Amer. Chem. Soc.* **98**, 8272-8273).

[24] Takano, N., Takeda, S. and Takeno, N. (1991) Asymmetric electrooxidation of sulfides on chiral conducting polymer coated electrode, *Intern. Sympos. Organic Reactions, Kyoto,* Aug. 19-21.

[25] Nonaka, T., Abe, S. and Fuchigami, T. (1983) Part 2. Electrochemical asymmetric reduction of citraconic and mesaconic acids on optically-active polyamino acid-coated electrodes, *Bull. Chem. Soc. Jpn,* **56**, 2778-2783.

[26] Gourley, R.N., Grimshaw, J. and Millar, P.C. (1967) Electrochemical reduction in the presence of tertiary amines an asymmetric synthesis of 3,4-dihydro-4-methylcoumarin, *J. Chem. Soc. Chem. Comm.* 1278-1279.

[27] Komori, T. and Nonaka, T. (1983) Stereochemical studies of the electrolytic reactions of organic compounds. Part 22. Electroorganic reactions on organic electrodes. 3. Electrochemical asymmetric oxidation of phenylcyclohexylsulfide on poly-L-Valine coated Platinum. *J. Amer. Chem. Soc.* **105**, 5690-5691.

[28] Fujihira, M., Yokozawa, A., Kinoshita, N. and Osa, T. (1982) Asymmetric synthesis by modified Raney Ni powder electrode, *Chem. Lett.* 1089-1092. (see also Osa T., and Matsue T. (1985) Asymmetric reduction of methyl acetoacetate on powder electrode modified Ni, *Denki Kagaku. Kaguo Butsuri Kagaku.* **53**, 104-108, *Chem. Abstr.* 1985, **103**, 13491f.

[29] Izumi, Y. and Tai, A. (1977) *Stereo-differentiating reactions*, Kodansha, Academic Press, N.Y.

30 Klabunovskii, E.I. and Vedenyapin, A.A. (1980) *Asymmetric catalysis, Hydrogenation on metals,* Nauka, Moscow, (russ.).

31 a) Baturova, M.D., Kuznetsova, T.I., Vedenyapin, A.A. and Klabunovskii, E.I. (1987) Electrohydrogenation of ethyl acetoacetate on modified Ni catalyst, *Izv. Acad. Sci. USSR (Ser. Khim.)* 2674–2677

b) Fish, M.J. and Ollis, D. (1978) Heterogeneous catalysis by optically selective surfaces, *Catal. Rev. Sci. Eng.* **18**, 259-295.

32 Giorgadze, N.G., Baturova, M.D., Murina, I.P., Kuznetsova, T.I., Areshidze, G.Kh. and Vedenyapin, A.A. (1986) Enantioselective hydrogenation, electrohydrogenation and adsorption on Ni and Ni-Pd catalysts, *First republic meeting on asymmetric reactions,* 25-27 May, Batumi, Gruz. SSR., Abstracts p. 12.

33 a) Kambe, S., Kawai, T., and Kawai, S. (1986) Photocatalytical asymmetric hydrogenation of methyl acetoacetate with CdS modified Raney Ni, *J. Chem. Soc. Jpn,* 1270-1274

b) Brack, A., Barbier, B., Bertrand, M, Chabin, A., and Westall, F. (2002) Polimerization of amino acids, thioesters on mineral surfaces in diluted solutions, in Lacoste H. (ed.) *Proc. Second European Workshop on exo / astrobiology,* Graz, Austria, 16-19 Sept., Publ. ESA Div., p.435 436.

34 Kawabata, S., Iwata, N. and Yoneyama, H. (2000) Asymmetric electrosynthesis of amino acids using an electrode modified with amino acid oxidase and electron mediator, *Chem. Lett.* 110-111.

35 Siemcu, N. (Institutul de cercetari chimico-farmaceutice, Bucuresti) (1979) Procedeu si instalatie pentru prepararea enantiomerului dextrogir sau levogir, *Romania inventiet No.* 67279, 30. 04. 1979.

36 a) Bonner, W.A. (1995) The quest for chirality, in *Physical origin of homochirality in life,* Conference Proceedings, 379. Santa Monica, Cal., American Institute of Physics Press, Woodbury, N.Y., p. 17-49

b) Bonner, W.A. (1990) Attempted asymmetric electrochemical reductions in magnetic-fields, *Origins of Life and Evolution of Biosphere* **20,** 1-13

c) Bonner, W.A. (1990) The electrochemical reduction of phenylglyoxylic acid in a magnetic-field, *Electrochem. Acta* **35**, 683-684.

d) Takahashi, F., Tomii, K. and Takahashi, H. (1986) The electrochemical asymmetric reduction of *alpha*-keto acids in the magnetic-fields, *Electrochem. Acta* **31**, 127-130.

Chapter 7

PRACTICAL ASYMMETRIC CATALYTIC REACTIONS

Abstracts

This chapter summarizes some of the most characteristic results obtained with the use of mainly homogeneous metal complex catalysts either in the industry or in processes recommended for practical use. These are large scale processes of asymmetric synthesis of the herbicide metolachlor, synthesis of optically pure menthol with the use of chiral iridium and rhodium phosphine complexes, consideration of the synthesis of ethyl 2-hydroxybutyrate as a monomer for the preparation of biodegradable polyesters with use of heterogeneous chiral modified nickel catalyst, the manufacturing of (*R*)-pantolactone by means of a possible catalytic systems for enantioselective hydrogenation of ketopantolactone, and catalytic systems for the preparation of other pharmaceuticals.

7.1. Asymmetric metal complex catalysts in industry

There are many pharmaceuticals, fragrants, and agrochemicals in which desirable biological properties are strongly related to a given configuration of an enantiomer included in the given compound. In many cases the pharmaceuticals are used as racemic compounds, where one enantiomer is an active component and another is an "isomeric ballast", but it is not always safe, as in the case of racemic thalidomide, in which the *S*-enantiomer proved to be a teratogen and therefore, a tragedy for mothers who took the racemic mixture. Therefore in practical manufacturing, it is desirable to synthesize only one enantiomer.

The preparation of enantiomerically pure compounds can be realized by methods of separation of racemic mixtures into enantiomers, by asymmetric synthesis using chiral auxiliaries, or better by using chiral homogeneous or heterogeneous catalysts. Below will be considered only the last two methods.

At first practical asymmetric catalysis used pure amino acids having biological activity. Thus the process elaborated at first by Monsanto Co. (St. Louis) (Knowles et al.[1ab]) consisted of the asymmetric synthesis of (*S*)-*DOPA* (see as <u>1</u> in Scheme 7.1.) served as a landmark in industrial synthesis based on the asymmetric catalytic hydrogenation of the precursor dehydroamino acid. The rhodium complex with the ligand *DIPAMP*, <u>2</u>, proved to be more effective as the catalyst in the asymmetric synthesis of this remedy against Parkinson disease. VEB ISIS Chemie Ltd.[2] elaborated this synthesis with a Rh-complex using ligand <u>3</u>.

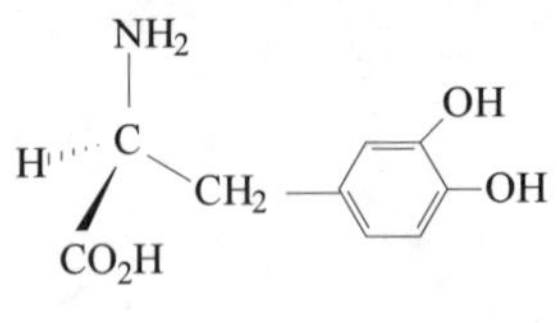

1

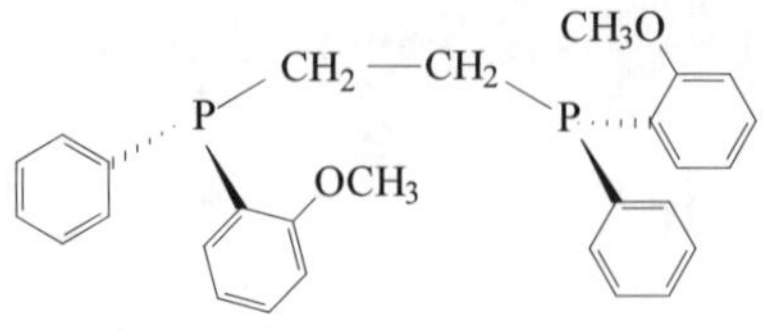

2

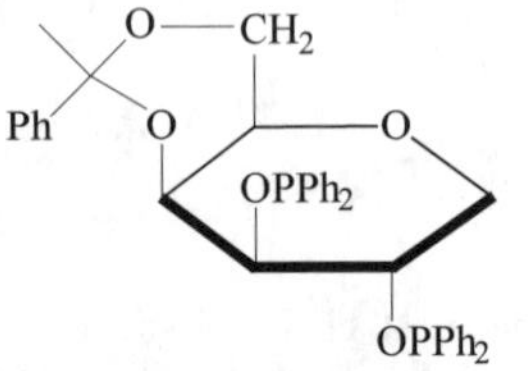

3

Scheme 7.1.

Another method of synthesis of (*S*)-*DOPA* and other *alpha*-amino acids was elaborated using hydrogenation of a precursor corresponding to the azlactone of substituted acrylic acids on a [$PdCl_2${(*S*)-(1-phenylethyl)-amine)}] complex that was prepared *in situ (*Karpeiskaya et al.[3]) (Scheme 7.2.).

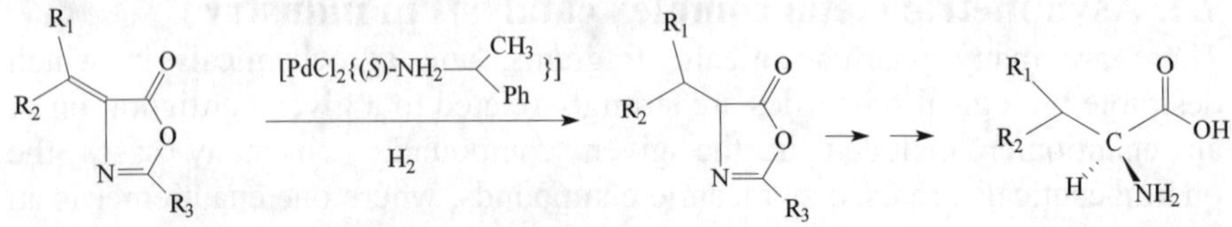

Scheme 7.2.

A lot of chiral diphosphines were used as ligands for Rh- and Ru-complexes as catalysts in the asymmetric hydrogenation of different dehydroamino acids as amino acid precursors and particularly in the almost optically pure *alpha*-phenylalanine for the manufacture of the sweetener *Aspartame* (Scheme 7.3.).

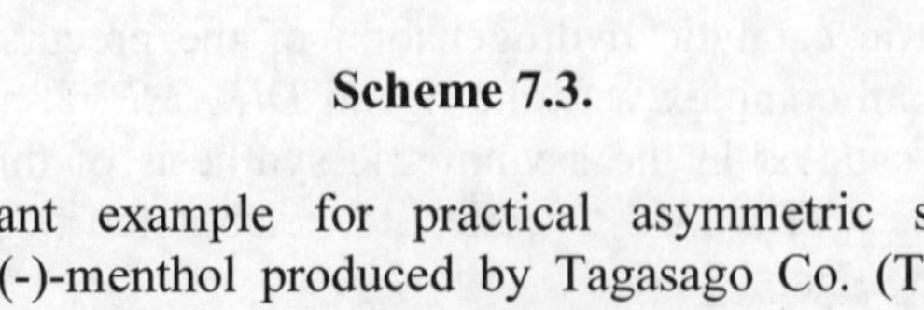

Scheme 7.3.

Another important example for practical asymmetric synthesis is the manufacture of (-)-menthol produced by Tagasago Co. (Tani et al.[4]) on a scale of more than 1000 tons per year according to process in Scheme 7.4. involving the asymmetric hydrogenation of *N,N*-diethylgeranylamine over a Rh-complex containing the chiral ligand, *BINAP* (Scheme 7.5.), first into (+)-citronellal and then into (-)-menthol with high chemical and optical yields [4].

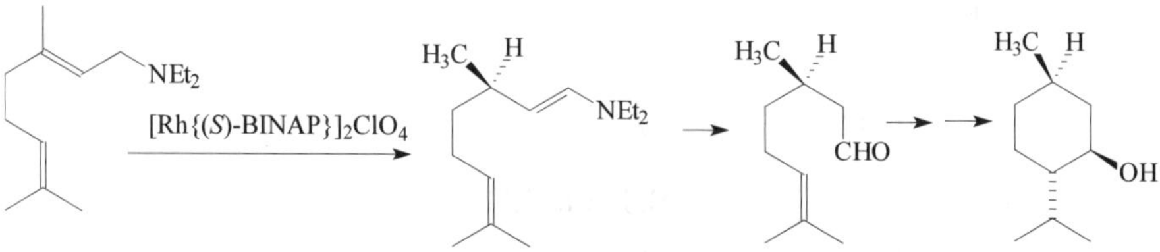

Scheme 7.4.

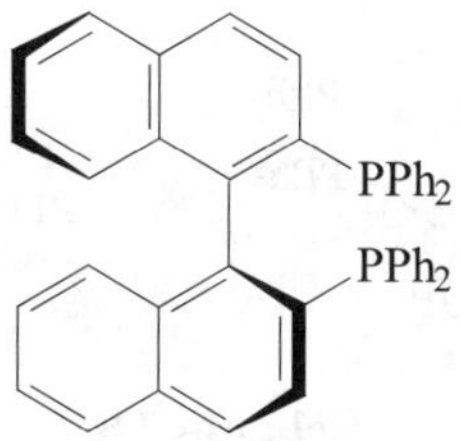

Scheme 7.5.

Rhodium and ruthenium complexes with chiral diphosphine ligands (L) are mainly active in hydrogenation of functionalized alkenes or ketones. In the hydrogenation of C=N bonds these complexes were less effective. And the first attempts to hydrogenate the imine precursor of the important herbicide *Metolachlor* (trade name DUAL) was not very successful (Scheme 7.6.).

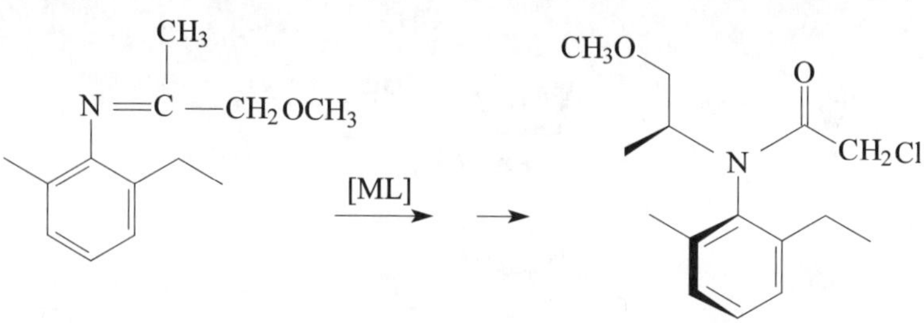

Scheme 7.6.

Rh complexes with chiral diphosphines (*DiPhos, NorPhos* and *CycPhos*) (Scheme 7.7.) were used to hydrogenate this imine precursor with an *ee* of not more than 50% [5].

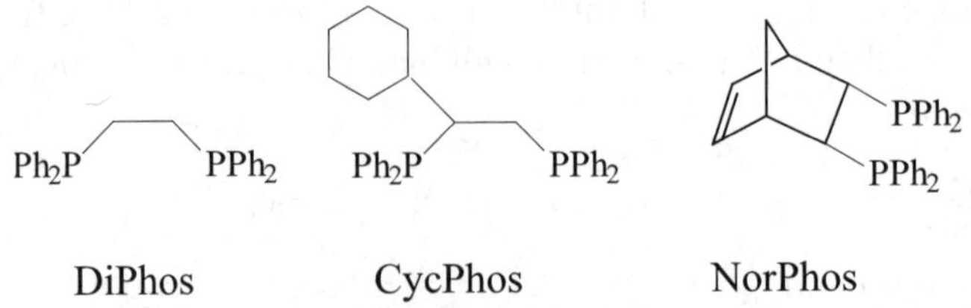

Scheme 7.7.

Iridium complexes of *DIOP* and *BDPP* ligands (Scheme 7.8.) proved to more effective catalysts. The highest optical yields were obtained with [Ir(bdpp)] catalysts.

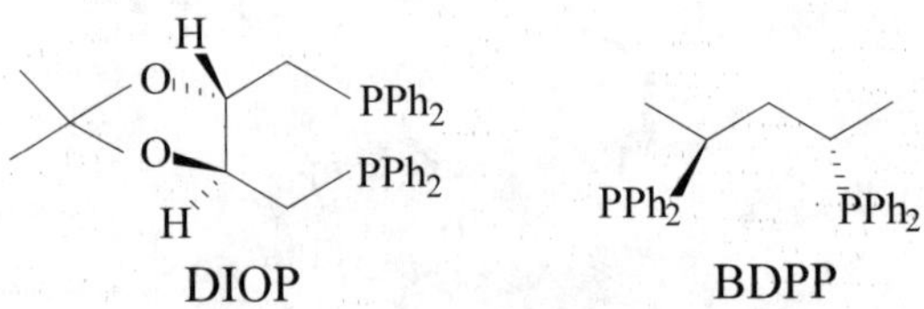

Scheme 7.8.

Even though [Ir(DIOP)] and [Ir(BDPP)] catalysts showed much higher activities than the best Rh complexes, they are were still far below the requirements [6,7]. The commercial production of the biologically active (*S*)-enantiomer was made possible by the Ir complexes of ferrocenyl-diphosphine ligands [7,8] (Scheme 7.9.). Especially the catalysts with R=C_6H_5 and R'= 3,5-diMeC_6H_5 groups gave very good results.

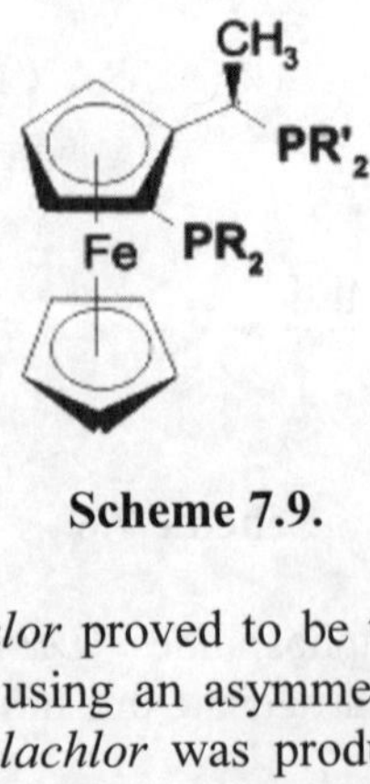

Scheme 7.9.

The preparation of *Metolachlor* proved to be the second example (after the synthesis of (-)-menthol) of using an asymmetric catalysis on a large scale 100,000 tons per year *Metolachlor* was produced in optically active form using this procedure.

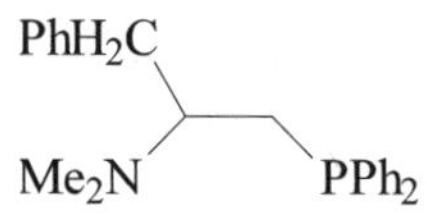

Scheme 7.10.

The Rh-complex of the chiral diphosphine ligand, (*S*)-*PhePhos* (Scheme 7.10.), was used as catalyst for the reductive hydrosilylation of the C=N bond of the imine precursor of the antidepressant, *Pyrazidol,* with an *ee* of 73%[9] (Scheme 7.11).

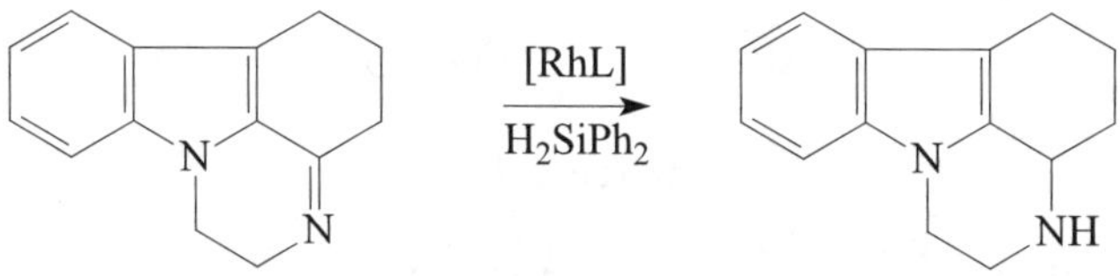

Scheme 7.11.

7.2. Heterogenized metal complex catalysts

The easy separation of a solid catalysts from the reaction mixture is one of the main advantages offered by heterogeneous catalysts. The cost of most chiral homogeneous metal complexes used as catalysts is very high; therefore, their heterogenization is highly desirable. But many attempts to obtain effective heterogenized complex catalysts have been unsuccessful [2]. The first successful step for the synthesis of reusable catalysts was the creation of water-soluble chiral ligands. High enantioselectivity was found for a water-soluble Ru-complex containing sulfonated chiral diphosphine *BINAP* (Scheme 7.12.).

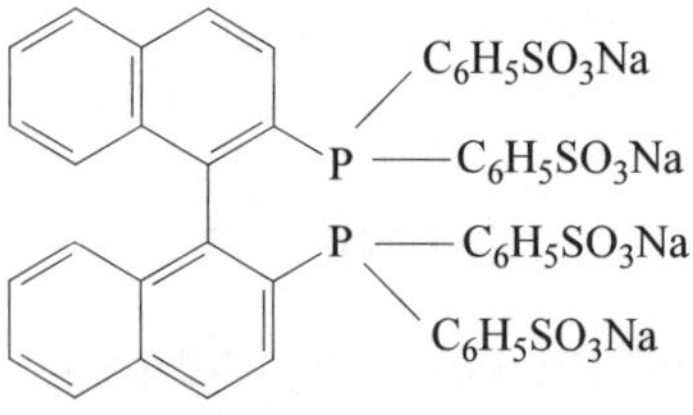

Scheme 7.12.

Using this homogeneous metal complex catalyst allowed hydrogenation in water-organic biphasic systems. In this case the catalyst was dissolved in water and could be used repeatedly, while the substrate remained in the organic solution [10]. Accordingly, the complex $[Rh(COD)Cl]_2$, with sulfonated diphosphine, *ChiraPhos* (Scheme 7.13.), is an effective catalyst for the hydrogenation of dehydro-*N*-acetylalanine into (*S*)-*N*-acetylalanine with an *ee* of 20-60% in water solution at 25^0C and 1-6 bar of hydrogen [11-14].

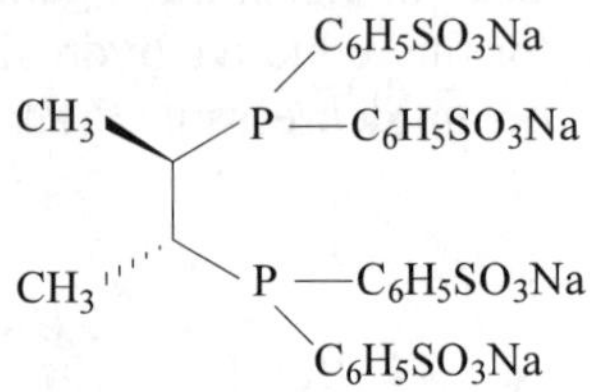

Scheme 7.13.

Davis et al.[13-14] described a new type of effective chiral catalysts, the so called "supported aqueous-phase catalysts". This hydrophilic complex is supported on a hydrophilic solid to create a large interface between the catalytic species and the organic reactants. The hydrophilicity of the ligands and the support creates an interaction sufficient to maintain immobilization of the sulfonized *BINAP* ligand (Scheme 7.12.) in a layer on the carrier.

The complex [Ru(*BINAP*-$4SO_3Na$)(benzene)Cl]Cl was supported on a solid carrier of controlled pore size porous glass [14]. In the process of heterogenization the complex was bounded covalently to the carrier. Sometimes such catalysts revealed enantioselectivity close to those of the homogeneous complex. In the case of the synthesis of the anti-inflammatory drug, *(S)-Naproxen,* the C=C bond of the precursor was hydrogenated (Scheme 7.14.) with high effectiveness.

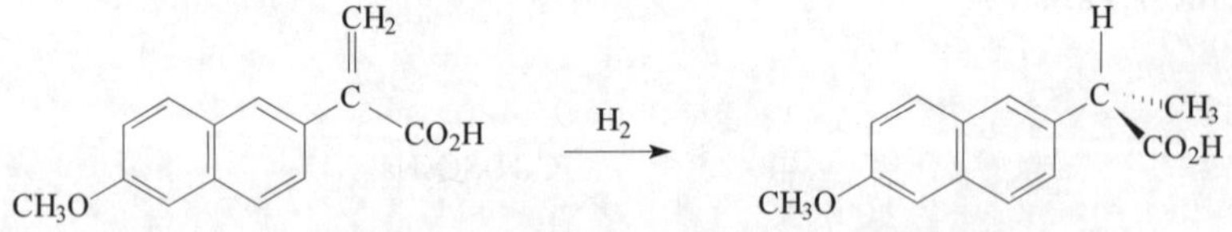

Scheme 7.14.

The homogeneous catalyst $[Ru(OAc)_2(BINAP)]$ also resulted in this synthesis with a chemical yield of 92% and an *ee* of 97%. Immobilization of Ru-complexes with phenyl-sulfonated groups on porous glass allowed for the synthesis of *Naproxen* in a MeOH-ethylene glycol solvent mixture at 20^0C and 100 bar with an *ee* of 89.1% and in an ethylene glycol solvent with an *ee* of 94.8% without decreasing enantioselectivity upon repeated use of catalyst.

Another example of the successful use of these catalysts is in the hydrogenation of butyrylactone (Scheme 7.15.) in ethylene glycol at 85 ^{0}C and 60 bar with an *ee* of 95% [13b] (in homogeneous hydrogenation the *ee* reached 97%).

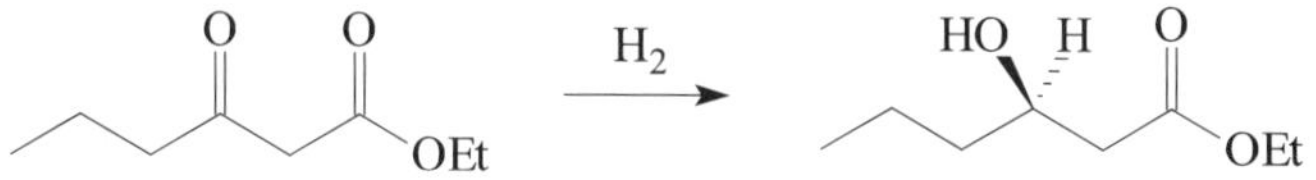

Scheme 7.15.

Comparison of this new type of catalyst with known catalytic heterogenized systems, for example, Ru-BINAP-catalyst deposited on zeolite-*beta* [12], showed that these catalysts are very effective, despite the metal complex not being covalently bonded to the carrier but distributed in the liquid film on the surface and thus representing a "hybrid genuine" homogeneous-heterogeneous catalytic system (Davis [13,14]).

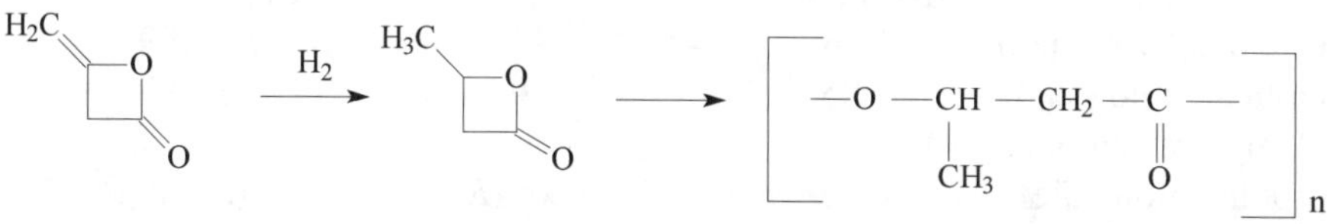

Scheme 7.16.

Hydrogenation of the C=O bond in diketene to its chiral lactone, followed by its polymerization into a polyhydroxybutyrate-type biodegradable polymer [15,16] (Scheme 7.16.) is a prospective process for the preparation of plastics that are produced by ICI with a process involving fermentation of glucose [17a]. It was found that over the complex [RhCl((*S*)-BINAP) (benzene)]Cl Net_3 in THF at 50 ^{0}C and 100 bar the product were produced with *ee*'s of 90-92% [15].

The preparation of optically active ethyl and methyl 3-hydroxybutyrates was considered in detail in Chapter 4. Here it will be compared with the effectiveness of production of these optically active esters using enzymatic, heterogeneous, and homogeneous catalysts [17].

Table 7.1. presents the main characteristics of the process involving the reduction of *beta*-keto butyrate, i) acted upon by Baker's yeast, ii) using a Raney Ni catalyst modified by (2*R*,3*R*)-tartaric acid + NaBr, and iii) in the presence of a chiral Ru complex containing chiral diphosphine *BINAP*.

As can be seen, all three methods ensure a high effectiveness of the process. The optical yields of 3-hydroxybutyrates (in best instances) reached 85-99%. But all three methods have their own advantages and disadvantages.

In spite of the low productivity of the enzymic process – the large volumes of solution with low concentration of product; the long duration of reaction and the need of additional purification, filtration, and extraction of product from reaction mixture; the instability of the enzyme; and the impossibility of recycling the enzyme - this process is still used in practice.

Characteristics	Enzymic reduction	Heterogeneous hydrogenation	Homogeneous hydrogenation
Substrate (g)	EAA, 40	MAA, 40	MAA, 40
Solvent (ml)	water, 2600	none	MeOH, 40
Catalyst (g)	Baker's yeast 200	Ni-TA-NaBr 2	Ru-BINAP 0.14
Reductant	Sucrose, 600 g	H_2, 75 bar	H_2, 20 bar
Temperature (°C)	20	100	20
Reaction time (h)	134	20	40
Product recovering	Filtration extraction distillation	Filtration distillation	Distillation
Chemical yield (%)	59-76	94	99
Optical yield (%)	85	90	99
Configuration of product molecule	(*S*)	*S*) or (*R*)	(*S*) or (*R*)
Productivity g / liter solution x h	0.07	40	12
Repeating of use of catalyst	none	5-30 times	none

Table 7.1. A comparison of various methods for the synthesis of 3-hydroxy-butyrates by the reduction of *beta*-keto esters.

In spite of the fact that metal-complexes are extremely efficient catalysts (*ee*'s above 91%, high productivity and high turnover numbers), the high price of the precious metal and the expensive synthesis of the chiral diphosphine ligand, like *BINAP* and the impossibility of recycling of the catalysts renders this method impractical for large scale production of butyrates.

Recently attempts to improve the process were described. A catalytic system obtained by *in situ* mixing $[RuCl_2(benzene)]_2$ + BINAP has been used in the hydrogenation of MAA with an *ee* of 95% [39]. The highest activity and enantioselectivity in the production of (*R*)-MHB were measured with the cationic catalyst [RuCl(iPr-Ph)]Cl-BINAP, when the reaction was carried out in methanol, ethanol or 2-propanol [40]. These solvents, which also acts as proton donors, accelerated the product release from the reaction intermediate. The presence of water in the reaction mixture was found to be detrimental for both activity and enantioselectivity. Despite the somewhat reduced optical yield, *ee* 94%, in comparison to the heterogeneous Ni catalysts modified with (2*R*,3*R*)-tartaric acid, the application of this catalyst is compensated by the high productivity of the process, by the low cost of the catalysts, their easy preparation, and by the possibility of its reuse many times without loss of catalytic activity and enantioselectivity. Therefore the catalyst of this type probably can be recommended for large scale production of alkyl butyrates as monomers for manufacturing biodegradable polyesters.

Another path of manufacture of practical catalysts is using immobilized chiral metal complexes. Thus, the complex [Rh-BINAP] was occluded [18] in an elastomeric type polydimethylsiloxane membrane, which gave a regenerable active membrane-catalyst with the same enantioselectivity as the homogeneous catalyst in the hydrogenation of acetoacetic acid ester into methyl (*R*)-(-)-3-hydroxybutyrate, that can be polymerized into polyester (Scheme 7.17.).

Scheme 7.17.

In contrast to earlier polymer-supported complex catalysts in which complexes were immobilized through electrostatic interaction, covalent bonds, or coordinative bonds, in this case the complex is captured in the elastomer network by occlusion in a dense polymer in the absence of any supplementary chemical bonding and only as result of steric restrictions. In the hydrogenation of methyl acetoacetate by this catalyst an *ee* of 70% was obtained in polyethyleneglycol solution at 60 ^{0}C. After regeneration of the catalyst and reuse, its activity and enantioselectivity were almost unchanged.

7.3. Asymmetric hydrogenation of *alpha*-keto esters on chiral metal complexes

Neutral Wilkinson type catalysts with chiral ligands are quite effective for the hydrogenation of C=O bonds in prochiral 2-oxocarboxylic acids or their esters and are applied to the asymmetric synthesis of 2-hydroxyesters with rather high *ee*'s of up to 80-95%. The reduction of *alpha*-keto esters, such as alkyl-pyruvates into alkyl-lactates (Scheme 7.18.) on chiral complexes is one of the most important methods of the synthesis of chiral synthones for practical use.

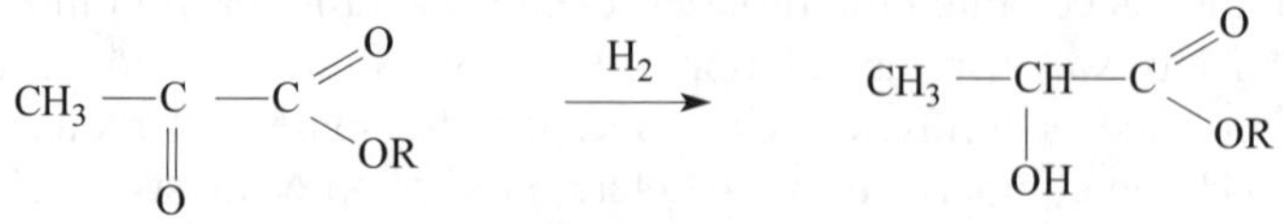

Scheme 7.18.

Thus, on the Rh complex of the *BPPM* ligand (Scheme 7.19.), isobutyl pyruvate was reduced into the lactate with an *ee* of 70% [19-20].

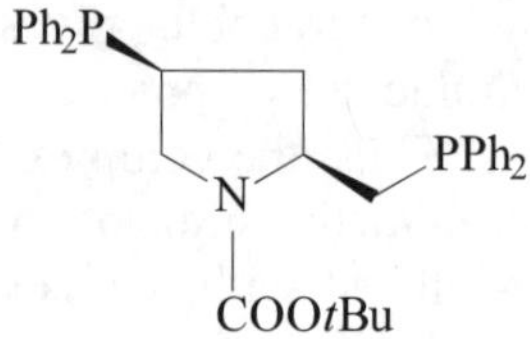

Scheme 7.19.

A comprehensive review of asymmetric hydrogenation, mainly 2-oxo-carboxylic acids and their esters, over heterogeneous catalysts of Pt-alumina modified with alkaloids was presented in Chapter 5 of this book. Here some practical aspects of these catalytic systems will be considered. Thus, ethyl (*R*)-4-phenyl-2-hydroxybutyrate is an important intermediate for the synthesis of the angiotensin-converting enzyme inhibitor *Benazepril* (Scheme 7.20.) [21], and other carboxyalkyl dipeptides like *Enalapril* (Scheme 7.21.) [17c].

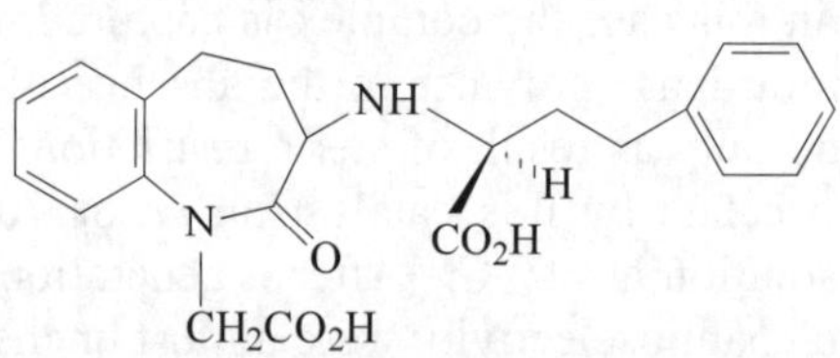

Scheme 7.20.

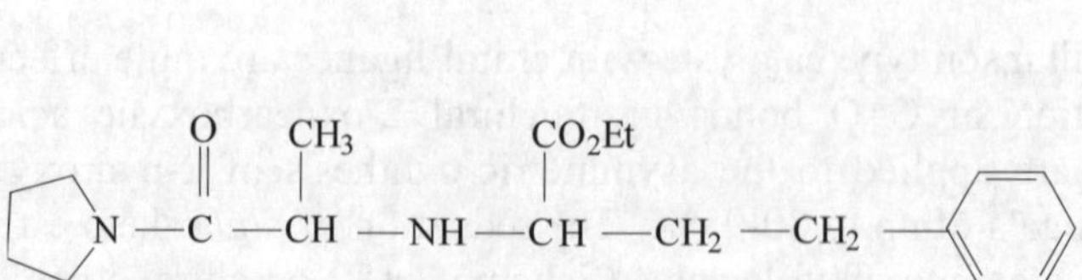

Scheme 7.21.

The ethyl (*R*)-4-phenyl-2-hydroxybutyrate intermediate was prepared by the enantioselective hydrogenation of a keto ester (Scheme 7.22.).

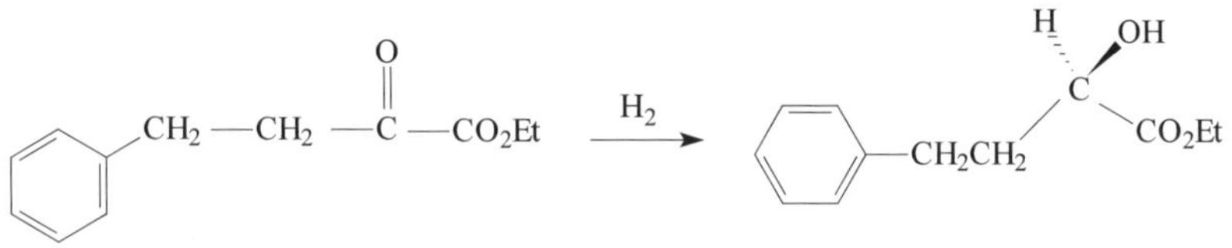

Scheme 7.22.

Using $[Rh(NBD)Cl]_2$ and *NorPhos* as chiral ligand, the 4-phenyl-2-oxobutyrate was hydrogenated with an *ee* of 96% [20]. (*R*)-4-phenyl-2-hydroxybutyrate can also be prepared with an *ee* of 73% on 5% Pt-alumina modified with the alkaloid MeODHCnd at 25^{0}C [21a]. This process has been developed by the Ciba Geigy AG. (Basel) and scaled up for production of up to 200 kg of *Benazepril* with an *ee* of 79-82% [21]. This catalyst gives products with (*R*)-configuration, but *Benazepril* must have (*S,S,S*)-configuration for medical use; therefore, the modifier cinchonidine must be changed to cinchonine giving (*S*)-configuration, but in this case the product will be obtained with a somewhat lower *ee*. The same problem exists in the synthesis of *Enalapril* in which the product must attain the (*S,S,S*)-configurations of all asymmetric centers [17c].

Very promising are supercritical fluids as solvents for asymmetric hydrogenations. Thus, ethyl pyruvate can be hydrogenated into ethyl lactate in supercritical ethane with a 3.5 times increased reaction rate and with the same *ee* as in the common solvent toluene [21b] (see Chapter 5.5 of this book).

7.4. Catalytic synthesis of pantolactone and other pharmaceuticals using chiral homogeneous metal complexes

The methods of preparation of pantolactone via asymmetric hydrogenation of ketopantolactone are of great interest [17a]. Here practical aspects of preparation of pantolactone using homogeneous chiral metal complex catalysts will be considered. The heterogeneous catalysts of this process were considered in Chapter 5.

Pantolactone (PL) [(*R*)-(-)-3-hydroxy-4,4-dimethyltetrahydofuran-2-on], **1** (Scheme 7.23.), is an intermediate in the preparation of several biologically important molecules such as D-(+)-pantothenic acid, **2**, which is a member of B vitamins (Vitamin B_5) and is an important constituent of coenzyme A. The biosynthesis of **2** involves the asymmetric reduction of ketopantolactone, **3**, (KPL) (4,4-dimethyltetrahydrofuran-2,3-dion) to *R*-(-)-PL, because only the *R*-(-)-enantiomer is biologically active.

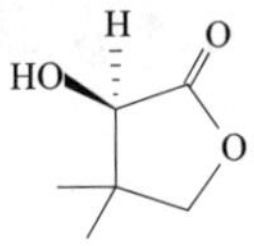

1

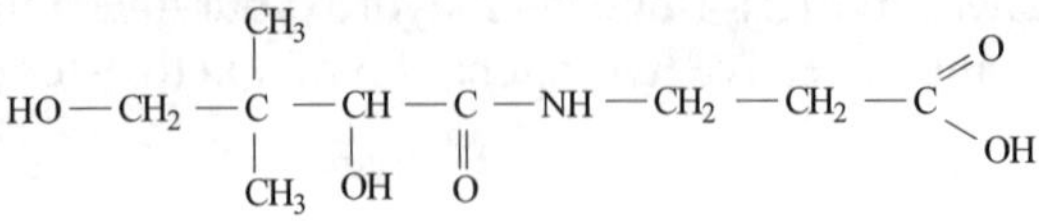

2

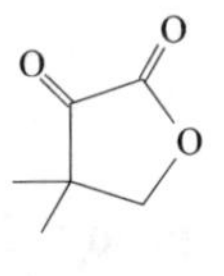

3

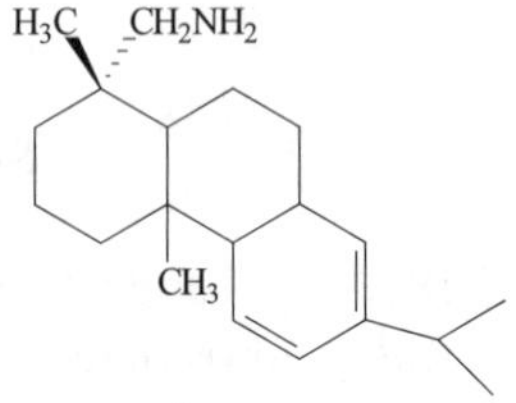

4

Scheme 7.23.

At this time in the industry, the preparation of *R*-(-)-PL is still based partly on the resolution of the racemic pantolactone into its two enantiomers, for example, with the use of dehydroabietylamine, 4 [22]. A more economical preparation of PL might be based on the asymmetric hydrogenation of KPL, using homogeneous or heterogeneous chiral catalysts. Along these lines, a rather effective hydrogenation of KPL to PL using chiral metal phosphine complexes has been realized. This asymmetric hydrogenation can be carried out in quantative yields with high optical purity (*ee* = 86.7%) under optimal conditions using a neutral Rh complex with the chiral phosphine ligand *BPPM* (Scheme 7.19.). The resulting product can be easily purified to pure *R*-(-)-pantolactone in 70-80% yield by recrystallization from a hexane-benzene mixture. The process is shown on Scheme 7.24.

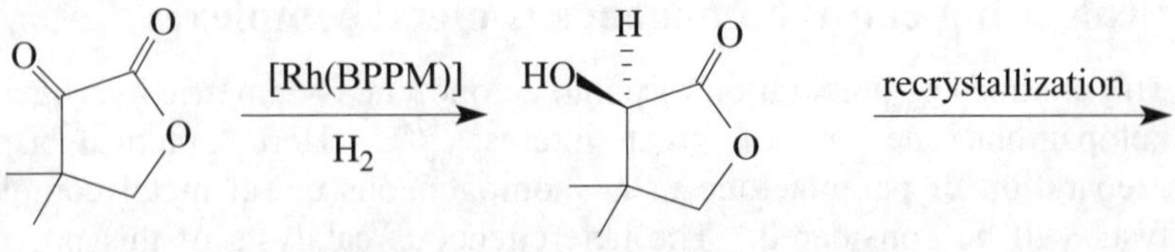

Scheme 7.24.

Reduction of ketopantolactone using Baker's yeast gives an *ee* of approximately 72% [23], but a chiral Rh catalyst has been shown to be superior to Baker's yeast in this process.

High optical yields were obtained using [Rh(hexa-1,5-diene)L] complexes, where L = *BPPM* (Scheme 7.19), or 1 and 2 in Scheme 7.25. [24].

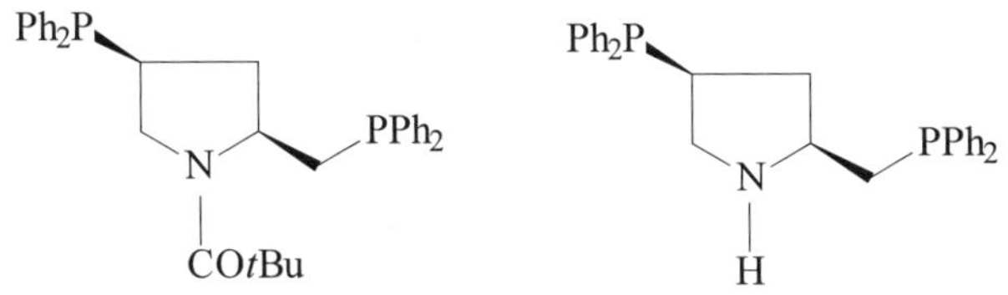

1 2

Scheme 7.25.

The reduction of KPL on [Rh(HD)L] catalysts with *BPPM* and **1** ligands lead to (*R*)-PL with *ee*'s of 54.6 and 59.2%, respectively, while the ligand **2** produces (*S*)-PL with an *ee* of 15.4%. The substituent at the *N*-atoms play an important role in the process because the sense of asymmetric induction depends on the *N*-substituents.

Using cycloocta-1,5-diene (COD) ligand instead of hexa-1,5-diene in the Rh-complexes, increased the *ee* from 54.6% *R* to 80.5% *R* [25] or 86.7% *R* [26] at 30^0C and 50 bar hydrogen pressure in benzene solution if the neutral [RhCl(COD)(BPPM)]Cl complex was used, while the cationic complex, [Rh(COD)(BPPM)]ClO_4, gave only 53.5% *R* [27]. In the case of the latter cationic complex, a remarkable solvent effect was observed. In MeOH, which is well known as a poorly coordinating solvent, the sense of asymmetric induction was opposite and (*S*)-(+)-pantolactone with an *ee* of 15.7% was produced [27].

Another type of ligand is the (-)-*Cy-DIOP* (Scheme 7.26.), which is based on ligand *DIOP* (Scheme 7.8.) [28,29]. The cyclohexyl analog is an effective ligand in the hydrogenation of ketopantolactone to a moderate *ee* of 45% [30]. On the other hand, including the cyclohexyl group instead of the phenyl into the *BPPM* produced the very effective *BCPM* ligand and the [Rh(COD)Cl]$_2$ (BCPM) produced (*R*)-(-)-PL with 100% yield and an *ee* of 92 % in THF and at a substrate to catalyst ratio of 1000 [31].

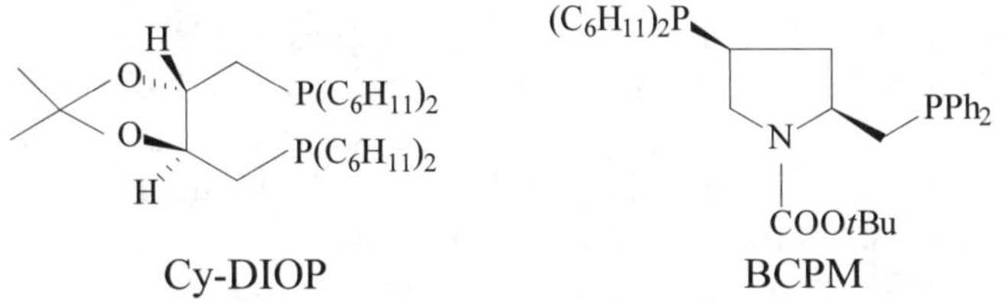

Cy-DIOP BCPM

Scheme 7.26.

In the hydrogenation of KPL at 50^0C and 50 bar in THF a markedly different effect of the dicyclohexyl-phosphino groups of the ligands at the C2 position was observed on the optical yields of PL.

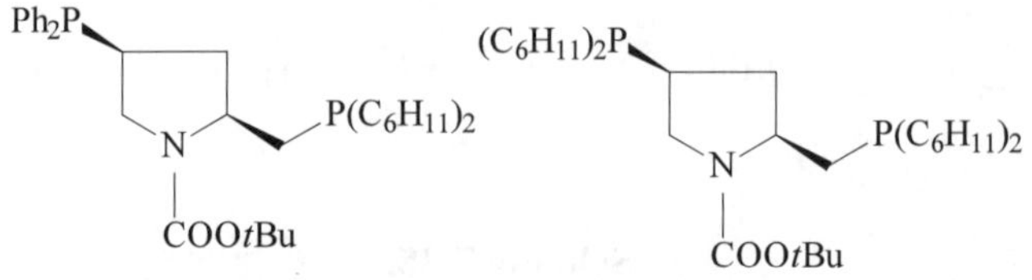

BCPP BCCP

Scheme 7.27.

The Rh-complexes with chiral ligands of *BPPM* (Scheme 7.19.), *BCPM* (Scheme 7.26.), *BCPP* and *BCCP* (Scheme 7.27.) were studied [32], and it was found that complexes with *BPPM, BCPM* and *BCPP* ligands gave (*R*)-PL with *ee*'s of 81%, 91% and 9%, respectively, while the *BCPP* ligand led to (*S*)-PL with an *ee* of 61%. This may be rationalized by the assumption that one of the Ph_2P-groups in the other ligands controls the chiral induction and accelerates the reaction rate.

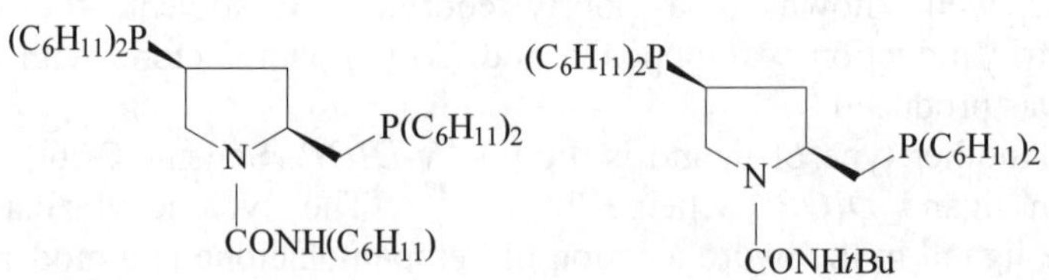

1 2

Scheme 7.28.

In the case of the Rh complexes with ligands <u>1</u> and <u>2</u> (Scheme 7.28.) was found that a neutral Rh-complexes of these ligands in THF gave pantolactone with an *ee*'s of 41% *S* and 60% *S*, respectively, while both cationic complexes gave an *ee* of 10% and *R*-configuration. Thus it was found [33] that the above Rh complexes are more effective catalysts than [Rh(Alk)(DIOP)] complexes in the asymmetric hydrogenation of ketopantolactone.

In 1990 Petit and the Mortreux's group [34] found that the $[Rh(COD)Cl]_2$ complex with rather simple chiral ligand containing one cyclopentyl and one Ph-groups (<u>1</u> in Scheme 7.29) revealed high enantioselectivity: *ee* 90.7% *R*, at 1 bar hydrogen, and with the relationship of substrate to Rh 400.

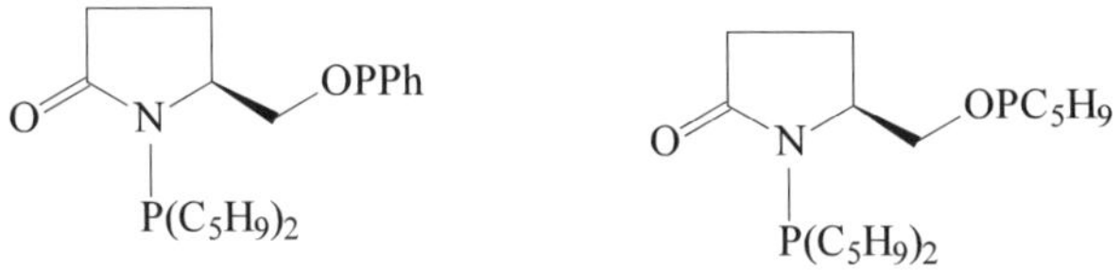

1 2

Scheme 7.29.

Recently, for the hydrogenation of KPL Mortreux et al.[35] found other more effective ligands for Rh-catalyst containing two cyclopentyl groups. They found very high *ee*'s up to 96.9% at 70 ^{0}C in the case of ligand 2 with a substrate to catalyst ratio of approximately 10 000. As the authors indicated "...to the best of our knowledge, the neutral Rh complex of this new chiral ligand is the most effective catalyst so far reported for the asymmetric hydrogenation of ketopantolactone into (*R*)-(-)-pantolactone..." [35].

As for the mechanism of this reaction, it was noted [33] that the studies revealed that:

i) the Schrock-Osborn mechanism [36] is not operative since no acceleration by the addition of water was observed

ii) there is a difference between cationic and neutral complexes in that the neutral complexes are much more enantioselective

iii) there is a remarkable solvent effect on the enantioselectivity of pantolactone formation.

A new Rh(I)-complex catalyst bearing two chiral ligands, the (*R,R*)-DIOP and an *N,N'*-co-ligand based on (*R*)- or (*S*)-phenylethylamines, pyrroleimines, or pyrroleoxazolines proved to be enantioselective in the hydrogenation of ketopantolactone into (*R*)-pantolactone at 50 ^{0}C and 50 bar in toluene solution at a relationship of substrate to Rh 200 [37].

In all cases, the investigated Rh-complexes with co-ligands, independently of their configuration, *R*- or *S*-, and their chemical structure, (*R*)-pantolactone was obtained with an *ee* of approximately 30%, while catalyst [Rh-(COD)Cl_2-(*RR*)-DIOP] resulted the (*R*)-pantolactone with an *ee* of 54%. Thus there was no matched-mismatched contributions from chirality of *N,N'*-co-ligands that was explained by formation of monodentate coordination of co-ligand with Rh only by the *N* of the pyrrole ring. Only in the case of a monodentate co-ligand with rigid structure (Scheme 7.30.) combined with the complex [Rh(*RR*-DIOP)Cl] was the (*R*)-pantolactone obtained with an *ee* of 50%.

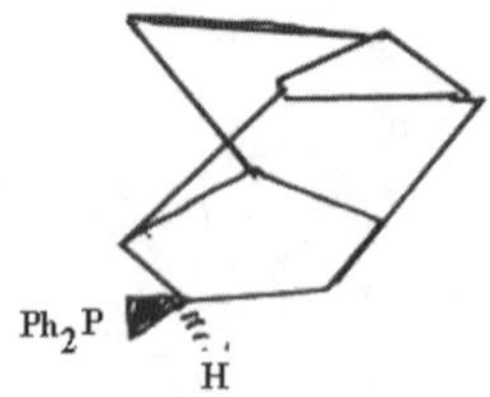

Scheme 7.30.

For preparation of (*R*)-(-)-pantolactone the enzymic method was applied. Under the action of *ketopantoyl lactone reductase* in Baker's yeast (*R*)-(-)-pantolactone was prepared with high chemical and optical yields [38].

References

1 a) Knowles, W.S, Sabacky, M.J. and Vineyard, B.D. (1972) Catalytic asymmetric hydrogenation, *J. Chem. Soc., Chem. Commun.* 10-11.
b) Knowles, W.S. (1986) Application of organometallic catalysis to the commercial production of L-DOPA, *J. Chem. Ed.*, **63**, 222-225.
c) Chirality in industry (1992) Collins, A.N., Sheldrake, G.N., and Crosby, J. (eds.), Wiley, N.Y., V.I. 409 pp.
d) Chirality in industry (1997) Collins, A.N., Sheldrake, G.N., and Crosby, J. (eds), Wiley, N.Y., V.II. 418 pp.

2 Kagan, H.B. (1988) Asymmetrric catalysis in organic synthesis with industrial perspectives, *Bull. Soc. Chim. Fr.*, N 5, 846-853.

3 Karpeiskaya, E.I., Levitina, E.S., Godunova, L.F. and Klabunovskii, E.I. (1986) Reaction of catalytic enantioselective aminolysis of *delta*-2-oxazolin-5-ones as a method of asymmetric synthesis of *alpha*-amino acids, *J. Mol. Catal. A: Chem.* **34**, 129-143.

4 Tani, K., Yamagata, T., Akutagawa, S., Kumobayashi, H., Taketomi, T., Takaya, H., Miyashita, A., Noyori, R. and Otsuka, S. (1984) Highly enantioselective isomerization of prochiral allylamines catalysed by chiral diphosphine Rh (I) complexes. Preparation of optically active enamines, *J. Amer. Chem. Soc.* **106**, 5208-5217.

5 a) Cullen, W.R., Frizuk, M.D., James, B.R., Kutney, J.P., Kang, G.J., Herb, G., Thorburn, J.S. and Spogliarich, R.S. (1990) Asymmetric homogeneous hydrogenation of imines using Rh-phosphine systems, *J. Mol. Catal. A. Chem.* **62**, 243-253.
b) Moser, H., Rihs, G. and Sauter, H (Ciba-Geigy Co.) (1982) Der Einfluss von Atropoisomerie und chiralen Zentrum auf die biologische Aktivitaet des Metolachlor, *Z. Natuforschung* **37b**, 451-462.

6 a) Spindler, F., Pugin, B., and Blaser, H.U. (1990) Novel diphosphino-iridium catalyst for the enantioselective hydrogenation of N-arylketimines, *Angew. Chem. Int. Ed. Engl.*, **29**, 558-559.
b) Spindler, F., Pugin, B., Jalett, H.-P., Buser, H.P., Pittelkow, U. and Blaser, H.U. (1996) A technically useful catalyst for the homogeneous enantioselective hydrogenation of N-arylimines: a case study, *Chemistry Industry* (*Catalysis of Organic Reactions*), (Dekker) **68,** 153-166.

7 Bader, H.B. and Blaser, H.U. (1997) Catalysis for agrochemicals: the case history of the DUAL Herbicide, *Stud. Surf. Sci. Catal.,* **108**, 17-29.

8 Togni, A., Breutel, C., Schnyder, A., Spindler, F., Landert, H. and Tijani, A. (1994) A novel easily accessible chiral ferrocenyl diphosphine for highly enantioselective hydroganation, allylic alkylation and hydroboration reactions, *J. Amer. Chem. Soc.,* **116,** 4062-4066.

9 Zhorov, E.Yu., Pavlov, V.A., Fedotova, O.A., Shvedov, V.I., Mistryukov, E.A., Platonov, D.N., Gorshkova, L.S. and Klabunovskii, E.I. (1991) Enantioselective metal-complex catalysis, *Comm. 7*., Asymmetric hydrosilylation of imines and oximes on catalyst $[Rh(COD)Cl]_2$ (*S*)-*Phephos, Izv. AN SSSR. Ser. Khim.* , 865-868.

10 Joo, F., and Katho, A. (2004) Hydrogenation, in *Aqueous Phase Organometallic Catalysis,* Herrmann, W.A. (ed.), Wiley-VCH, Weinheim, 2nd ed., 429-463.

11 Amrani, Y., Lecomte, L., and Sinou, D. (1989) Chiral sulfonated phosphines, syntheses and use as ligands in asymmetric hydrogenation using an aqueous-organic two phase solvent system, *Organometallics*, **8**, 542-547.

12 Tas, D., Rane, V.H., Parton, R.F. and Jacobs, P.A (1997) Immobilized Noyori catalyst on beta-zeolite in hydrogenation of beta-ketoester, *15th North American Catalysis Society Meeting,* Chicago, IL. Abstracts.

13 a) Wan, K.-T. and Davis, M.E. (1995) Asymmetric synthesis of naproxen by a new heterogeneous catalyst, *J. Catal.* **152**, 25-30.
b) Davis, M.E. and Wan, K.-T. (1997) A route to the rational design of heterogeneous catalysts, in *"The Robert A. Welch Foundation Conference on Chemical Research. XXXX. Chemistry on the Nanometer Scale".* Houston, 21 Oct. 1996, pp. 1-13.

14 a) Wan, K.-T. and Davis, M.E. (1994) Design and synthesis of a heterogeneous asymmetric catalysts, *Nature*, **370**, 449-450
b) Wan, K.-T. and Davis, M.E. (1993) Asymmetric hydrogenation in water by a Rhodium complex of sulfonated 2,2-bis(diphenylphosphine)-1,1-binaphthyl (BINAP), *J. Chem. Soc., Chem. Comm.* 1262-1264.

15 Ohta, T., Miyake, T. and Takaya, H. (1992) An efficient synthesis of optically active methyloxetan-2-one: asymmetric hydrogenation of diketene catalysed by BINAP-Ru complexes, *J. Chem. Soc., Chem. Comm.* 1725-1726.

16 Chiellini, E. and Solaro, R. (1996) Biodegradable polymeric materials, *Adv. Mater.*, **8**, 305 ; *Chem. Abstr.* 1996, **125**, 60185.

17 a) Sheldon, R.A. (1993) *Chirotechnology. Industrial synthesis of optically active compounds,* Marcel Dekker Inc., N.Y., Basel, Hong Kong, 423 pp.
b) Klabunovskii, E.I. (1996) Catalytic asymmetric synthesis of beta-hydroxyacids and their esters, *Russian Chem. Rev.* **65**, 329-344
c) Merck Co. Inc., SU 1287756.A3. KRKA, SU 1757471.A3.

18 Vankelecom, I.F.J., Tas, D., Parton, R.F., Van de Vyver, V. and Jacobs, P.A. (1996) Chiral catalytic membranes, *Angew. Chem. Int. Ed. Engl.* **35**, 1346-1348.

19 Ojima, I., Kogure, T. and Achiwa, K. (1977) Effective homogeneous hydrogenation of alpha-ketoesters catalysed by neutral phosphine ligands and application to the asymmetric synthesis of lactates, *J. Chem. Soc., Chem. Comm.* 428-430.

20 Spindler, F., Pittelkow, U. and Blaser, H.U. (1991) A highly enantioselective Rh-catalyst for the hydrogenation of aliphatic alpha-keto esters, *Chirality,* **3**, 370-375.

21 a) Blaser, H.U. and Jalett, H.-P. (1993) Enantioselective hydrogenation of alpha-ketoacids using Pt-catalysts modified with cinchona alkaloids, *Stud. Surf. Sci. Catal.,* **78,** 139-146.
b) Baiker, A. (1999) Supercritical fluids in heterogeneous catalysis, *Chem. Rev.* **99**, 453-473.

22 Hoffmann-LaRoche, US 3.884.966, 20.05.1975, *Chem. Abstr.* **84**, 16758 (1976)

23 Ojima, I. (1984) Homogeneous asymmetric catalysis by means of chiral rhodium complexes, *Pure Appl. Chem.*, **56**, 99-110.

24 Achiwa, K., Kogure, T. and Ojima, I. (1977) Catalytic asymmetric synthesis of *R*(-)- and (*S*)-(+)-pantolactone, *Tetrahedron Lett.*, 4431-4432.

25 Achiwa, K., Kogure, T. and Ojima, I. (1978) Highly effective catalytic asymmetric synthesis of (*R*)-(-)-pantolactone, *Chem. Lett.* 297-298.

26 Ojima, I., Kogure, T. and Terasaki, S. (1978) Effective biomimetic route to D-(+)-pantothenate using asymmetric hydrogenation catalyzed by a chiral Rh complex in the key steps, *J. Org. Chem.* **43**, 3444-3446.

27 Ojima, I., and Kogure, T. (1980) The asymmetric hydrogenation of alpha-keto esters catalyzed by Rh complexes with chiral diphosphine ligands. On the catalytic cycles and the mechanism of asymmetric induction, *J. Organomet. Chem.*, **195**, 239-248.

28 Weishaeupl, J. (1984) DIOP als Ligand in der asymmetrischen Katalyse mit Uebergangsmetall-Phosphin-Komplexen, *Oesterreichische Chemie-Zeitschrift* **85,** H 1, 1-5.

29 Kagan, H.B. (1982) Asymmetric synthesis using organometallic catalysis, in Wilkinson G. (ed.) *Comprehensive organometallic chemistry*, Pergamon Press, Oxford, **8**, 463-498.

30 Yamamoto, K. and Saeed-ur-Rahman (1984) New entry to the preparation of chiral bis(dicyclohexylphosphino)alkane derivatives. Use for Rh-catalyzed hydrogenation of carbonyl compounds, *Chem. Lett.* 1603-1606.

[31] Takahashi, H., Hattori, M., Chiba, M., Morimoto, T. and Achiwa, K. (1986) Preparation of new chiral pyrrolidinebisphosphines as highly effective ligands for catalytic asymmetric synthesis of (*R*)-(-)-pantolactone, *Tetrahedron Lett.*, **27**, 4477-4480.

[32] Morimoto, T., Takahashi, H., Fujii, K., Chiba, M. and Achiwa, K. (1986) Synthesis of a new chiral pyrrolidine ligand bearing two different types of phosphino groups and their effects on the asymmetric hydrogenation of ketopantolactone, *Chem. Lett.*, 2061-2064.

[33] Tani, K., Suwa, K., Tanigawa, E., Ise, T., Yamagata, T., Tatsuno, Y., and Otsuka, S. (1989) Preparation of optically active peralkyldiphosphines and their use, as the Rh complex, in the asymmetric catalytic hydrogenation of ketones, *J. Organomet. Chem.* **370,** 203-221.

[34] Hatat, C., Karim, A., Kokel, N., Mortreux, A. and Petit, F. (1990) Hydrogenation asymetrique de cetones activees par des complexes rhodium alkyl-AMPP: optimisation et etude du mechanisme, *New J. Chem.* **14**, 141-152.

[35] Roucoux, A., Agboussou, F., Mortreux, A. and Petit, F. (1993) New alkylarylamido phosphine phosphinites as chiral diphosphines for asymmetric hydrogenation of acilated keto compounds, *Tetrahedron Asymm.* **4**, 2279-2282.

[36] Schrock, R.R., and Osborn, J.A. (1970) Rhodium catalysts for the homogeneous hydrogenation of ketones, *J. Chem. Soc., Chem. Comm.* 667-668.

[37] Brunner, H., and Tracht, T. (1998) Asymmetric catalysis. 128. Diastereomeric Rh(I) complexes in the enantioselective hydrogenation of ketopantolactone, *Tetrahedron Asymm.* **9,** 3773-3780.

[38] Shimizu, S., Hata, H., and Yamada, H. (1984) Reduction of ketopantolactone to D-(-)-pantolactone by microorganisms, *Agr. Biol. Chem.* **48,** 2285-2291.

[39] Matteoli, U., Beghetto, V. and Serivanti, A. (1999) Asymmetric hydrogenation by an in situ prepared (*S*)-BINAP-Ru(II) catalytic system, *J. Mol. Catal. A. Chem.* **140,** 131-137.

[40] Wolfson, A., Vankelekom, I.F.J., Geresh, S. and Jacobs, P.A. (2003) The role of the solvent in the asymmetric hydrogenation of beta-keto esters with Ru-BINAP, *J. Mol. Catal. A. Chem.* **198,** 39-45.

INDEX

Catalysis by Metal Complexes

Series Editors:
R. Ugo, *University of Milan, Milan, Italy*
B.R. James, *University of British Colombia, Vancouver, Canada*

1. F.J. McQuillin: *Homogeneous Hydrogenation in Organic Chemistry.* 1976 ISBN 90-277-0646-8
2. P.M. Henry: *Palladium Catalyzed Oxidation of Hydrocarbons.* 1980 ISBN 90-277-0986-6
3. R.A. Sheldon: *Chemicals from Synthesis Gas.* Catalytic Reactions of CO and H_2. 1983 ISBN 90-277-1489-4
4. W. Keim (ed.): *Catalysis in* C_1 *Chemistry.* 1983 ISBN 90-277-1527-0
5. A.E. Shilov: *Activation of Saturated Hydrocarbons by Transition Metal Complexes.* 1984 ISBN 90-277-1628-5
6. F.R. Hartley: *Supported Metal Complexes.* A New Generation of Catalysts. 1985 ISBN 90-277-1855-5
7. Y. Iwasawa (ed.): *Tailored Metal Catalysts.* 1986 ISBN 90-277-1866-0
8. R.S. Dickson: *Homogeneous Catalysis with Compounds of Rhodium and Iridium.* 1985 ISBN 90-277-1880-6
9. G. Strukul (ed.): *Catalytic Oxidations with Hydrogen Peroxide as Oxidant.* 1993 ISBN 0-7923-1771-8
10. A. Mortreux and F. Petit (eds.): *Industrial Applications of Homogeneous Catalaysis.* 1988 ISBN 90-2772-2520-9
11. N. Farrell: *Transition Metal Complexes as Drugs and Chemotherapeutic Agents.* 1989 ISBN 90-2772-2828-3
12. A.F. Noels, M. Graziani and A.J. Hubert (eds.): *Metal Promoted Selectivity in Organic Synthesis.* 1991 ISBN 0-7923-1184-1
13. L.I. Simándi (ed.): *Catalytic Activation of Dioxygen by Metal Complexes.* 1992 ISBN 0-7923-1896-X
14. K. Kalyanasundaram and M. Grätzel (eds.): *Photosensitization and Photocatalysis Using Inorganic and Organometalic Compounds.* 1993 ISBN 0-7923-2261-4
15. P.A. Chaloner, M.A. Esteruelas, F. Joó and L.A. Oro: *Homogeneous Hydrogenation.* 1994 ISBN 0-7923-2474-9
16. G. Braca (ed.): *Oxygenates by Homologation or CO Hydrogenation with Metal Complexes.* 1994 ISBN 0-7923-2628-8
17. F. Montanari and L. Casella (eds.): *Metalloporphyrins Catalyzed Oxidations.* 1994 ISBN 0-7923-2657-1

Catalysis by Metal Complexes

18. P.W.N.M. van Leeuwen, K. Morokuma and J.H. van Lenthe (eds.): *Theoretical Aspects of Homogeneous Catalisis.* Applications of *Ab Initio* Molecular Orbital Theory. 1995 ISBN 0-7923-3107-9
19. T. Funabiki (ed.): *Oxygenases and Model Systems.* 1997 ISBN 0-7923-4240-2
20. S. Cenini and F. Ragaini: *Catalytic Reductive Carbonylation of Organic Nitro Compounds.* 1997 ISBN 0-7923-4307-7
21. A.E. Shilov and G.P. Shul'pin: *Activation and Catalytic Reactions of Saturated Hydrocarbons in the Presence of Metal Complexes.* 2000 ISBN 0-7923-6101-6
22. P.W.N.M. van Leeuwen and C. Claver (eds.): *Rhodium Catalyzed Hydroformylation.* 2000 ISBN 0-7923-6551-8
23. F. Joó: *Aqueous Organometallic Catalysis.* 2001 ISBN 1-4020-0195-9
24. R.A. Sánchez-Delgado: *Organometallic Modeling of the Hydrodesulfurization and Hydrodenitrogenation Reactions.* 2002 ISBN 1-4020-0535-0
25. F. Maseras and A. Lledós (eds.): *Computational Modeling of Homogeneous Catalysis.* 2002 ISBN 1-4020-0933-X
26. L.I. Simándi (ed.): *Advances in Catalytic Activation of Dioxygen by Metal Complexes.* 2003 ISBN 1-4020-1074-5
27. A. Sen (ed.): *Catalytic Synthesis of Alkene-Carbon Monoxide Copolymers and Cooligomers.* 2003 ISBN 1-4020-1129-6
28. A. Kotschy and G. Timári: *Heterocycles from Transition Metal Catalysis.* Formation and Functionalization. 2005 ISBN 1-4020-3624-8
29. P.J. Dyson and T.J. Geldbach: *Metal Catalysed Reactions in Ionic Liquids.* 2005 ISBN 1-4020-3914-X
30. R.P. Tooze (ed.): *Separation, Recovery and Recycling in Homogeneous Catalysis.* 2005 ISBN 1-4020-4086-5
31. E. Klabunovskii, G.V. Smith and Á. Zsigmond: *Heterogeneous Enantioselective Hydrogenation.* Theory and Practice. 2006 ISBN 1-4020-4294-9

springer.com

* *Volume 1 is previously published under the Series Title: Homogeneous Catalysis in Organic and Inorganic Chemistry*